Optimal control of degenerate parabolic equations in image processing

Optimal control of degenerate parabolic equations in image processing. Analysis of evolution equations with variable degeneracy and associated minimization problems

von Kristian Bredies

Dissertation
zur Erlangung des Grades eines Doktors der Naturwissenschaften
– Dr. rer. nat. –

Vorgelegt im Fachbereich 3 (Mathematik & Informatik)
der Universität Bremen
im Juli 2007

Bibliografische Information der Deutschen Nationalbibliothek

Die Deutsche Nationalbibliothek verzeichnet diese Publikation in der
Deutschen Nationalbibliografie; detaillierte bibliografische Daten sind
im Internet über http://dnb.d-nb.de abrufbar.

ISBN 978-3-8325-1833-2

Logos Verlag Berlin GmbH
Comeniushof, Gubener Str. 47,
10243 Berlin
Tel.: +49 30 42 85 10 90
Fax: +49 30 42 85 10 92
INTERNET: http://www.logos-verlag.de

Datum des Promotionskolloquiums: 24. Oktober 2007

Gutachter: Prof. Dr. Peter Maaß, Universität Bremen
           Prof. Dr. Arnd Rösch, Universität Duisburg-Essen

Abstract

This dissertation examines an optimal control and parameter identification problem in image processing involving a class degenerate parabolic partial differential equations for which the degeneracy is controlled.

The main contribution is the mathematical analysis of the considered partial differential equations and associated optimal control problems, especially with respect to the coefficients of the partial differential equation. Motivated by an interpolation problem from medical image processing, a minimization problem is suggested which is constrained by a class of degenerate parabolic equations, where the degeneracy, controlled by a parameter, leads to preservation of edges and consequently to non-smooth solutions. The minimization problem tries to find an optimal parameter with respect to certain objectives.

Existence and uniqueness of solutions for the class of degenerate parabolic equations in spaces of non-smooth functions which vary with the control parameter is proven. The mathematical tool of weighted and directional Sobolev spaces is introduced to obtain analytical properties of the solution operator with respect to its degenerate coefficients. One main result is the characterization of the solution spaces in terms of special weak weighted and directional derivatives. Such notions allow to describe the non-smooth functions, the solution operator of the partial differential equations produces, in a common framework. Moreover, weak closedness properties with respect to the varying parameter are obtained. Combining these ingredients, the existence of minimizers under appropriate regularization is shown, the main result regarding the associated optimal control problems. Furthermore, first-order optimality conditions are derived where differentiation of the objective functional is possible.

A numerical method for finding stationary points for the minimization problem is suggested. Computations illustrating the outcome of the proposed algorithm for the image-processing problem are carried out, confirming its appropriateness and feasibility.

iv

Acknowledgments

I would like to thank Prof. Dr. Peter Maaß for supervision and motivation during the completion of this dissertation. His great support has made this research possible.

Thanks go to the people from my group at the "Zentrum für Technomathematik" for a very nice and pleasant working environment, with special thanks to Malte Peter, Lutz Justen, Dirk Lorenz, Sebastian Meier and Bettina Suhr for innumerable discussions, support and advice – both mathematically and personally. Moreover, I am greatly in debt to Malte Peter, Sebastian Meier, Lutz Justen, Dirk Lorenz and Rike Grotmaack for the careful reading of this work and helpful hints and suggestions.

I would like to thank Prof. Dr. Arnd Rösch for his interest in my work and agreeing to examine my dissertation.

Finally, I also wish to acknowledge the support of the DFG Priority Program 1114 "Mathematical methods for time series analysis and digital image processing" which helped me to do the research to produce this dissertation.

Contents

1 Introduction   1

2 Problem formulation   7
  2.1 A class of degenerate parabolic equations   9
  2.2 The optimal control problem   12
  2.3 Related work   14
    2.3.1 Image processing 14
    2.3.2 Optimal control of PDEs   15
    2.3.3 Parameter identification   16
    2.3.4 Degenerate partial differential equations 17

3 Linear degenerate parabolic equations   19
  3.1 The weak formulations and solution spaces 20
    3.1.1 The weak formulations   20
    3.1.2 Abstract stationary solution spaces 23
    3.1.3 Abstract time-variant solution spaces 27
  3.2 Existence and Uniqueness   32
  3.3 Weak compactness results   43
    3.3.1 Spaces of weakly continuous functions 43
    3.3.2 Relative weak compactness in $\mathcal{C}^*\big(0,T;L^2(\Omega)\big)$ 47

4 Weighted and directional Sobolev spaces   53
  4.1 The space $H^{1,\infty}$ as a dual space 54
    4.1.1 Sequential weak*-convergence and compactness 58
    4.1.2 Superposition operators in Y_k^* 60
  4.2 Weak weighted and directional derivatives 64
    4.2.1 The weak weighted derivative 65
    4.2.2 The weak directional derivative 67
  4.3 Weighted and directional Sobolev spaces 76
    4.3.1 Closedness of the weak weighted gradient 77

    4.3.2   Closedness of the weak directional derivative   81
4.4   Density of smooth functions   84
    4.4.1   Calculus rules in $H^r_{w,\partial q_1,\ldots,\partial q_K}$ 97

5   Time-variant spaces and application to degenerate PDEs   101
5.1   Time-variant weights and directions 102
    5.1.1   Superposition operators in $\mathcal{Y}^*_k$ 104
5.2   Time-variant weighted and directional spaces 107
    5.2.1   The time-variant weak weighted and directional derivative 108
    5.2.2   Density of smooth functions 111
5.3   Application to degenerate parabolic PDEs 115
5.4   Boundedness and additional regularity 130
    5.4.1 L^∞-estimates 130
    5.4.2   Additional regularity with respect to time 132

6   Analysis of the optimization problem   141
6.1   Existence of optimal solutions 142
6.2   First-order optimality conditions 153
    6.2.1   Differentiability of the control-to-state mapping 154
    6.2.2   The adjoint equations 158
    6.2.3   First-order necessary conditions 163

7   Numerical realization   175
7.1   Discretization of the PDE with finite differences 176
7.2   The discrete optimization problem 182
7.3   A gradient descent algorithm 189
7.4   Numerical examples 191

8   Summary   203

A   Notation   209
A.1   Symbols 209
A.2   Functions, operators and sets 210
A.3   Vector and function spaces 212

Bibliography   215

Chapter 1

Introduction

In this work, an optimal control and parameter identification problem in image processing involving degenerate parabolic partial differential equations is examined. The considered problem and its analysis is settled between different fields of applied mathematics: on the one hand, it has its motivation in image processing, where methods based on PDEs which are enhancing the structure of the image are in the focus of research and have been applied successfully for quite some time. On the other hand, the problem can be interpreted as an optimal control problem with a partial differential equation as constraint, but also as a parameter identification problem for the coefficients of a PDE, which in turn is often considered as an inverse problem. And, needless to say, the theory of weak solutions for PDEs, especially for degenerate parabolic equations, is playing in important role in this work, since it is the aspect common to the mathematical fields it is located in.

The motivation for the optimal control problem we are going to deal with is located in the area mathematical image processing. It originates from the problem of generating an image sequence or movie which interpolates two given images with similar features on different scales in a natural way. This problem arises, for example, if one tries to support a medical expert in the diagnosis of cancer by optimizing the representation of mammography images. The characteristics of natural moving pictures are in general dominated by the movement of several objects at different directions and speed without the objects themselves changing too much. This observation can be mathematically described by the optical flow constraint. Recovering the optical flow for an image sequence, or, quite similar, finding a mapping which matches one image with another (image matching or registration), is an important task in image processing, see [HS81, ADK99] a for optical flow and [Bro81, Mod04] for image registration, for example.

1

A different situation occurs if one wants to describe the evolution of a still image which possibly changes contrast, sharpness and features in time. Our motivation is the imagination that this evolution should happen in a natural way, which can be often modeled by some kind of underlying physical process from which we think that it acts more like diffusion and is influenceable by sources or drains. In mathematical image processing, such diffusion processes are often used to smooth images (see, for instance, [AK02] and [CS05]). In particular, special attention is paid to non-linear partial differential equations which realize anisotropic diffusion and are able to smooth the image while the edges are preserved. Perhaps the best-known equation of this type is the Perona-Malik equation [PM90] which is generally ill-posed. Many well-posed modifications, e.g. [CLMC92] and enhancements, e.g. [Wei98] were proposed since then. Such equations are also able to describe contrast changes by simply adding a 0-th order term which models sources or drains.

However, in most cases, well-posedness is only achieved by strong regularization which often also enforces a regularity on the solution which does not allow for jumps across edges. On the other hand, the Perona-Malik equation is known to exist until a shock occurs. Numerically, one can nevertheless compute a discretized version of the equation beyond the shock which, by experience, forms jumps across edges. This suggests to develop a framework in which the existence of solutions of anisotropic diffusion equations is guaranteed even in the presence of discontinuities (shocks) across hypersurfaces. In this work, we introduce a class of diffusion equations for which such existence can be proven. Its solutions have the property that real discontinuities are able to emerge; there are no smoothing steps involved.

The thesis is devoted to the study of an optimal control problem associated with the above-mentioned image interpolation problem, for which a more detailed description will be given in the following. First, the class of evolution equations under consideration is modeling diffusion which can be completely suppressed in one direction; a process described by the equation

$$\frac{\partial y}{\partial t} = \operatorname{div}(D_p^2 \nabla y^{\mathrm{T}}) + u \quad \text{in }]0, T[\times \Omega \,,$$
$$D_p^2 \nu \cdot \nabla y = 0 \qquad\qquad \text{on }]0, T[\times \partial\Omega \,,$$
$$y(0) = y_0 \qquad\qquad\quad \text{on } \{0\} \times \Omega$$

with

$$D_p^2 = \left(I - \sigma(|p|) \frac{p}{|p|} \otimes \frac{p}{|p|} \right) \,,$$
$$p :]0, T[\times \Omega \to \mathbb{R}^d \quad, \quad |p| \le 1 \quad, \quad \sigma : [0, 1] \to [0, 1] \,.$$

The vector field p indicates the direction and amount of which the diffusion is suppressed. The distinctive feature of this equation is the degeneracy for $|p| = 1$, i.e. $z \cdot D_p^2 z \geq 0$ for $z \in \mathbb{R}^d$ with 0 attained for $z = p$, in contrast to the usual uniform-ellipticity assumption $z \cdot D_p^2 z \geq c|z|^2$ for all $z \in \mathbb{R}^d$. These degenerate parabolic equations are the underlying constraint for the joint optimal control and parameter identification problem

$$\min_{u,|p|\leq 1} \sum_{n=1}^{N} \frac{\|y(t_n) - y_{\Omega,n}\|_2^2}{2} + \Phi_u(u) + \Phi_p(p) \ ,$$

$$0 < t_1 < \ldots < t_N = T \quad , \quad y_{\Omega,n} : \Omega \to \mathbb{R} \ .$$

Such a minimization problem not only controls the sources/sinks u of a diffusion equation but also tries to find an optimal diffusion parameter p. Usual parameter identification problems restrict the diffusion coefficients such that the equation remains uniformly elliptic with respect to the space-variable. In contrast to that, the parameter p can be chosen such that the underlying equation degenerates on arbitrary subsets of the time/space domain. Again, it is notable that the class of degenerate parabolic equations allows weak solutions which have proper discontinuities across hypersurfaces in the space domain. The positions of these discontinuities are implicitly determined through the parameter p of the above minimization problem.

The degenerate parabolic equations are analyzed for existence and uniqueness of solutions which can be established, in special solution spaces depending on the parameter p, by using Lions' projection theorem, a generalization of the well-known Lax-Milgram theorem, as well as Galerkin approximations. Usually, parabolic PDEs with a fixed degeneracy with respect to the space-variable (and smooth coefficients) are studied in terms of elliptic regularization, that is replacing the original equation by a sequence of approximate equations which are uniformly elliptic in the space-variable (for instance, by adding $\varepsilon \Delta y$) and examining convergence. For the study of the solution spaces associated with variable degeneracy, however, more effort is necessary. To deal with this situation, notions of weak weighted and directional differentiability and associated Sobolev spaces are introduced. It is presented how these notions extend the classical weak derivatives and that the smooth function are dense in the associated spaces. In turn, this leads to a characterization of all weak solution spaces for p with a certain smoothness and thus giving an analog to $H = W$ for degenerate parabolic equations.

Furthermore, the notion of weighted and directional Sobolev spaces is the essential tools for analyzing the solution operator with respect to the parameters and, implicitly, the varying solution spaces. Existence of solutions for the optimal control problem can then be proven. The problem, however, is non-convex due to the highly non-linear structure of the solution operator. Furthermore, differentiability can only be obtained for the non-degenerate case, making it difficult to formulate first-order necessary conditions. We show in which situations one is able to derive first-order necessary conditions for optimality. Eventually, the numerical realization with finite differences is discussed and experiments are performed.

The thesis can be outlined as follows. In Chapter 2, a more detailed description and derivation of the problem discussed in this work is given. It also contains references to related works and tries to put the work in context with the existing literature. Chapters 3–5 are mainly devoted to the analysis of the class of degenerate parabolic PDEs with respect to both the data (u, y_0) and the parameter p. In particular, Chapter 3 introduces appropriate weak solution spaces, constitutes existence and uniqueness of weak solutions of the PDE in a certain sense and shows weak compactness properties with respect to all p and bounded (u, y_0).

The topic of Chapters 4 and 5 is to develop a framework which captures the essential properties of the weak solution spaces. In Chapter 4, notions of weak weighted and directional derivatives for the stationary case as well as associated Sobolev spaces are introduced. In opposition to multiplying the classical weak gradient by a weight, the way weighted Sobolev spaces are usually defined, these spaces base on the weak weighted and directional derivatives which are defined by a variational formulation. It is moreover shown that these notions extend the classical weak derivatives by pointing out connections. Additionally, we prove the density of smooth functions which will establish the equivalence of weak solution spaces for the degenerate parabolic PDE. Finally, weak (and weak*) closedness results are derived which are the key for gaining insight into the behavior of the weak solution operator of the considered PDE with respect to varying edge fields p and become important when proving existence of optimal solutions in Chapter 6.

Chapter 5 deals with transferring of the results from Chapter 4 to the time-variant case. It is shown which properties of the time-invariant notions and spaces are still valid for the time-variant case. Basically, most properties can be transferred, but since one dimension is added, certain weak* sequential continuity may fail. To give a remedy for this defect, the notion of bounded semi-variation is applied which seems to give the required amount of compactness under minimal assumptions. The results are then applied to characterize, for p fulfilling some smoothness conditions, the actual solution spaces for the class of degenerate parabolic PDEs. Moreover, some further applications are pointed out, such as L^∞-estimates, for instance.

The subsequent chapter, Chapter 6, focuses on the analysis of the optimization problem. This is where the results of the preceding chapters are applied to show existence of optimal solutions, consisting of the pair (u^*, p^*) representing an optimal right-hand side of the PDE and an optimal edge field. The difficulty here is the treatment of the parameter p. It turns out that, with the appropriate weak closedness properties, existence for arbitrary dimensions can basically be obtained for edge fields which are smooth enough. The smoothness assumptions are on the one hand necessary to apply the equivalence of the solution spaces to time-variant weighted and directional Sobolev spaces (see Chapter 5). On the other hand, a lack of smoothness may only imply some convergence (weak*-convergence in L^∞, for instance) for which the application of the subsequent nonlinear operations fail to be continuous. The optimal control problem is also analyzed with respect to optimality conditions. First-order necessary conditions can be derived in some cases, in particular where the derivative of the weak solution operator of the PDE with respect to p exists. Unfortunately, this can only be ensured for the non-degenerate case where $\|p\|_\infty < 1$. Nevertheless, the corresponding adjoint equations as well as the derivative the solution operator of the PDE are presented and the corresponding optimality system is derived.

Finally, the optimal control problem is also treated numerically in Chapter 7. There, the problem in two dimensions is discretized with finite differences and analyzed with respect to the numerical implementation. This includes discussing the discretization of the PDE as well as the optimization problem and then deriving the corresponding discrete optimality system associated with discrete first-order necessary conditions. Based on this optimality system, a gradient-descent algorithm is implemented which is able to find stationary points. Examples are shown which depict the outcome of this algorithm for some sample images.

A summary and outlook can eventually be found in Chapter 8. In Appendix A, for the reader's convenience, the notation used throughout this work is explained in the hope to reduce the possibility of ambiguity.

Readers with specific interest may take the following suggestions into consideration: The reader interested in the analysis of PDEs and function-space theory can concentrate on the Chapters 3–5, while the reader interested in optimal control problems and the corresponding necessary optimality conditions can be recommended to skip parts of Chapters 3–5 and to focus on Chapter 6. Readers with most interest in image processing and the actual application may skip Chapters 3–5 as well as most parts of Chapter 6 in the first reading and take a closer look at the results of Chapter 7.

Chapter 2

Problem formulation

The optimal control problem examined in this dissertation is motivated by the following image-processing problem. Imagine two or more images showing essentially the same object but with different scales emphasized. This can for example be two versions of a mammogram with coarse and fine structure enhanced, respectively, see Figure 2.1. The problem now is to create a transition (movie) from these images which provides a reasonable interpolation with respect to the visual perception of humans. This is, for instance, important for the diagnosis of breast cancer, when the latter-mentioned versions of the mammograms are presented to a medical expert who has to form a decision. In order to support the expert, it is necessary to show a smooth transition between the different versions. Simply switching between the representations, however, is likely to confuse the expert rather than to support him. Additionally, the human visual perception is very strong in recognizing features at different scales, hence we require that the interpolation fades along intermediate scales when going from a coarse-scale image to a fine-scale image.

To summarize, we impose the following conditions on the interpolating image sequence:

1. The image sequence should interpolate the given images.

2. The movie should contain no jumps with respect to time.

3. The interpolation between two neighboring images should be direct, i.e. there should not be the impression of a third image in the interpolation of two images.

4. The transition should interpolate between fine and coarse scales by showing intermediate scales.

Mammogram   Natural image   Artificial image

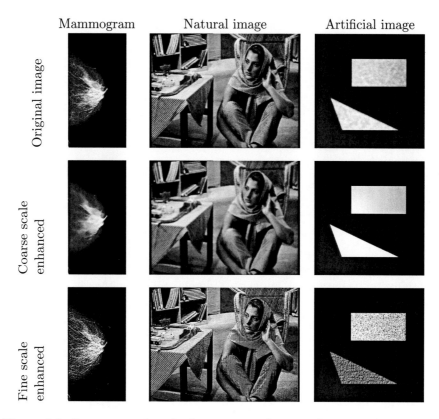

Figure 2.1: Some examples of enhancement of certain features of images.

5. The image structure should be preserved, i.e. existing edges should not
   be blurred.

Note that a key point here is the requirement 4, i.e. the interpolation should
reveal intermediate scales. Omitting this condition, one can verify that simple
linear interpolation satisfies the remaining conditions. However, interpolating
linearly between a coarse-scale image and a fine-scale image instantly reveals
all fine details.

   In the following, we introduce an optimal control problem which respects
all these conditions. The idea behind this is the following. We model the
interpolating image sequence as one solution of a special class of parabolic
partial differential equations which incorporates some of the above conditions
and also depends on certain parameters. The optimal control problem now tries
to minimize over all parameters of this PDE in order to fulfill the remaining
conditions. The solution associated with the optimal set of parameters then is

the desired interpolating movie.

Let y_n, where $n = 0, \ldots, N$, be the collection of images which has to be interpolated by the movie $y(t)$. Each $y(t)$ is a gray-scale image which we model as a function $y(t) : \Omega \to \mathbb{R}$ where $\Omega \subset \mathbb{R}^d$ is a domain (usually, $d = 2$ and $\Omega =]0, 1[^2$ for images). We also prescribe the times $0 = t_0 < \ldots < t_n < \ldots < t_N = T$ at which the y should interpolate the y_n. We will now describe a PDE which is suitable to model transitions between images with edge preservation as well as multiscale fading.

2.1 A class of degenerate parabolic equations

To motivate the resulting partial differential equation, let us first take a look at the well-known heat equation.

$$
\begin{aligned}
\frac{\partial y}{\partial t} &= \Delta y \quad \text{in }]0, T[\times \Omega , \\
\frac{\partial y}{\partial \nu} &= 0 \quad \text{on }]0, T[\times \partial\Omega , \\
y &= y_0 \quad \text{on } \{0\} \times \Omega .
\end{aligned}
\tag{2.1}
$$

Its solution describes the isotropic diffusion of the initial heat distribution y_0 over time t. The underlying physical process suggests that this generates a smooth transition between y_0 and some $y(t_n)$ (which is given in this context and cannot be prescribed). Moreover, $y(t)$ represent the intermediate scales of y_0 and $y(t_n)$. Hence one can say this model meets the second, third and fourth of the above conditions. Indeed, regarded in the whole space $\mathbb{R}^d$, $y(t)$ is Gaussian scale-space representation of y_0. It is well-known that intermediate scales are represented through $y(t)$, see [Lin94] and [SNFJ97], for example.

Of course, we will not be able to interpolate between images with (2.1) since y_0 determines all $y(t)$. One obvious extension therefore is to introduce a right-hand side u which represents heat sources or drains in the physical interpretation. Thus, y should be governed by the equation

$$
\begin{aligned}
\frac{\partial y}{\partial t} &= \Delta y + u \quad \text{in }]0, T[\times \Omega , \\
\frac{\partial y}{\partial \nu} &= 0 \quad \text{on }]0, T[\times \partial\Omega , \\
y &= y_0 \quad \text{on } \{0\} \times \Omega .
\end{aligned}
\tag{2.2}
$$

Here, we assume that u is a function $u \in L^2\big(0, T; L^2(\Omega)\big)$ rather than a distribution. A non-zero u allows to introduce contrast changes which, depending on the time of occurrence, have more or less time to diffuse. This allows to fade one smooth image into another in a way which does not create jumps. Thus, the input u can be understood as the source leading to actual changes in the contrast or the features.

One drawback of this model is that it does not account for one important property of images: the edges. The heat equation produces images which always have a certain smoothness. In most cases, this smoothness prevents that the resulting images have edges, so they have to be blurred. Such effects are not always desired since they cannot be undone without enormous effort and may essentially change the impression of the image. Moreover, the human visual perception heavily relies on the recognition of edges. Thus, it is important to take them into account when modeling the way one image evolves into another. One idea to overcome this problem is to change the parameters of the partial differential equation.

The background behind this is the observation that the disappearance of the edges is caused by diffusion along all directions. In the extreme case, such as evolution equation (2.2), the amount of diffusion is independent of the direction. A different situation occurs if we deal with anisotropic diffusion which means that there are certain directions in which there is less diffusion. So if one prevents diffusion over the edge, it will be preserved. Mathematically, such behavior is described by the diffusion tensor, which is the identity in the case of Equation (2.2). The diffusion tensor determines how the gradient of y is connected with the direction in which the diffusion actually takes place. So if one wants to prevent diffusion along a certain direction, one has to reduce or eliminate this direction from the gradient. A possibility to do this is to multiply the gradient by the following diffusion tensor:

$$D_p^2 = I - \sigma(|p|)\frac{p}{|p|} \otimes \frac{p}{|p|} \qquad \text{with } p \in \mathbb{R}^d \text{ and } |p| \leq 1 \qquad (2.3)$$

where σ is a suitable function mapping $[0,1]$ injectively and strictly monotonically onto $[0,1]$ (the exact requirements are stated in Chapters 3 and 5). The Laplacian of y in the evolution equation (2.2) is then replaced by $\text{div}(D_p^2 \nabla y^T)$. To understand what effect this modification has, it makes sense to examine the mapping $\nabla y \mapsto D_p^2 \nabla y^T$ which can be interpreted as the reduction of the direction p by the factor $\sigma(|p|)$ from the gradient, leaving perpendicular directions invariant. So if $\sigma(|p|) = 1$, then diffusion along p is completely blocked (see Figure 2.2(a) for an illustration). In the case where there is an edge in the image with direction p, there will be no diffusion across the edge. Consequently, the edge will not blur due to diffusion (see Figure 2.2(b)).

Hence, in addition to u, we also introduce a vector field p which may vary in time and space (with $|p| \leq 1$) and a suitable edge-intensity function $\sigma : [0,1] \rightarrow$

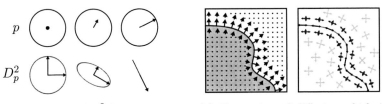

(a) Illustration of D_p^2 for some p

(b) Illustration of diffusion which is blocked over the edges

Figure 2.2: (a) Illustration of the diffusion tensor D_p^2 for some p. In the top row, the vector p is depicted while in the bottom row, you can see the image of one corresponding set of orthogonal eigenvectors as well as the image of the unit sphere under D_p^2. (b) Illustration of diffusion which is blocked over the edges. On the left, a vector field p which is aligned to an edge is depicted. The image on the right indicates the diffusion process associated with this vector field.

$[0,1]$. The above considerations lead to the partial differential equation

$$
\begin{aligned}
\frac{\partial y}{\partial t} &= \operatorname{div}\left(D_p^2 \nabla y^{\mathrm{T}}\right) + u & &\text{in }]0,T[\, \times \Omega \ , \\
\frac{\partial y}{\partial_p \nu} &= 0 & &\text{on }]0,T[\, \times \partial\Omega \ , \\
y &= y_0 & &\text{on } \{0\} \times \Omega \ .
\end{aligned}
\tag{2.4}
$$

where

$$
\frac{\partial y}{\partial_p \nu} = \partial_{D_p^2 \nu} y = \nu \cdot D_p^2 \nabla y^{\mathrm{T}}
$$

denotes the directional derivative of y with respect to the possibly reduced outer normal $D_p^2 \nu$ on $\partial\Omega$.

Unfortunately, although being linear in u, this class of parabolic equations does not fit any more into the usual framework of uniformly elliptic evolution equations. The reason is that the diffusion tensor is allowed to degenerate, i.e. if $|p| = 1$ we only have $z \cdot D_p^2 z \geq 0$ for $z \in \mathbb{R}^d$ and 0 attained if $z = p$. In contrast to that, one can assume in the uniformly elliptic case that there is a $c > 0$ such that $z \cdot D_p^2 z \geq c|z|^2$ for all $z \in \mathbb{R}^d$. Such an estimate plays an important part when proving existence and uniqueness of associated evolution equations. Therefore, there is need to analyze existence and uniqueness properties for the degenerate case as well as the nature of the solution operator. This will be the topic of Chapter 3.

To summarize so far, Equation (2.4) provides a model for smooth transitions between images which respects the human visual perception. It is, however, not

clear how this equation can be used to solve the interpolation problem. The idea
here is to formulate the interpolation condition as an optimal control problem.

2.2   The optimal control problem

The interpolation condition (condition 1 in the above list), can formally be
expressed by

$$y(t_n) = y_n \quad , \quad n = 0, \ldots, N . \tag{2.5}$$

Here, we suppose that the given images y_n are generally non-smooth, i.e. $y_n \in$
$L^2(\Omega)$. While the interpolation constraint can always be fulfilled for y_0, it is
not clear if $y(t_n) = y_n$ can be established for $n = 1, \ldots, N$. Indeed, in optimal
control theory this is the problem of (approximate) controllability which has
been studied for the heat equation (2.2) in [Lio68], with a negative answer. With
this equation, it is only possible to reach a dense subset of smooth $y_n \in L^2(\Omega)$
with a control $u \in L^2(0, T; L^2(\Omega))$. It suggests that we have the same situation
for the each degenerate parabolic equation of the form (2.4), too. Hence, if we
replace the exact interpolation condition (2.5) by an approximation condition
such as

$$\min \sum_{n=1}^{N} \|y(t_n) - y_n\|^2 \tag{2.6a}$$

where y is the solution of Equation (2.4) with parameters u and p, this problem
is potentially not well-posed in the sense that the minimum is 0 but there
is no pair (u, p) for which the minimum is attained. Additionally, the above
functional can become arbitrarily small regardless of p, by just varying $u \in$
$L^2(0, T; L^2(\Omega))$. Thus, too many solutions are allowed in a certain sense. Hence,
some regularization as well as further constraints are necessary. The idea here
is to add regularization terms which also have an interpretation in terms of the
modeled problem.

First recall the role of u. By varying this parameter, it is possible to change
the brightness in some regions of the image. This datum affects the equation
linearly, hence large u also lead to large contrast changes. Since we assume that
the images which have to be interpolated are similar, it makes sense to prefer
small u. In terms of optimal control, one penalizes u, for example, with

$$\Phi_u(u) = \lambda_u \int_0^T \int_\Omega |u(t, x)|^2 \; \mathrm{d}x \; \mathrm{d}t \tag{2.6b}$$

where $\lambda_u > 0$ is a regularization parameter.

Moreover, the choice of p is also open. On the one hand, it should adapt to
the edges and block diffusion such that they are not blurred, but on the other

hand, this is not necessary for the smooth parts. To obtain a good transition, one wants as few edges as necessary. The amount of edges can for example be measured in terms of

$$\Phi_p(p) = \lambda_p \int_0^T \int_\Omega \sigma\big(|p(t,x)|\big) \; \mathrm{d}x \; \mathrm{d}t \; , \tag{2.6c}$$

again with some regularization parameter $\lambda_p > 0$.

Hence, finding an interpolating y amounts to solving the optimal control problem

$$\min_{\substack{u \in L^2 \\ |p| \leq 1}} \sum_{n=1}^{N} \frac{\|y(t_n) - y_n\|^2}{2} + \Phi_u(u) + \Phi_p(p) \tag{2.7}$$

where y is the solution of Equation (2.4) associated with u and p. This is a combination of a distributed degenerate parabolic optimal control problem and a parameter identification problem for a class of degenerate parabolic equations. It will turn out later that further regularization terms are needed to solve the optimal control and parameter identification problem (2.7). This is not too surprising: It is well-known that problems of the above type (i.e. no additional compactness is required) are generally not well-posed even if the degenerate case is excluded, see [Mur77]. We will see in Chapter 6, however, under which conditions the above problem turns out to be well-posed. For now, we write the additional regularization as $\tilde{\Phi}_p$. The full optimal control/parameter identification problem then reads as

$$\boxed{\begin{aligned}
&\min_{\substack{u \in L^2 \\ |p| \leq 1}} \sum_{n=1}^{N} \frac{\|y(t_n) - y_n\|^2}{2} + \Phi_u(u) + \Phi_p(p) + \tilde{\Phi}_p(p) \; , \\
&\frac{\partial y}{\partial t} = \mathrm{div}\big(D_p^2 \nabla y^{\mathrm{T}}\big) + u \quad \text{in }]0,T[\times \Omega \; , \\
&\frac{\partial y}{\partial_p \nu} = 0 \qquad\qquad\qquad\quad \text{on }]0,T[\times \partial\Omega \; , \\
&y = y_0 \qquad\qquad\qquad\qquad \text{on } \{0\} \times \Omega \; .
\end{aligned}} \tag{2.8}$$

The analysis of this problem is strongly connected with the analysis of the solution operator for the associated equation (2.4). Special attention has to be paid to the degenerate case. Here, a framework is necessary which deals with the fact that the natural solution spaces of the PDE vary with p. Such a framework is developed in the subsequent chapters.

In Chapter 3, the existence and uniqueness of weak solutions of a slightly more general version of (2.4) in appropriate spaces depending on p is established.

Moreover, it is shown that all solutions with bounded data u and all possible p are contained in a weakly compact set of $C^*\big(0,T;L^2(\Omega)\big)$, which is the space common to all solutions. Chapters 4 and 5 then analyze the solution spaces more deeply. In Chapter 4, the notions of weak weighted and directional derivatives and associated function spaces as a generalization of (weighted) Sobolev spaces are introduced. Density results are derived showing the equivalence to the stationary spaces introduced in Chapter 3. Furthermore, weak closedness properties with respect to both y and p are discussed, which will turn out to be an important tool analyzing the weak solution operator of (2.4). The topic of Chapter 5 is to transfer the notion of weighted and directional Sobolev spaces to the time-variant setting and to apply the results to the class of degenerate parabolic PDEs introduced in Chapter 3, allowing to describe the behavior of the weak solution operator with respect to both the data u and the degeneracy associated with p.

Finally, in Chapter 6 the results are collected to obtain existence results for the optimal control problem (2.8). Moreover, it will turn out that the derivative of the objective functional (i.e. the functional to be minimized) does not exist for all p. We show in which situations this is nevertheless the case and formulate first-order necessary conditions. These conditions also turn out to be useful in the numerical treatment in Chapter 7.

2.3 Related work

Since this work is intended to be settled between different fields of mathematics, many relations to the existing literature can be seen. The following is an attempt to give a rough overview over the subjects which are close to the topics discussed in the thesis and to highlight some particular works, without the aim of being complete.

2.3.1 Image processing

The optimal control problem (2.8) can be related to many ideas in image processing. First, partial differential equations are now widely used for solving particular image processing tasks. For example, scale-space theory, which aims at analyzing features of an image on different scales, is based on the solution of the heat equation where the initial heat distribution is represented by the image [Wit83, Lin94, SNFJ97]. Another application is image denoising, where the image is plugged into the initial value of a smoothing instationary PDE, resulting in denoised versions after a suitable amount of time. Like the parabolic equation examined in this work, the smoothing PDEs are often non-linear anisotropic diffusion processes like the ill-posed Perona-Malik equation [PM90] or well-posed regularizations which do not degenerate [CLMC92, Wei98].

The minimization of suitable functionals is also a general approach often used in mathematical image processing. For instance, a denoised image can be interpreted as a minimizing argument of a functional which penalizes both noise and inaccuracy to the original image. The literature comprises a variety of models and methods, for instance total-variation [CL97] or Besov/wavelet penalty-terms [CDLL98] and, recently, also combinations with norms measuring oscillations [VO04, DT05]. Beside denoising, plenty of other problems in mathematical image processing base on minimization of suitable functionals. The already-mentioned tasks of determining optical flow or registration of images fall into this class [HS81, ADK99, Bro81, Mod04], as do many others [AK02]. To a certain extend, the problem considered in this work can be regarded as a registration problem, but the focus here lies on more on providing a smooth multiscale interpolation rather than identifying the displacement of an image with respect to another.

Finally, a main part of this work is devoted to the analysis of a special class of function spaces which seems to be appropriate for describing images. Finding mathematical models for natural images is also subject to investigation in mathematical image processing. Such models often try to capture the important features: images can be thought of consisting of objects which are clearly defined by edges and may possess a texture and shading. A well-known approach to representing smooth images with edges is utilizing the total-variation seminorm [ROF92], while textures can be modeled by certain dual norms [Mey01]. Another famous approach of modeling smooth images with edges was developed by Mumford and Shah [MS85, MS89], which can also be related to the function spaces introduced here, see Chapter 3.

2.3.2 Optimal control of PDEs

Influencing systems which are governed by a partial differential equations in an "optimal way" such that given objectives are reached is the goal of optimal control of PDEs. There is an extensive research in this field of applied mathematics since its foundations were laid in the 1960ies. Mathematically, posing optimal control problems with PDEs lead to minimization problems in infinite-dimensional spaces. When analyzing well-posedness and necessary as well as sufficient conditions for optimality based on differentiation, one has to combine methods from partial differential equations with optimization techniques in infinite-dimensional spaces.

Problems with the control entering linearly in the partial differential equation are quite well-analyzed [Lio68]. More recent works deal with the control of all kinds of non-linear PDEs including semi-linear and quasi-linear parabolic equations (see [DR00, AQ06, CFY95], to name a few works, there are many more). The treatment of optimal control problems with non-linear PDEs is, besides the theory for non-linear PDEs, much more complicated. To name some examples, it is not immediate that non-linear operations usually do possess weak sequential continuity, closedness or differentiability properties with respect to the function spaces the problem is set in. Such results are crucial for proving existence of minimizers and deriving first-order necessary conditions, so additional effort is required. Moreover, second-order sufficient conditions can usually only be established with respect to a second norm, which results in the so-called two-norm discrepancy (see [Trö05] for an introduction into the topic and further references). The minimization problem considered in this work can be interpreted as an optimal control problem with partial differential equations. It is linear in one variable, u, and non-linear in the other variable, p, thus resulting in a non-convex objective functional and falling into the class of non-linear optimal control problems with PDE-constraint. Therefore, some ideas from optimal control theory with partial differential equations can be borrowed, for example when necessary conditions are derived: Such conditions usually result in an optimality system with an adjoint equation which is a PDE as well as equations which relate the solutions of the PDE and its adjoint to the optimal control. A system with this structure emerges in the analysis of first-order necessary conditions of (2.8).

Furthermore, regarding degenerate equations, it seems like that there is only little literature with respect to optimal control. A few works deal with parabolic equations with degeneracy [LY95, Bel05], but, to the best knowledge of the author, there is only little literature in which the degeneracy is actually controlled. In [LR91], for instance, a parabolic equation is considered for which the time-derivative is allowed to degenerate. There, the coefficients with respect to the space-variable are still uniformly elliptic, so it does not cover the case considered in (2.4).

2.3.3   Parameter identification

The problem of identifying the parameters of a given model or PDE arises, for instance, when real-life situations have to be simulated on a computer. Often, these parameters, such as diffusion coefficients or reaction coefficients, are not precisely known for the particular situation or the model involves parameters which cannot be measured directly. Thus, the problem of parameter identification of a "forward" model is often referred to as an inverse problem.

Mathematically, there are different ways of looking at parameter identification problems. One possibility is to assume that the model is available as a black box which can be fed with sufficiently many inputs. So, for instance, in the context of PDEs, essential parts of the corresponding forward operator (such as a Dirichlet-to-Neumann mapping) are considered as given. The question now arises if and how it is possible to recover the parameters of the model [KV84, KV85]. Another view is in context of inverse problems. Here, one tries to solve the parameter-to-measurement mapping for some given measurement, or to minimize some corresponding regularized functional. Such an approach is similar to optimal control of partial differential equations, where one seeks data for the PDE which reaches a given objective [CER90].

The control problem (2.8) can be more identified with the latter. Here, one uses the parameter p to control the diffusion tensor such that, together with the source term u, the objectives formulated in the beginning of this chapter are reached. The underlying class of partial differential equation does not model a physical process which has given fixed parameters, it can more be interpreted as an underlying process which needs the proper adjustments via the edge field p.

Some famous results on the control of coefficients of elliptic partial differential equations are motivated in the field of homogenization: One wants to find an optimal mixture of two materials with diffusivity α and β, respectively, such that the corresponding solution of the Dirichlet problem minimizes a certain energy. Using a special type of convergence, the H-convergence, one can show that the energy is minimized by the solution of a limit problem where the diffusivity is no longer a single number, but a full tensor [CK97]. This type of control of the coefficients of a PDE involves no penalty term for the coefficients themselves. If one introduces a cost functional with respect to the parameters, the situation changes. The resulting problems are not necessarily well-posed unless some compactness is introduced, such as a total-variation or $H^{1,r}$-term [BK89, Gut90, CKP98]. The regularization of the parameter identification problem discussed in this work is similar in nature, it can roughly be interpreted as an $H^{1,\infty}$-term plus some compactness. The additional order of differentiability originates from the fact that degeneracies which lead to lower regularity of the solution have to be identified.

2.3.4 Degenerate partial differential equations

There is much literature concerning degenerate partial differential equations of all kinds, so we can only mention some examples to give a rough overview. In many situations, PDEs model a physical process which is known to produce

smooth or, at least, continuous results. The equation itself, however, turns out
to be degenerate in the sense that potentially, some non-linearities can lead to a
degeneracy or coefficients vanish at the boundary or some designated points of
the considered domain. In the former case, it often turns out that the points of
degeneracy are linked with the solution itself and some smoothness can nevertheless be established [DiB93]. The latter case usually produces solutions which
are smooth except, in some situations, for some points, e.g. at the boundary
[FY99].

The setting is different for degeneracies which occur within the domain and
not only on hypersurfaces (or Lebesgue null-sets). A general approach to these
kinds of equation is the theory developed for second-order equations with non-
negative characteristic form [OR73]. A common technique in this field is the
application of elliptic regularization and passage to the limit (also known as
vanishing viscosity). Variational techniques which are related to the approach
in this thesis, however, have also been employed.

The deeper examination of degenerate equations and its solutions can also
motivate the study of weighted Sobolev spaces. In this context, there is also
assumed that the weights, which correspond to the degeneracies in a way, admit
a special structure, i.e. vanish only at the boundary and with a certain power
[Kuf80, Tri95], for instance. Another approach utilizes the class of Muckenhoupt A_p-weights to prove properties of the weighted spaces such as inclusion
properties and analogs to $H = W$, see [Tur00, Kil94, Kil97]. In this work, the
class of degenerate equations is also analyzed with the help of certain weighted
(and directional) Sobolev spaces. But unlike the above approaches, which are
all based on the classical notion of weak or distributional derivatives, a variational formulation is used to define weak weighted and directional derivatives
and to introduce associated Sobolev spaces. It seems that such a technique does
not exist in the literature so far.

Chapter 3

Linear degenerate parabolic equations

This chapter is devoted to the analysis of linear degenerate parabolic equations of the form

$$\frac{\partial y}{\partial t} = \operatorname{div}\left(D^2_{p_1,\dots,p_d} \nabla y^{\mathrm{T}}\right) + u \quad \text{in }]0, T[\times \Omega\ ,$$

$$\frac{\partial y}{\partial_p \nu} = 0 \qquad\qquad\qquad \text{on }]0, T[\times \partial\Omega\ , \qquad\qquad (3.1)$$

$$y = y_0 \qquad\qquad\qquad \text{on } \{0\} \times \Omega$$

where

$$D^2_{p_1,\dots,p_d} = I - \sum_{i=1}^{d} \sigma\big(|p_i|\big) \tfrac{p_i}{|p_i|} \otimes \tfrac{p_i}{|p_i|}\ , \qquad\qquad (3.2\text{a})$$

with $p_1, \dots, p_d$ being vector fields on $]0, T[\times \Omega$ which satisfy

$$p_i \cdot p_{i'} = 0 \quad \text{if}\ \ i \neq i'\ , \quad \text{and}\quad |p_i| \leq 1 \text{ for } i = 1, \dots, d \qquad (3.2\text{b})$$

as well as a suitable $\sigma : [0,1] \to [0,1]$.

This class of parabolic equations is slightly more general than the class associated with the control problem (2.8), since diffusion is potentially reduced in more than one direction. As we will see in this chapter, such a generalization introduces no difficulties and requires only little additional effort when proving existence and uniqueness of solutions in a certain sense. Observe that because of the degeneracy, Equation (3.1) does not fit any more into the usual framework of weak solutions for parabolic equations which are uniformly elliptic with respect to the space-variable. In particular, consideration of solutions in $L^2\big(0, T; H^1(\Omega)\big)$ with time derivative in $L^2\big(0, T; H^1(\Omega)^*\big)$ is not sufficient.

Adapted solution spaces which depend on the parameters $p_1, \ldots, p_d$ have to be introduced. This is of particular importance since one of these parameters is varied in the optimal control problem (2.8) and consequently has an influence on the solution spaces. In this chapter, it is first assumed that the vector fields $p_1, \ldots, p_d$ are fixed. The analysis of the solution operator with respect to these parameters will then be carried out subsequently and in Chapters 4–5.

In the following sections, a notion of weak solutions for (3.1) for a fixed set of vector fields $p_1, \ldots, p_d$ which is measurable and satisfies (3.2b) almost everywhere will be introduced. It is shown that unique weak solutions exist in a suitable space and depend continuously on the data (y_0, u) in $L^2(\Omega) \times L^2(0, T; L^2(\Omega))$. The solution space depends on $p_1, \ldots, p_d$, is embedded in $L^2(0, T; L^2(\Omega))$ and contains functions which may be discontinuous across hypersurfaces. Existence is obtained with Lions' projection-theorem, while uniqueness will be established with the help of Galerkin approximations. Unfortunately, the uniqueness does not hold in the largest space in which weak solutions make sense, we will, however, resolve this issue in Chapter 5.

Moreover, a-priori estimates uniform with respect to all $p_1, \ldots, p_d$ are derived and the weak compactness of solutions for bounded data (y_0, u) and all suitable $(p_1, \ldots, p_d)$ in an appropriate metric space is established.

3.1 The weak formulations and solution spaces

3.1.1 The weak formulations

In the following, let Ω be a domain with Lipschitz boundary. We need two conditions on the parameters of the diffusion tensor $D^2_{p_1, \ldots, p_d}$ to make the subsequent constructions well-defined.

Condition 3.1. *Let* $\sigma : [0, 1] \to [0, 1]$ *be strictly monotone, continuous with* $\sigma(0) = 0$ *as well as* $\sigma(1) = 1$.

Condition 3.2. *Let* $p_1, \ldots, p_d \in L^\infty(]0, T[\times \Omega, \mathbb{R}^d)$ *such that* (3.2b) *holds almost everywhere., i.e.* $\|p_i\|_\infty \le 1$ *for* $i = 1, \ldots, d$ *and*

$$p_i(t, x) \cdot p_{i'}(t, x) = 0 \quad if \quad i \ne i'$$

for almost every $(t, x) \in {]0, T[} \times \Omega$.

We will later restrict ourselves to the case with only one direction of degeneracy, i.e. where $p_1 = p$ and $p_2 = \ldots = p_d = 0$, and pose more conditions on σ. But for now, Conditions 3.1 and 3.2 ensure that the diffusion tensor $D^2_{p_1, \ldots, p_d}$ can be well-defined and is positive semi-definite almost everywhere.

Lemma 3.3. *Let σ satisfy Condition 3.1 and let $p_1, \ldots, p_d$ such that Condition 3.2 holds. Then $D^2_{p_1,\ldots,p_d}$ is measurable, symmetric and positive semi-definite with eigenvalues in $[0,1]$ for almost every $(t,x) \in [0,T] \times \Omega$.*

Proof. According to Condition 3.1, the mapping

$$p \mapsto \begin{cases} \sigma(|p|)\frac{p}{|p|} \otimes \frac{p}{|p|} & p \neq 0 \\ 0 & p = 0 \end{cases}$$

for $p \in \mathbb{R}^d$ with $|p| \leq 1$ is continuous and, of course, bounded. Thus, the superposition operator (or Nemytskii operator) associated with this operation (that is applying the above mapping pointwise almost everywhere) maps measurable $p_i \in L^\infty(]0,T[\times \Omega, \mathbb{R}^d)$ to bounded measurable matrix-functions, i.e. into the space $L^\infty(]0,T[\times \Omega, \mathbb{R}^{d \times d})$, see also [AZ90].

The symmetry of $D^2_{p_1,\ldots,p_d}(t,x)$ can immediately be seen by looking at the defining equation (3.2a). If some $p_i(t,x) \neq 0$ then

$$D^2_{p_1,\ldots,p_d}(t,x)p_i(t,x) = p_i(t,x) - \sum_{i'=1}^{d} \sigma(|p_{i'}(t,x)|) \frac{p_{i'}(t,x)}{|p_{i'}(t,x)|} \otimes \frac{p_{i'}(t,x)}{|p_{i'}(t,x)|} p_i(t,x)$$

$$= (1 - \sigma(|p_i(t,x)|))p_i(t,x)$$

due to the orthogonality in Condition 3.2. Hence $p_i(t,x)$ is an eigenvector with eigenvalue $1 - \sigma(|p_i(t,x)|) \in [0,1]$. If for some $p \neq 0$ we have $p \cdot p_i(t,x) = 0$ for all $i = 1, \ldots, d$, then, by similar arguments

$$D^2_{p_1,\ldots,p_d}(t,x)p = p \ .$$

This considerations lead to d eigenvectors for almost every (t,x) with eigenvalues in $[0,1]$, establishing the positive semi-definiteness. Finally, the above applies to almost every $(t,x) \in]0,T[\times \Omega$, hence the result follows. ◄

It is also useful to know the square root of the diffusion tensor $D^2_{p_1,\ldots,p_d}$.

Corollary 3.4. *In the situation of Lemma 3.3, the square root of $D^2_{p_1,\ldots,p_d}$ is given by*

$$\sqrt{D^2_{p_1,\ldots,p_d}} = D_{p_1,\ldots,p_d} = I - \sum_{i=1}^{d} \left(1 - \sqrt{1 - \sigma(|p_i|)}\right) \frac{p_i}{|p_i|} \otimes \frac{p_i}{|p_i|} \qquad (3.3)$$

Proof. With the proof of Lemma 3.3 in mind, one can easily see that $D^2_{p_1,\ldots,p_d}$ and the right-hand side of (3.3) have the same eigenspaces. Taking the root of the eigenvalues and comparing almost everywhere, i.e.

$$\sqrt{1 - \sigma(|p_i(t,x)|)} = 1 - \left(1 - \sqrt{1 - \sigma(|p_i(t,x)|)}\right) \ ,$$

yields the desired result. ◄

In the following, we assume that $y_0 \in L^2(\Omega)$, $u \in L^2\big(0,T;L^2(\Omega)\big)$ and y is smooth enough for the operations applied to the function to make sense. To derive a weak formulation for (3.1), the usual way is to test with a smooth function $z \in C^\infty([0,T] \times \overline{\Omega})$ with $z(T,x) = 0$ for all $x \in \Omega$. In the following, we will make use of the abbreviations $y(t) = y(t, \cdot)$ and $z(t) = z(t, \cdot)$ etc. Applying the divergence theorem and taking the boundary condition $\frac{\partial y}{\partial_p \nu} = 0$ into account yields

$$\int_0^T \Big\langle \frac{\partial y}{\partial t}(t),\, z(t) \Big\rangle_{L^2} \, dt$$

$$= -\int_0^T \Big\langle \Big(\Big(I - \sum_{i=1}^d \sigma(|p_i(t)|)\frac{p_i(t)}{|p_i(t)|} \otimes \frac{p_i(t)}{|p_i(t)|}\Big)\nabla y(t)^{\mathrm{T}}\Big),\, \nabla z(t)^{\mathrm{T}}\Big\rangle_{L^2} \, dt$$

$$+ \int_0^T \langle u(t),\, z(t)\rangle_{L^2} \, dt$$

which leads to, after integration by parts with respect to t,

$$-\int_0^T \Big\langle y(t),\, \frac{\partial z}{\partial t}(t) \Big\rangle_{L^2} \, dt$$

$$+ \int_0^T \Big\langle \Big(I - \sum_{i=1}^d \sigma\big(|p_i(t)|\big)\frac{p_i(t)}{|p_i(t)|} \otimes \frac{p_i(t)}{|p_i(t)|}\Big)\nabla y(t)^{\mathrm{T}},\, \nabla z(t)\Big\rangle_{L^2} \, dt$$

$$= \int_0^T \langle u(t),\, z(t)\rangle_{L^2} \, dt + \langle y_0,\, z(0)\rangle_{L^2} \,.$$

By taking the square root of the diffusion tensor which is given by (3.3), we obtain the weak formulation

$$\boxed{\begin{aligned}
\int_0^T -\Big\langle y(t),\, \frac{\partial z}{\partial t}(t) \Big\rangle_{L^2} + \big\langle D_{p_1,\dots,p_d}(t)\nabla y(t)^{\mathrm{T}},\, D_{p_1,\dots,p_d}(t)\nabla z(t)^{\mathrm{T}}\big\rangle_{L^2} \, dt \\
= \int_0^T \langle u(t),\, z(t)\rangle_{L^2} \, dt + \langle y_0,\, z(0)\rangle_{L^2} \\
\text{for all } z \in C^\infty([0,T]\times\overline{\Omega}) \text{ with } z(T) = 0
\end{aligned}}$$

(3.4)

Introducing the operator $\mathcal{A}(t) : H^1(\Omega) \to H^1(\Omega)^*$ defined by

$$\mathcal{A}(t)(y)(z) = \int_\Omega D_{p_1,\dots,p_d}(t,x)\nabla y(x)^{\mathrm{T}} \cdot D_{p_1,\dots,p_d}(t,x)\nabla z(x)^{\mathrm{T}} \, dx \,,$$

the weak formulation can also we written as

$$y(0) = y_0 \qquad \text{in } L^2(\Omega)$$

$$\frac{\partial y}{\partial t}(t) + \mathcal{A}(t)\big(y(t)\big) = u(t) \quad \text{in } H^1(\Omega)^* \text{ for a.e. } t \in \,]0,T[\,.$$

(3.5)

Remark 3.5. Linear equations of the type (3.5) with diffusion tensors $D_{p_1,\dots,p_d}$ which are uniformly elliptic are quite well understood (see [DL90] and [LSU68], for example). This corresponds to $|p_i(t,x)| \leq 1 - \varepsilon$ almost everywhere with a given $\varepsilon \in]0, 1]$. The situation is getting more complicated for the general case $|p_i(t,x)| \leq 1$ almost everywhere, since it involves other solution spaces than

$$W(0,T) = \left\{ y \in L^2(0,T;H^1(\Omega)) \ \Big| \ \frac{\partial y}{\partial t} \in L^2(0,T;H^1(\Omega)^*) \right\},$$

$$\|y\|_{W(0,T)} = \sqrt{\int_0^T \|y(t)\|_{H^1}^2 + \left\| \frac{\partial y}{\partial t}(t) \right\|_{(H^1)^*}^2 \ \mathrm{d}t} \ .$$

(3.6)

The ideas from linear functional analysis, however, can still be applied.

Remark 3.6. The theory we present here to solve this degenerate partial differential equation is a variational method for the solution of evolution problems. This theory fits well when further properties of the solution operator such as continuity and differentiability have to be analyzed. A different, but also suitable approach would be to apply the semi-group method with time-variant operators $\mathcal{A}(t)$ to obtain existence and uniqueness.

3.1.2 Abstract stationary solution spaces

In the following, we introduce solution spaces for (3.4) which turn out to be not necessarily a sub- or superset of $W(0,T)$. We follow the general approach which has been developed in [OR73] for this kind of equations. There, existence of certain weak solutions of (3.1) with coefficients which additionally satisfy $p_1,\dots,p_d \in \mathcal{C}^2([0,T] \times \overline{\Omega}, \mathbb{R}^d)$ as well as smooth initial condition y_0 has been established. Here, by using the special structure of the diffusion tensor with respect to x and embedding the ideas in the theoretical framework of evolution equations, we can extend the results to coefficients $p_1,\dots,p_d \in L^\infty(]0,T[\times \Omega, \mathbb{R}^d)$ and initial data $y_0 \in L^2(\Omega)$.

We first study the function space which is associated with weak solutions of (3.1) with time-invariant coefficients $p_1,\dots,p_d$.

Definition 3.7. Let $p_1,\dots,p_d \in L^\infty(\Omega,\mathbb{R}^d)$ be given such that Condition 3.2 (with the time-variable t omitted) is satisfied.

The space $V_{p_1,\dots,p_d}$ is defined as the closure of $H^1(\Omega)$ under the norm

$$\|y\|_{V_{p_1,\dots,p_d}} = \sqrt{\int_\Omega |y|^2 \ \mathrm{d}x + \int_\Omega |D_{p_1,\dots,p_d} \nabla y^{\mathrm{T}}|^2 \ \mathrm{d}x} \ .$$

(3.7a)

The associated scalar product is formally

$$\langle y, z \rangle_{V_{p_1,\dots,p_d}} = \int_\Omega yz \ \mathrm{d}x + \int_\Omega D_{p_1,\dots,p_d} \nabla y^{\mathrm{T}} \cdot D_{p_1,\dots,p_d} \nabla z^{\mathrm{T}} \ \mathrm{d}x \ .$$

(3.7b)

In the case $p_1 = p$, $p_2, \ldots, p_d = 0$, we abbreviate $V_p = V_{p_1,\ldots,p_d}$.

Remark 3.8.

(a) By the positive semi-definiteness of $D_{p_1,\ldots,p_d}$ (see Lemma 3.3) it is immediate that $\|\cdot\|_{V_{p_1,\ldots,p_d}}$ is a well-defined norm, hence $V_{p_1,\ldots,p_d}$ is indeed a Hilbert space.

(b) It is clear by construction that for each representing Cauchy sequence $\{y_k\}$ in $H^1(\Omega)$ with respect to (3.7a) and associated with an element $y \in V_{p_1,\ldots,p_d}$, the sequence $\{D_{p_1,\ldots,p_d}\nabla y_k^{\mathrm{T}}\}$ also forms a Cauchy sequence (in $L^2(\Omega, \mathbb{R}^d)$) and converges to a limit we write formally as $D_{p_1,\ldots,p_d}\nabla y^{\mathrm{T}}$, hence the expression

$$\int_\Omega |D_{p_1,\ldots,p_d}(x)\nabla y(x)^{\mathrm{T}}|^2 \, \mathrm{d}x$$

also makes sense. However, it is important to remark that, in opposition to what the notation $D_{p_1,\ldots,p_d}\nabla y^{\mathrm{T}}$ suggests, it is not clear that it has anything to do with the gradient of a function. For such statements, an appropriate notion of weak differentiability has to be introduced for which its connection to the above setting has to be proven. This is, among other things, topic of Chapter 4.

Likewise, with an analog argumentation and some abuse of notation, we can associate with each $y \in V_{p_1,\ldots,p_d}$ a function $y \in L^2(\Omega)$. Remember that by definition, $V_{p_1,\ldots,p_d}$ contains equivalence classes of Cauchy sequences. We will therefore say that a $y \in L^2(\Omega)$ belongs to $V_{p_1,\ldots,p_d}$ if there is a sequence $\{y_k\}$ in $H^1(\Omega)$ which is a Cauchy sequence with respect to (3.7a) and $y_k \to y$ in $L^2(\Omega)$.

(c) Furthermore, since $H^1(\Omega) \hookrightarrow L^2(\Omega)$ and $\|y\|_2 \le \|y\|_{V_{p_1,\ldots,p_d}}$, we also have $V_{p_1,\ldots,p_d} \hookrightarrow L^2(\Omega)$ and the elements can be identified with L^2-functions. By the usual identification of $L^2(\Omega) = L^2(\Omega)^*$ we additionally have the evolution triple

$$V_{p_1,\ldots,p_d} \hookrightarrow L^2(\Omega) \hookrightarrow V_{p_1,\ldots,p_d}^* \, .$$

Note that, since $V_{p_1,\ldots,p_d}$ is a Hilbert space, both embeddings are dense (see [GGZ74], for instance).

The following example shows that an important model for images is contained in V_p with p depending on the edges.

Proposition 3.9. *Consider a function $y \in L^2(\Omega)$ with the property that there exists a relatively closed null-set $\Gamma \subset \Omega$ and $y \in H^1(\Omega \backslash \Gamma)$.*

Then there exists a $p \in L^\infty(\Omega)$ with $|p| \leq 1$ almost everywhere such that $y \in V_p$ (for each σ fulfilling Condition 3.1).

Proof. Fix a σ satisfying Condition 3.1 which has to be bijective and consider the distance function $v(x) = \inf_{\xi \in \Gamma} \{|x - \xi|\}$ which is always Lipschitz continuous with Lipschitz constant 1, thus there exists a measurable gradient $|\nabla v| \leq 1$ almost everywhere in Ω (by Rademacher's theorem, see [EG92]). Consequently, the edge field

$$p(x) = \sigma^{-1}\left(\max\left\{0, (1 - v(x)^2)\right\}\right)\nabla v(x)^{\mathrm{T}} \qquad (*)$$

satisfies Condition 3.2 and it makes sense to define the associated V_p.

Now choose a smooth cutoff function $\rho : \mathbb{R} \to [0,1]$, i.e. ρ is arbitrarily smooth and satisfies

$$\rho^{(k)}(s) = 0 \quad \text{for all } s \leq 1, k \geq 0$$
$$\rho(s) = 1 \quad \text{for all } s \geq 2$$
$$\rho^{(k)}(s) = 0 \quad \text{for all } s \geq 2, k \geq 1 .$$

Define for each $\varepsilon > 0$ the functions

$$y_\varepsilon(x) = \rho\big(v(x)/\varepsilon\big)y(x)$$

which satisfy $y_\varepsilon \in H^1(\Omega)$ since the support of y_ε has a positive distance from Γ.

We now want to show that $\{y_\varepsilon\}$ is a Cauchy sequence with respect to (3.7a) approximating y as $\varepsilon \to 0$. We only show this for the term involving the gradient since L^2-convergence follows from pointwise convergence and uniform L^2-boundedness of y_ε. First note that the weak derivatives of y_ε are given by

$$\nabla y_\varepsilon(x) = \rho\big(v(x)/\varepsilon\big)\nabla y(x) + \varepsilon^{-1}y(x)\rho'\big(v(x)/\varepsilon\big)\nabla v(x) \qquad (**)$$

and observe that the second term vanishes outside of $\{x \in \Omega \mid \varepsilon < v(x) < 2\varepsilon\}$.

Choose an $\frac{1}{2} > \varepsilon_0 > 0$ and some $0 < \varepsilon_1, \varepsilon_2 \leq \varepsilon_0$. Consider the first term of $(**)$ for ε_1 and ε_2 and estimate

$$\int_\Omega \big|\rho(v/\varepsilon_1) - \rho(v/\varepsilon_2)\big|^2 \big|D_p\nabla y^{\mathrm{T}}\big|^2 \, \mathrm{d}x \leq 4 \int_{\{v \leq \varepsilon_0\}\backslash\Gamma} |\nabla y|^2 \, \mathrm{d}x .$$

Note here, that $|\nabla y|^2 \in L^1(\Omega\backslash\Gamma)$ and $|\{v \leq \varepsilon_0\}| \to 0$ if $\varepsilon_0 \to 0$ (this follows from the assumption that Γ is a relatively closed null-set), so the integral on the right-hand side also becomes arbitrarily small.

Next, evaluate, for x with $v(x) \leq 1$,

$$D_p \nabla v^{\mathrm{T}} = \left(I - \left(1 - \sqrt{1 - \sigma(|p|)}\right)\nabla v^{\mathrm{T}} \otimes \nabla v^{\mathrm{T}}\right)\nabla v^{\mathrm{T}}$$

$$= \sqrt{1 - \sigma\left(\sigma^{-1}(1 - v^2)\right)}\nabla v^{\mathrm{T}} = v\nabla v^{\mathrm{T}}$$

using $(*)$ and keeping in mind that $|\nabla v| = 1$ almost everywhere in Ω. Consequently, the second term of $(**)$ can be estimated by

$$\int_\Omega \left|\varepsilon_k^{-1}\rho'(v/\varepsilon_k)y D_p\nabla v^{\mathrm{T}}\right|^2 \, \mathrm{d}x \leq \|\rho'\|_\infty \int_{\{\varepsilon_k < v < 2\varepsilon_k\}} \left|\varepsilon_k^{-1}vy\right|^2 \, \mathrm{d}x$$

$$\leq 4\|\rho'\|_\infty \int_{\{\varepsilon_k < v < 2\varepsilon_k\}} |y|^2 \, \mathrm{d}x$$

$$\leq 4\|\rho'\|_\infty \int_{\{v < \varepsilon_0\}} |y|^2 \, \mathrm{d}x$$

for $k = 1, 2$. Likewise, the integral on the right-hand side vanishes as $\varepsilon_0 \to 0$.

Putting the above estimates together yields

$$\int_\Omega \left|D_p\nabla(y_{\varepsilon_1} - y_{\varepsilon_2})^{\mathrm{T}}\right|^2 \, \mathrm{d}x \leq 2 \int_\Omega \left|\rho(v/\varepsilon_1) - \rho(v/\varepsilon_2)\right|^2 \left|D_p\nabla y^{\mathrm{T}}\right|^2 \, \mathrm{d}x$$

$$+ 4 \sum_{k=1}^{2} \int_\Omega \left|\varepsilon_k^{-1}\rho'(v/\varepsilon_k)y D_p\nabla v^{\mathrm{T}}\right|^2 \, \mathrm{d}x$$

$$\leq C \int_{\{v < \varepsilon_0\}\backslash\Gamma} |y|^2 + |\nabla y|^2 \, \mathrm{d}x$$

with a suitable $C > 0$. This shows that $\{y_\varepsilon\}$ is a Cauchy sequence and one can say that $y \in V_p$.   ◄

Example 3.10. Let $\Omega' \subset \Omega$ be a relatively open subset with Lipschitz boundary and $y = \chi_{\Omega'}$. Then $y \in H^1(\Omega \backslash \partial\Omega')$ and Proposition 3.9 can be applied with $\Gamma = \partial\Omega'$. Hence, $y \in V_p$ with a properly chosen p. See Figure 3.1 for an illustration of y as well as the construction used in the proof of Proposition 3.9.

Remark 3.11. The image model which is used in Proposition 3.9, i.e. y is in a Sobolev space except for the edges Γ, is corresponding to the model known from the Mumford-Shah minimization problem for image segmentation [MS85, MS89]. See, for example, [Dav05] for a comprehensive study of the problem and its solutions.

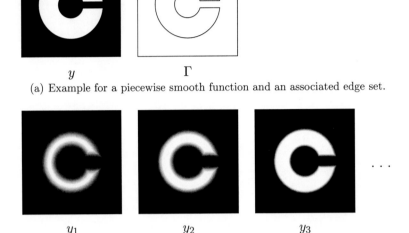

(a) Example for a piecewise smooth function and an associated edge set.

y_1 y_2 y_3
(b) An associated approximating Cauchy sequence.

Figure 3.1: (a) An example for a $y \in V_p$ according to Example 3.10. On the left-hand side the characteristic function of a Lipschitz domain in $\Omega =]0,1[^2$ is depicted. The set Γ, where y has jumps, is shown on the right-hand side. The construction of the edge field p is based on the distance function associated with Γ. (b) Illustration of an approximating sequence for y as constructed in Proposition 3.9. The sequence $\{y_k\}$ in $H^1(\Omega)$ forms a Cauchy sequence with respect to (3.7a).

3.1.3   Abstract time-variant solution spaces

In the usual setting of evolution equations, one would now define, for time invariant $p_1, \ldots, p_d$,

$$W_{p_1,\ldots,p_d}(0,T) = \left\{ y \in L^2\left(0,T;V_{p_1,\ldots,p_d}\right) \ \middle| \ \frac{\partial y}{\partial t} \in L^2\left(0,T;V_{p_1,\ldots,p_d}^*\right) \right\},$$

$$\|y\|_{W_{p_1,\ldots,p_d}} = \sqrt{\int_0^T \|y(t)\|_{V_{p_1,\ldots,p_d}}^2 + \left\|\frac{\partial y}{\partial t}(t)\right\|_{V_{p_1,\ldots,p_d}^*}^2 \, \mathrm{d}t}$$

and get an analog for (3.6) adapted to the degenerate parabolic equation (3.1).

Here, we also want to deal with time-dependent $p_1, \ldots, p_d$, thus we generalize the space $L^2\left(0,T;V_{p_1,\ldots,p_d}\right)$ accordingly. Observe that $H^1(\Omega)$ is, by construction, dense in almost every $V_{p_1(t),\ldots,p_d(t)}$ which might vary along t. Thus we introduce the following space.

Definition 3.12. Let σ and $p_1, \ldots, p_d \in L^\infty(]0, T[\times \Omega, \mathbb{R}^d)$ satisfy Condition 3.1 and Condition 3.2, respectively.

The space $\mathcal{V}_{p_1, \ldots, p_d}$ is defined as the closure of $L^2(0, T; H^1(\Omega))$ under the norm

$$\|y\|_{\mathcal{V}_{p_1, \ldots, p_d}} = \sqrt{\int_0^T \|y(t)\|_{V_{p_1(t), \ldots, p_d(t)}}^2 \, dt} \tag{3.8a}$$

The associated scalar product is, at least formally,

$$\langle y, \, z \rangle_{\mathcal{V}_{p_1, \ldots, p_d}} = \int_0^T \langle y(t), \, z(t) \rangle_{V_{p_1(t), \ldots, p_d(t)}} \, dt \, . \tag{3.8b}$$

Analogously, the case $p_1 = p, \, p_2, \ldots, p_d = 0$ is denoted by $\mathcal{V}_p = \mathcal{V}_{p_1, \ldots, p_d}$.

Remark 3.13.

(a) Again, the definition of the norm makes sense for each measurable $y :$ $]0, T[\rightarrow H^1(\Omega)$ if $p_1, \ldots, p_d$ are measurable in $]0, T[\times \Omega$ and satisfying Condition 3.2. One can verify that $\mathcal{V}_{p_1, \ldots, p_d}$ is a Hilbert space in this situation.

(b) For each $y \in \mathcal{V}_{p_1, \ldots, p_d}$ we have that $y(t) \in V_{p_1(t), \ldots, p_d(t)}$ for almost every $t \in]0, T[$. This can be seen with an argument analog to the one utilized in the proof of the Fischer-Riesz theorem. For a representing Cauchy sequence $\{y_k\}$ in $L^2(0, T; H^1(\Omega))$ we know that a subsequence (not relabeled) satisfies $\|y_{k+1} - y_k\|_{\mathcal{V}_{p_1, \ldots, p_d}} \leq 2^{-k}$. With

$$u_k(t) = \sum_{l=1}^k \|y_{l+1}(t) - y_l(t)\|_{V_{p_1(t), \ldots, p_d(t)}}$$

it follows that

$$\int_0^T \lim_{k \to \infty} u_k(t)^2 \, dt \leq \liminf_{k \to \infty} \int_0^T u_k(t)^2 \, dt$$

$$\leq \left(\sum_{l=1}^\infty \|y_{l+1}(t) - y_l(t)\|_{\mathcal{V}_{p_1, \ldots, p_d}} \right)^2 < \infty$$

which in turn implies $\lim_{k \to \infty} u_k(t) < \infty$ for almost every $t \in]0, T[$. Consequently, by definition of u_k, $\{y_k(t)\}$ has to be a Cauchy sequence with respect to the $V_{p_1(t), \ldots, p_d(t)}$-norm almost everywhere. Moreover, each representing Cauchy sequence $\{y_k\}$ with respect to the $\mathcal{V}_{p_1, \ldots, p_d}$-norm yields equivalent Cauchy sequences $\{y_k(t)\}$ with respect to the $V_{p_1(t), \ldots, p_d(t)}$-norm almost everywhere, hence one can say that $y(t) \in V_{p_1(t),\ldots,p_d(t)}$ for almost every $t \in \,]0, T[$ as well as

$$\|y\|^2_{\mathcal{V}_{p_1,\ldots,p_d}} = \int_0^T \|y(t)\|^2_{V_{p_1(t),\ldots,p_d(t)}} \, dt \, .$$

(c) By construction, we have the dense embeddings

$$L^2(0, T; H^1(\Omega)) \hookrightarrow \mathcal{V}_{p_1,\ldots,p_d} \hookrightarrow L^2(0, T; L^2(\Omega))$$

with constant 1, so it immediately follows that the dual space $\mathcal{V}^*_{p_1,\ldots,p_d}$ satisfies

$$L^2(0, T; L^2(\Omega)) \hookrightarrow \mathcal{V}^*_{p_1,\ldots,p_d} \hookrightarrow L^2(0, T; H^1(\Omega)^*)$$

with dense embeddings. With the above result, it is then clear that $\mathcal{V}^*_{p_1,\ldots,p_d}$ is given by the closure of $L^2(0, T; L^2(\Omega))$ under the norm

$$\|y\|_{\mathcal{V}^*_{p_1,\ldots,p_d}} = \sqrt{\int_0^T \|y(t)\|^2_{V^*_{p_1(t),\ldots,p_d(t)}} \, dt} \, ,$$

leading to the "evolution triple"

$$\mathcal{V}_{p_1,\ldots,p_d} \hookrightarrow L^2(0, T; L^2(\Omega)) \hookrightarrow \mathcal{V}^*_{p_1,\ldots,p_d} \, .$$

As a side result, we also have for each $y \in \mathcal{V}^*_{p_1,\ldots,p_d}$ that $y(t) \in V^*_{p_1(t),\ldots,p_d(t)}$ for almost every $t \in \,]0, T[$.

With this in mind we can introduce the natural space associated with weak solutions of (3.1).

Definition 3.14. Let $p_1, \ldots, p_d$ given as in Definition 3.12. Define the space $W_{p_1,\ldots,p_d}(0, T)$ as the closure

$$W_{p_1,\ldots,p_d}(0, T) = \overline{\left\{ y \in \mathcal{AC}(0, T; H^1(\Omega)) \ \middle| \ \|y\|_{W_{p_1,\ldots,p_d}} < \infty \right\}}^{\|\cdot\|_{W_{p_1,\ldots,p_d}}},$$

$$\|y\|_{W_{p_1,\ldots,p_d}} = \sqrt{\|y\|^2_{\mathcal{V}_{p_1,\ldots,p_d}} + \left\| \frac{\partial y}{\partial t} \right\|^2_{\mathcal{V}^*_{p_1,\ldots,p_d}}} \, .$$

$$(3.9)$$

Moreover, define $\bar{W}_{p_1,\ldots,p_d}(0, T)$ as the space

$$\bar{W}_{p_1,\ldots,p_d}(0, T) = \left\{ y \in \mathcal{V}_{p_1,\ldots,p_d} \ \middle| \ \frac{\partial y}{\partial t} \in \mathcal{V}^*_{p_1,\ldots,p_d} \right\}$$

also equipped with the $W_{p_1,\ldots,p_d}(0, T)$-norm.

As before, $W_p(0, T)$ and $\bar{W}_p(0, T)$ denote the spaces associated with $p_1 = p$ and $p_2 = \ldots = p_d = 0$, respectively.

Remark 3.15.

The statement that $\frac{\partial y}{\partial t} \in \mathcal{V}^*_{p_1,\dots,p_d}$ has to be understood in the following sense: The distributional derivative $\frac{\partial y}{\partial t}$ is the element in $\mathcal{V}^*_{p_1,\dots,p_d}$ satisfying

$$-\int_0^T \langle y(t),\, z_2\rangle_{L^2} \frac{\partial z_1}{\partial t}(t)\, \mathrm{d}t = \left\langle \frac{\partial y}{\partial t},\, z\right\rangle_{\mathcal{V}^* \times \mathcal{V}}$$

for each $z(t,x) = z_1(t)z_2(x)$ with $z_1 \in C_0^\infty(]0,T[)$ and $z_2 \in H^1(\Omega)$, abbreviating the dual pairing between $\mathcal{V}^*_{p_1,\dots,p_d}$ and $\mathcal{V}_{p_1,\dots,p_d}$ by $\langle\,\cdot\,,\,\cdot\,\rangle_{\mathcal{V}^*\times\mathcal{V}}$.

Note that, by the density of the embedding $\mathcal{AC}(0,T;H^1(\Omega)) \hookrightarrow \mathcal{V}_{p_1,\dots,p_d}$, the above choice of test functions z is sufficient to determine $\frac{\partial y}{\partial t}$ uniquely.

The construction of $W_{p_1,\dots,p_d}(0,T)$ allows us to draw conclusions which are familiar analogs to well-known statements regarding solution spaces for parabolic equations.

Proposition 3.16. *We have*

1. $W_{p_1,\dots,p_d}(0,T) \hookrightarrow \mathcal{C}(0,T;L^2(\Omega))$ *with an embedding constant which is uniformly bounded for all* $p_1,\dots,p_d$,

2. *for all* $y,z \in W_{p_1,\dots,p_d}(0,T)$ *that the function* $t \mapsto \langle y(t),\, z(t)\rangle_{L^2}$ *is absolutely continuous with*

$$\frac{\partial}{\partial t}\langle y(t),\, z(t)\rangle_{L^2} = \left\langle \frac{\partial y}{\partial t}(t),\, z(t)\right\rangle_{V^*(t)\times V(t)} + \left\langle \frac{\partial z}{\partial t}(t),\, y(t)\right\rangle_{V^*(t)\times V(t)}$$

a.e. in $]0,T[$, *where* $\langle\,\cdot\,,\,\cdot\,\rangle_{V(t)^*\times V(t)}$ *denotes the dual pairing between* $V^*_{p_1(t),\dots,p_d(t)}$ *and* $V_{p_1(t),\dots,p_d(t)}$.

Proof. The arguments for proving the statements for $W_{p_1,\dots,p_d}(0,T)$ do not differ too much from the ones used in the case $W(0,T)$ according to (3.6), see [Sho96], for example, but we repeat them here for convenience.

First extend the vector fields $p_1,\dots,p_d$ and function $y \in W_{p_1,\dots,p_d}(0,T)$ to $[-T,T]$ by

$$\bar{y}(t) = \begin{cases} \frac{t+T}{T}y(-t) & \text{if } t < 0 \\ y(t) & \text{if } t \geq 0, \end{cases} \qquad \bar{p}_i(t) = \begin{cases} p_i(-t) & \text{if } t < 0 \\ p_i(t) & \text{if } t \geq 0. \end{cases}$$

The spaces $\mathcal{V}_{\bar{p}_1,\dots,\bar{p}_d}$ and $W_{\bar{p}_1,\dots,\bar{p}_d}(-T,T)$ can then be defined analogously to (3.8) and (3.9), respectively. The extension mapping $y \mapsto \bar{y}$ is of course linear with norm estimates

$$\|\bar{y}\|^2_{\mathcal{V}_{\bar{p}_1,\dots,\bar{p}_d}} \le 2\|y\|^2_{\mathcal{V}_{p_1,\dots,p_d}}$$

$$\left\|\frac{\partial \bar{y}}{\partial t}\right\|^2_{\mathcal{V}^*_{\bar{p}_1,\dots,\bar{p}_d}} \le \left\|\frac{\partial y}{\partial t}\right\|^2_{\mathcal{V}^*_{p_1,\dots,p_d}} + \int_0^T \left\|\frac{y(t)}{T} + \frac{T-t}{T}\frac{\partial y}{\partial t}(t)\right\|^2_{V^*_{p_1(t),\dots,p_d(t)}} dt$$

$$\le 3\left\|\frac{\partial y}{\partial t}\right\|^2_{\mathcal{V}^*_{p_1,\dots,p_d}} + \frac{2}{T^2}\|y\|^2_{\mathcal{V}_{p_1,\dots,p_d}}.$$

Moreover, it is well-defined mapping $W_{p_1,\dots,p_d}(0,T) \to W_{\bar{p}_1,\dots,\bar{p}_d}(-T,T)$ on the dense subset $\{y \in W_{p_1,\dots,p_d}(0,T) \mid y \in \mathcal{AC}(0,T;H^1(\Omega))\}$. It follows that the extension is continuous on the whole space.

Now we can obtain an estimate for $\|y(t)\|^2_2$ for all $t \in]0,T[$. Let $y \in \mathcal{AC}(0,T;H^1(\Omega))$ with $\|y\|_{W_{p_1,\dots,p_d}} < \infty$. Then $t \mapsto \|\bar{y}(t)\|^2_2$ is absolutely continuous and, consequently, differentiable almost everywhere in $[-T,T]$ with derivative

$$\frac{\partial}{\partial t}\|\bar{y}(t)\|^2_2 = 2\langle\frac{\partial \bar{y}}{\partial t}(t), \bar{y}(t)\rangle_{L^2}. \tag{$*$}$$

Furthermore, for $t \in [0,T]$,

$$\frac{1}{2}\|y(t)\|^2_2 \le \int_{-T}^t \left|\langle\frac{\partial \bar{y}}{\partial t}(s), \bar{y}(s)\rangle_{L^2}\right| ds$$

$$\le \int_{-T}^T \left\|\frac{\partial \bar{y}}{\partial t}(s)\right\|_{V^*_{\bar{p}_1(s),\dots,\bar{p}_d(s)}} \|\bar{y}(s)\|_{V_{\bar{p}_1(s),\dots,\bar{p}_d(s)}} ds$$

$$\le \frac{1}{2}\left(\left\|\frac{\partial \bar{y}}{\partial t}\right\|^2_{\mathcal{V}^*_{\bar{p}_1,\dots,\bar{p}_d}} + \|\bar{y}\|^2_{\mathcal{V}_{\bar{p}_1,\dots,\bar{p}_d}}\right)$$

$$\le C\|y\|^2_{W_{p_1,\dots,p_d}}$$

with the embedding constant obtained above. Note that C does not depend on $p_1,\dots,p_d$. Taking the maximum over all $t \in [0,T]$ leads to a norm estimate for the embedding $W_{p_1,\dots,p_d}(0,T) \hookrightarrow \mathcal{C}(0,T;L^2(\Omega))$. Finally observe that the above y form a dense subset of $W_{p_1,\dots,p_d}(0,T)$, establishing the first statement.

The second statement follows analogously to $(*)$ with a density argument. Note that if $y, z \in \mathcal{AC}(0,T;H^1(\Omega)) \cap W_{p_1,\dots,p_d}(0,T)$, then $\langle y(t), z(t)\rangle_{L^2}$ is also absolutely continuous with derivative

$$\frac{\partial}{\partial t}\langle y(t), z(t)\rangle_{L^2} = \langle\frac{\partial y}{\partial t}(t), z(t)\rangle_{L^2} + \langle\frac{\partial z}{\partial t}(t), y(t)\rangle_{L^2}$$

$$= \langle\frac{\partial y}{\partial t}(t), z(t)\rangle_{V(t)^* \times V(t)} + \langle\frac{\partial z}{\partial t}(t), y(t)\rangle_{V(t)^* \times V(t)},$$

almost everywhere in $]0, T[$. For general $y, z \in W_{p_1,\ldots,p_d}(0, T)$, the above identity can be extended by density. Note that, due to Remark 3.13, the right-hand side still makes sense. Also, $\frac{\partial}{\partial t}\langle y_k(\cdot), z_k(\cdot)\rangle_{L^2}$ converges in $L^1(]0, T[)$ if $y_k, z_k \in W_{p_1,\ldots,p_d}(0, T)$ converge, hence the limit function is still absolutely continuous in $[0, T]$.   ◀

Finally, with the notion of the solution space $\mathcal{V}_{p_1,\ldots,p_d}$, it is possible to derive a weak formulation in a functional-analytic setting.

Definition 3.17. Let σ and $p_1, \ldots, p_d$ fulfill Condition 3.1 and Condition 3.2, respectively. Let $u \in \mathcal{V}_{p_1,\ldots,p_d}^*$ and $y_0 \in L^2(\Omega)$ be given.

Then a $y \in \mathcal{V}_{p_1,\ldots,p_d}$ is called a *weak solution* of problem (3.1), if for each $z \in W_{p_1,\ldots,p_d}(0, T)$ with $z(T) = 0$ the equation

$$-\left\langle \frac{\partial z}{\partial t}, y \right\rangle_{\mathcal{V}^* \times \mathcal{V}} + \langle D_{p_1,\ldots,p_d} \nabla y^{\mathrm{T}}, D_{p_1,\ldots,p_d} \nabla z^{\mathrm{T}} \rangle_{L^2}$$
$$= \langle u, z \rangle_{\mathcal{V}^* \times \mathcal{V}} + \langle y_0, z(0) \rangle_{L^2} \quad (3.10)$$

is satisfied. Here, $\langle \cdot, \cdot \rangle_{\mathcal{V}^* \times \mathcal{V}}$ denotes the dual pairing between $\mathcal{V}_{p_1,\ldots,p_d}^*$ and $\mathcal{V}_{p_1,\ldots,p_d}$.

Remark 3.18. As one can easily see, each weak solution in the above sense has a distributional derivative $\frac{\partial y}{\partial t}$ in $\mathcal{V}_{p_1,\ldots,p_d}$, hence $y \in \bar{W}_{p_1,\ldots,p_d}(0, T)$. It is, however, not clear at the moment, whether $y \in W_{p_1,\ldots,p_d}(0, T)$ since we only have $W_{p_1,\ldots,p_d}(0, T) \hookrightarrow \bar{W}_{p_1,\ldots,p_d}(0, T)$.

This has consequences for the uniqueness of solutions. As it will be shown in the following, uniqueness can only be obtained for $W_{p_1,\ldots,p_d}(0, T)$ (since we are able to apply Proposition 3.16), so if the spaces do not coincide, we cannot exclude the possibility of the existence of a weak solution in the sense of Definition 3.17 in the set $\bar{W}_{p_1,\ldots,p_d}(0, T) \backslash W_{p_1,\ldots,p_d}(0, T)$. This issue can be resolved by proving appropriate density theorems for $\bar{W}_{p_1,\ldots,p_d}(0, T)$, an approach which is carried out in Chapter 5.

3.2   Existence and Uniqueness

This section is devoted to prove existence and uniqueness of solutions of (3.1) in the weak sense of Definition 3.17. Basically, in the context of the variational method, there are two different approaches for solving linear evolution equations. One is the Galerkin method, i.e. construct a sequence of discretized solutions which are finite-dimensional with respect to the space-variable and show convergence. The other method is based on variational or weak formulations such as (3.10). We will use the latter approach to prove existence, but employ the Galerkin method to obtain uniqueness in a certain sense. For existence, a generalized version of the Lax-Milgram theorem, called Lions' projection-theorem, see [Lio61], will be applied. Recall that in the situation of parabolic equations, the Lax-Milgram theorem cannot be applied: The associated bilinear form is not both continuous and coercive with respect to both variables in a common Hilbert space. For "inversion", however, one can drop the continuity assumption with respect to one variable. This is roughly what Lions' projection-theorem states. Having proven the existence, the uniqueness of the solution then follows by an adaptation of the usual monotonicity argument.

Let us state, for convenience, a version of Lions' projection-theorem for solutions of a linear equation, the proof of which can also be found in [Sho96].

Theorem 3.19. *Let H be a Hilbert space, X be a normed space and $E : H \times X \to \mathbb{R}$ a bilinear form such that $E(\cdot, z) \in H^*$ for each $z \in X$.*

1. *Then, for each $f \in X^*$, there exists a solution $y \in H$ of the variational problem*

$$E(y, z) = f(z) \quad \text{for all } z \in X ,$$

if and only if there is a $c > 0$ such that

$$\inf_{\|z\|_X = 1} \sup_{\|y\|_H \leq 1} |E(y, z)| \geq c .$$

2. *There is a solution y of the variational problem which depends linearly and continuously on f with the a-priori estimate*

$$\|y\|_H \leq c^{-1} \|f\|_{X^*} .$$

In order to apply this theorem, we will make use of the following corollary.

Corollary 3.20. *If, in the situation of Theorem 3.19, $X \hookrightarrow H$ with constant c_1 and E is X-coercive, i.e.*

$$E(z, z) \geq c_2 \|z\|_X^2 \quad \text{for all } z \in X ,$$

then Theorem 3.19 can be applied with constant $c = c_2/c_1$.

Remark 3.21. Note that the particular solution in Theorem 3.19 which fulfills $\|y\|_H \leq c^{-1}\|f\|_{X^*}$ is also the minimum-norm solution in H.

Remember that for the proof of Theorem 3.19, an injective operator $A : X \to H$ is constructed by the definition $\langle y, Az \rangle_H = E(y, z)$. Its inverse A^{-1} is continuous and can be extended to $\overline{\text{rg}(A)} \subset H$, yielding the particular solution y by the construction $y = \left(\overline{A^{-1}P_{\overline{\text{rg}(A)}}}\right)^* f$.

Now observe that each $z \in \text{rg}(A)^\perp$ satisfies

$$\langle y, z \rangle_H = \langle \left(\overline{A^{-1}P_{\overline{\text{rg}(A)}}}\right)^* f, z \rangle_H = \langle f, \overline{A^{-1}P_{\overline{\text{rg}(A)}}}z \rangle_{X^* \times X} = 0$$

meaning that $y \in \overline{\text{rg}(A)}$. Moreover, $\bar{y} \in \text{rg}(A)^\perp$ is equivalent to

$$\langle \bar{y}, Az \rangle_H = E(\bar{y}, z) = 0 \quad \text{for all } z \in X \ ,$$

thus, two solutions for the same data f only differ in $\text{rg}(A)^\perp$. But implies that y constructed above is the solution of the variational problem with the minimal H-norm.

The existence of a weak solution in the space $\bar{W}_{p_1,\dots,p_d}(0, T)$ can now be proven.

Proposition 3.22. *Let $\sigma, p_1, \dots, p_d, u, y_0$ be given according to Definition 3.17.*
Then there exists a weak solution y of (3.1) in the sense of Definition 3.17. Moreover, $y \in \bar{W}_{p_1,\dots,p_d}(0, T)$ and there exists a $C > 0$, only depending on T, such that

$$\|y\|^2_{\bar{W}_{p_1,\dots,p_d}} \leq C\left(\|u\|^2_{\mathcal{V}^*_{p_1,\dots,p_d}} + \|y_0\|^2_2\right) \ .$$

Proof. The usual approach to show existence of a weak solution is to switch to the exponentially shifted problem. We therefore pose the modified variational problem: Set $\tilde{u}(t) = e^{-t}u(t)$ and find a $\tilde{y} \in \mathcal{V}_{p_1,\dots,p_d}$ such that for each $z \in W_{p_1,\dots,p_d}(0, T)$ with $z(T) = 0$ the equation

$$-\langle \frac{\partial z}{\partial t}, \tilde{y} \rangle_{\mathcal{V}^* \times \mathcal{V}} + \langle \tilde{y}, z \rangle_{L^2} + \langle D_{p_1,\dots,p_d}\nabla \tilde{y}^{\mathrm{T}}, D_{p_1,\dots,p_d}\nabla z^{\mathrm{T}} \rangle_{L^2}$$
$$= \langle \tilde{u}, z \rangle_{\mathcal{V}^* \times \mathcal{V}} + \langle y_0, z(0) \rangle_{L^2} \ , \quad (*)$$

is satisfied.

Introduce the normed space

$$X = \{z \in W_{p_1,\dots,p_d}(0, T) \mid z(T) = 0\}$$
$$\|z\|^2_X = \|z\|^2_{\mathcal{V}_{p_1,\dots,p_d}} + \|z(0)\|^2_2$$

and the Hilbert space $H = V_{p_1,\ldots,p_d}$. It is immediate that $X \hookrightarrow H$ with embedding constant $c_1 = 1$. Let us check that the bilinear form $E : H \times X \to \mathbb{R}$ defined by

$$E(\tilde{y}, z) = -\left\langle \frac{\partial z}{\partial t}, \tilde{y} \right\rangle_{V^* \times V} + \langle \tilde{y}, z \rangle_{V_{p_1,\ldots,p_d}}$$

is X-coercive. But this is clear since

$$E(z, z) = -\int_0^T \left\langle \frac{\partial z}{\partial t}(t), z(t) \right\rangle_{V(t)^* \times V(t)} \mathrm{d}t$$

$$+ \int_0^T \|z(t)\|_2^2 + \|D_{p_1(t),\ldots,p_d(t)} \nabla z(t)^{\mathrm{T}}\|_2^2 \, \mathrm{d}t$$

$$= \frac{\|z(0)\|^2 - \|z(T)\|^2}{2} + \|z\|_{V_{p_1,\ldots,p_d}}^2 \geq \frac{\|z\|_X^2}{2}$$

using that $z(T) = 0$ as well as the formula for $\frac{1}{2}\frac{\partial}{\partial t}\|z(t)\|_2^2$ derived in Proposition 3.16. Note that the coercivity constant is $c_2 = 1/2$.

At last, it is necessary to see that the right-hand side of $(*)$ is in X^*. By definition,

$$f(z) = \langle \tilde{u}, z \rangle_{V^* \times V} + \langle y_0, z \rangle_{L^2},$$

hence

$$|f(z)| \leq \left(\|\tilde{u}\|_{V^*}^2 + \|y_0\|_2^2\right)^{1/2} \left(\|z\|_V^2 + \|z(0)\|_2^2\right)^{1/2} = C_1 \|z\|_X$$

with $C_1 = \left(\|\tilde{u}\|_{V^*}^2 + \|y_0\|_2\right)^{1/2}$. Corollary 3.20 then yields the existence of a $\tilde{y} \in H = V_{p_1,\ldots,p_d}$ such that for each $z \in X$ the equation

$$E(\tilde{y}, z) = f(z)$$

is satisfied. By definition, $\tilde{y}$ also solves $(*)$. Note that we can assume that the a-priori norm estimate holds which reads as

$$\|\tilde{y}\|_{V_{p_1,\ldots,p_d}} \leq \frac{c_1}{c_2}\|f\|_{X^*} \leq 2C_1 = 2\left(\|\tilde{u}\|_{V^*}^2 + \|y_0\|_2^2\right)^{1/2}.$$

Moreover, the solution $\tilde{y}$ is an element of $\bar{W}_{p_1,\ldots,p_d}(0,T)$. To see this, test with a $z \in W_{p_1,\ldots,p_d}(0,T)$ whose support with respect to t is compact in $]0,T[$ and use the variational formulation $(*)$ to verify that $\frac{\partial \tilde{y}}{\partial t} \in V_{p_1,\ldots,p_d}^*$:

$$\left| -\left\langle \frac{\partial z}{\partial t}, \tilde{y} \right\rangle_{V^* \times V} \right| = \left| \langle \tilde{u}, z \rangle_{V^* \times V} - \langle \tilde{y}, z \rangle_{V_{p_1,\ldots,p_d}} \right|$$

$$\leq \sqrt{2}\left(\|\tilde{u}\|_{V_{p_1,\ldots,p_d}^*}^2 + \|\tilde{y}\|_{V_{p_1,\ldots,p_d}}^2\right)^{1/2} \|z\|_{V_{p_1,\ldots,p_d}}$$

$$\leq \sqrt{2}\left(5\|\tilde{u}\|_{V_{p_1,\ldots,p_d}^*}^2 + 4\|y_0\|_2^2\right)^{1/2} \|z\|_{V_{p_1,\ldots,p_d}}$$

This also shows the a-priori estimate

$$\|\tilde{y}\|_{W_{p_1,\dots,p_d}} \le \left(14\|\tilde{u}\|_{\mathcal{V}^*_{p_1,\dots,p_d}}^2 + 12\|y_0\|_2^2\right)^{1/2} \le \sqrt{14}\left(\|\tilde{u}\|_{\mathcal{V}^*_{p_1,\dots,p_d}}^2 + \|y_0\|_2^2\right)^{1/2}$$

The next step is to prove that $y(t) = e^t \tilde{y}(t)$ solves (3.10). Note the identity

$$-\int_0^T \langle e^t \tilde{y}(t), \frac{\partial z}{\partial t}(t)\rangle_{L^2} \, dt = -\int_0^T \langle \tilde{y}(t), \frac{\partial}{\partial t}(e^t z(t))\rangle_{L^2} \, dt$$
$$+\int_0^T \langle \tilde{y}(t), e^t z(t)\rangle_{L^2} \, dt$$

which gives, by denoting $\tilde{z}(t) = e^t z(t)$,

$$-\langle \frac{\partial z}{\partial t}, y\rangle_{\mathcal{V}^* \times \mathcal{V}} = -\langle \frac{\partial \tilde{z}}{\partial t}, \tilde{y}\rangle_{\mathcal{V}^* \times \mathcal{V}} + \langle \tilde{y}, \tilde{z}\rangle_{L^2}$$
$$= \langle \tilde{u}, \tilde{z}\rangle_{\mathcal{V}^* \times \mathcal{V}} + \langle y_0, \tilde{z}(0)\rangle_{L^2} - \langle \tilde{y}, \tilde{z}\rangle_{\mathcal{V}_{p_1,\dots,p_d}} + \langle \tilde{y}, \tilde{z}\rangle_{L^2}.$$

Reordering this and using $\langle \tilde{y}, \tilde{z}\rangle_{\mathcal{V}_{p_1,\dots,p_d}} = \langle y, z\rangle_{\mathcal{V}_{p_1,\dots,p_d}}$ as well as $\langle \tilde{y}, \tilde{z}\rangle_{L^2} = \langle y, z\rangle_{L^2}$ yields the desired equation

$$-\langle \frac{\partial z}{\partial t}, y\rangle_{\mathcal{V}^* \times \mathcal{V}} + \langle D_{p_1,\dots,p_d}\nabla y^{\mathrm{T}}, D_{p_1,\dots,p_d}\nabla z^{\mathrm{T}}\rangle_{L^2} = \langle u, z\rangle_{\mathcal{V}^* \times \mathcal{V}} + \langle y_0, z(0)\rangle_{L^2}.$$

Taking the exponential shifts $u \mapsto \tilde{u}$ and $\tilde{y} \mapsto y$ into account moreover leads to the a-priori estimates

$$\|y\|_{\mathcal{V}_{p_1,\dots,p_d}} \le 2e^T \left(\|u\|_{\mathcal{V}^*_{p_1,\dots,p_d}}^2 + \|y_0\|_2^2\right)^{1/2},$$
$$\left\|\frac{\partial y}{\partial t}\right\|_{\mathcal{V}^*_{p_1,\dots,p_d}} \le \sqrt{2}e^T \left(14\|u\|_{\mathcal{V}^*_{p_1,\dots,p_d}}^2 + 12\|y_0\|_2^2\right)^{1/2},$$
$$\|y\|_{\bar{W}_{p_1,\dots,p_d}} \le \sqrt{32}e^T \left(\|u\|_{\mathcal{V}^*_{p_1,\dots,p_d}}^2 + \|y_0\|_2^2\right)^{1/2},$$

the latter being the claimed statement.   ◀

As already suggested in Remark 3.18, the weak solution is not necessarily unique in $\bar{W}_{p_1,\dots,p_d}(0,T)$. It is nevertheless possible to establish uniqueness in the potentially smaller space $W_{p_1,\dots,p_d}(0,T)$. For this purpose, the tool of Galerkin approximation is employed. It will moreover turn out that the sequence of Galerkin approximations converges weakly to the solution given by Proposition 3.22.

Lemma 3.23. *In the situation of Proposition 3.22, there exists a Galerkin basis* $\varphi_1, \varphi_2, \ldots \in H^1(\Omega)$ *and a sequence* $\{y_l\}$ *in* $W_{p_1,\ldots,p_d}(0,T)$ *such that for each* $l \geq 1$ *the function* $y_l : [0,T] \to \text{span}\{\varphi_1, \ldots, \varphi_l\}$ *satisfies*

$$\left\langle \frac{\partial y_l}{\partial t}(t), z \right\rangle_{L^2} = -\langle D_{p_1(t),\ldots,p_d(t)} \nabla y_l(t)^{\mathrm{T}}, D_{p_1(t),\ldots,p_d(t)} \nabla z^{\mathrm{T}} \rangle_{L^2}$$

$$+ \langle u(t), z \rangle_{V(t)^* \times V(t)} \tag{3.11}$$

$$\langle y_l(0), z \rangle_{L^2} = \langle y_0, z \rangle_{L^2}$$

for each $z \in \text{span}\{\varphi_1, \ldots, \varphi_l\}$ *and almost every* $t \in [0,T]$. *Moreover, all* y_l *obey the estimate*

$$\|y_l\|_{W_{p_1,\ldots,p_d}} \leq C\big(\|u\|^2_{V^*_{p_1,\ldots,p_d}} + \|y_0\|^2_2\big)^{1/2}$$

with a suitable $C > 0$ *and, for a subsequence,* $y_l \rightharpoonup y$ *in* $W_{p_1,\ldots,p_d}(0,T)$ *with* y *being a weak solution of* (3.1) *with data* u *and* y_0.

Proof. First, we choose an orthogonal basis $\varphi_1, \varphi_2, \ldots$ of $H^1(\Omega)$ which is simultaneously an orthonormal basis of $L^2(\Omega)$. Such a basis exists [Eva98] and is also a Galerkin basis, since by construction, $L^2\big(0,T;H^1(\Omega)\big)$ is dense in $\mathcal{V}_{p_1,\ldots,p_d}$ and consequently

$$\overline{\bigcup_{l \geq 1} L^2\big(0,T;\text{span}\{\varphi_1, \ldots, \varphi_l\}\big)}\bigg|_{\|\cdot\|_{\mathcal{V}_{p_1,\ldots,p_d}}} = \mathcal{V}_{p_1,\ldots,p_d}.$$

Furthermore,

$$\overline{\bigcup_{l \geq 1} \mathcal{AC}\big(0,T;\text{span}\{\varphi_1, \ldots, \varphi_l\}\big) \cap W_{p_1,\ldots,p_d}(0,T)}\bigg|_{\|\cdot\|_{W_{p_1,\ldots,p_d}}} = W_{p_1,\ldots,p_d}(0,T)$$

again by construction, also see Definition 3.14. This density ensures that it is sufficient to test (3.11) with each φ_m and let $l \to \infty$ as we will see later.

In prior to this, we note that the approximate problems (3.11) have solutions as claimed. Writing

$$y_l(t) = \sum_{m=1}^{l} \zeta^l_m(t)\varphi_m,$$

the l-th problem becomes

$$\frac{\partial \zeta^l}{\partial t} = -M^l(t)\zeta^l + \bar{u}^l(t), \quad \zeta^l(0) = \bar{y}^l_0$$

where

$$M^l(t)_{m,m'} = \langle D_{p_1(t),\ldots,p_d(t)} \nabla \varphi_m^{\mathrm{T}}, D_{p_1(t),\ldots,p_d(t)} \nabla \varphi_{m'}^{\mathrm{T}} \rangle_{L^2},$$

$$\bar{u}^l(t)_m = \langle u(t), \varphi_m \rangle_{V(t)^* \times V(t)}, \quad \bar{y}^l_{0,m} = \langle y_0, \varphi_m \rangle_{L^2},$$

which is a finite-dimensional time-variant, inhomogeneous linear ordinary differential equation with $L^1(]0,T[,\mathbb{R}^l)$-data. In particular, one can estimate the coefficients of the matrix $M^l(t)$ as follows:

$$|M^l(t)_{m,m'}| \leq \|D_{p_1(t),...,p_d(t)}\nabla\varphi_m^{\mathrm{T}}\|_{L^2}\|D_{p_1(t),...,p_d(t)}\nabla\varphi_{m'}^{\mathrm{T}}\|_{L^2}$$

$$\leq \|\varphi_m\|_{V_{p_1(t),...,p_d(t)}}\|\varphi_{m'}\|_{V_{p_1(t),...,p_d(t)}} \leq \|\varphi_m\|_{H^1}\|\varphi_{m'}\|_{H^1} .$$

Hence, a unique solution ζ^l exists in the sense of Carathéodory which is absolutely continuous with values in $\mathbb{R}^l$, leading to $y_l \in AC(0,T;H^1(\Omega))$ since y_l is a linear combination of $H^1(\Omega)$-functions depending on t.

Moreover, we can deduce a $W_{p_1,...,p_d}(0,T)$-estimate which only depends on the data u and y_0. This is done in analogy to Proposition 3.22, but this time, we also have to make sure that the estimates are independent of l. An estimate for all $\|y_l\|^2_{V_{p_1,...,p_d}}$ then follows from the following: First, observe that we are able to test (3.11) with time-variant functions $\bar{z} \in L^2(0,T;\mathrm{span}\{\varphi_1,...,\varphi_l\})$ instead of $z \in \mathrm{span}\{\varphi_1,...,\varphi_l\}$ since (3.11) obviously holds for simple functions on $]0,T[$ with values in $\mathrm{span}\{\varphi_1,...,\varphi_l\}$ which form a dense subset.

In particular, defining $\tilde{y}_l(t) = \mathrm{e}^{-t}y_l(t)$ and testing yields

$$\left\langle\frac{\partial\tilde{y}_l}{\partial t}(t), \tilde{y}_l(t)\right\rangle_{L^2} = \left\langle\mathrm{e}^{-t}\frac{\partial y_l}{\partial t}(t), \tilde{y}_l(t)\right\rangle_{L^2} - \langle\mathrm{e}^{-t}y_l(t), \tilde{y}_l(t)\rangle_{L^2}$$

$$= -\|D_{p_1(t),...,p_d(t)}\nabla\tilde{y}_l(t)^{\mathrm{T}}\|_2^2 - \|\tilde{y}_l(t)\|_2^2$$

$$+ \langle\mathrm{e}^{-t}u(t), \tilde{y}_l(t)\rangle_{V(t)^*\times V(t)}$$

$$\leq -\tfrac{1}{2}\|\tilde{y}_l(t)\|^2_{V_{p_1(t),...,p_d(t)}} + \tfrac{1}{2}\|\mathrm{e}^{-t}u(t)\|^2_{V^*_{p_1(t),...,p_d(t)}}$$

for almost every $t \in]0,T[$, with the help of standard estimates and Young's inequality. The usual integration with respect to t and usage of $\|\tilde{y}_l(T)\|_2^2 \geq 0$ then gives

$$\tfrac{1}{2}\left(\|\tilde{y}_l(T)^2\|_2^2 - \|y_0\|_2^2\right) \leq \tfrac{1}{2}\left(-\|\tilde{y}_l(t)\|^2_{V_{p_1,...,p_d}} + \|\mathrm{e}^{-t}u\|^2_{V^*_{p_1,...,p_d}}\right)$$

$$\Rightarrow \qquad \|\tilde{y}_l\|^2_{V_{p_1,...,p_d}} \leq \|\mathrm{e}^{-t}u\|^2_{V^*_{p_1,...,p_d}} + \|y_0\|_2^2$$

and going back to $y_l(t) = \mathrm{e}^t\tilde{y}_l(t)$ leads to

$$\|y_l\|^2_{V_p} \leq \mathrm{e}^{2T}\left(\|u\|^2_{V^*_{p_1,...,p_d}} + \|y_0\|_2^2\right) .$$

The $\mathcal{V}^*_{p_1,\ldots,p_d}$-norm of $\frac{\partial y_l}{\partial t}$ is then bounded uniformly in l by

$$
\left| \left\langle \frac{\partial y_l}{\partial t}(t),\, z(t) \right\rangle_{V(t)^* \times V(t)} \right|
$$
$$
\leq \left| \left\langle D_{p_1(t),\ldots,p_d(t)} \nabla y_l(t)^{\mathrm{T}},\, D_{p_1(t),\ldots,p_d(t)} \nabla z(t)^{\mathrm{T}} \right\rangle_{L^2} \right|
$$
$$
+ \left| \langle u(t),\, z(t) \rangle_{V(t)^* \times V(t)} \right|
$$
$$
\leq \sqrt{2} \left(\|y_l(t)\|^2_{V_{p_1(t),\ldots,p_d(t)}} + \|u(t)\|^2_{V^*_{p_1(t),\ldots,p_d(t)}} \right)^{1/2} \|z(t)\|_{V_{p_1(t),\ldots,p_d(t)}}
$$

for all $z \in L^2\big(0,T; \mathrm{span}\{\varphi_1,\ldots,\varphi_l\}\big)$ and almost every t. Integration with respect to t and the estimate on $\|y_l\|^2_{\mathcal{V}_{p_1,\ldots,p_d}}$ then yields

$$
\left\| \frac{\partial y_l}{\partial t} \right\|^2_{\mathcal{V}^*_{p_1,\ldots,p_d}} \leq 2(\mathrm{e}^{2T}+1)\big(\|u\|^2_{\mathcal{V}^*_{p_1,\ldots,p_d}} + \|y_0\|^2_2 \big)
$$
$$
\Rightarrow \quad \|y_l\|^2_{W_{p_1,\ldots,p_d}} \leq (3\mathrm{e}^{2T}+2)\big(\|u\|^2_{\mathcal{V}^*_{p_1,\ldots,p_d}} + \|y_0\|^2_2 \big) .
$$

Hence, each y_l is in $W_{p_1,\ldots,p_d}(0,T)$.

Moreover, it is obvious that $W_{p_1,\ldots,p_d}(0,T)$ is reflexive, hence there exists a subsequence of y_l (not relabeled) which converges weakly to a $y \in W_{p_1,\ldots,p_d}(0,T)$. This means in particular that $y_l \rightharpoonup y$ in $\mathcal{V}_{p_1,\ldots,p_d}$, $\frac{\partial y_l}{\partial t} \rightharpoonup \frac{\partial y}{\partial t}$ in $\mathcal{V}^*_{p_1,\ldots,p_d}$ as well as $y_l(0) \rightharpoonup y(0)$ in $L^2(\Omega)$ due to the continuous embedding $W_{p_1,\ldots,p_d}(0,T) \hookrightarrow \mathcal{C}\big(0,T; L^2(\Omega)\big)$, see Proposition 3.16. Now, for each $z = \rho\varphi_m$ where $\rho \in \mathcal{C}^1([0,T])$ with $\rho(T)=0$, there is an l_0 such that (3.11) is satisfied for $l \geq l_0$ hence, "integration by parts" with respect to t and passage to the limit as $l \to \infty$, gives

$$
- \left\langle y,\, \frac{\partial z}{\partial t} \right\rangle_{\mathcal{V}^* \times \mathcal{V}} + \left\langle D_{p_1,\ldots,p_d} \nabla y^{\mathrm{T}},\, D_{p_1,\ldots,p_d} \nabla z^{\mathrm{T}} \right\rangle_{L^2}
$$
$$
= \langle u,\, z \rangle_{\mathcal{V}^* \times \mathcal{V}} + \langle y_0,\, z(0) \rangle_{L^2}
$$

which means that $y \in W_{p_1,\ldots,p_d}(0,T)$ is a weak solution of (2.4) with data u and initial value y_0, since the above z span a dense subset in $\mathcal{AC}\big(0,T; H^1(\Omega)\big) \cap W_{p_1,\ldots,p_d}(0,T)$ and consequently also in $W_{p_1,\ldots,p_d}(0,T)$. ◀

Lemma 3.24. *The solution* $\bar{y} \in \bar{W}_{p_1,\ldots,p_d}(0,T)$ *obtained in Proposition 3.22 and the solution obtained from Galerkin approximation* $y \in W_{p_1,\ldots,p_d}(0,T)$ *in Lemma 3.23 coincide whenever the data u and y_0 is the same.*

Moreover, y is the unique weak solution in $W_{p_1,\ldots,p_d}(0,T)$.

Proof. In order to prove $y = \bar{y}$, observe that, since the data (u, y_0) coincides for both solutions, it is sufficient (by Remark 3.21) to verify that the exponentially shifted Galerkin solution $\tilde{y}(t) = e^{-t}y(t)$ satisfies $\tilde{y} \in \overline{\mathrm{rg}(A)} \subset \mathcal{V}_{p_1,\dots,p_d}$ with A being the linear operator associated with the bilinear form E chosen in Proposition 3.22. Assume the opposite, i.e. that

$$\langle \tilde{y}, P_{\mathrm{rg}(A)^\perp} \tilde{y} \rangle_{\mathcal{V}_{p_1,\dots,p_d}} = \| P_{\mathrm{rg}(A)^\perp} \tilde{y} \|^2_{\mathcal{V}_{p_1,\dots,p_d}} > 0 \ .$$

Since $\langle \tilde{y}(t), P_{\mathrm{rg}(A)^\perp} \tilde{y}(t) \rangle_{V_{p_1(t),\dots,p_d(t)}}$ is a measurable non-negative function in t, it is possible to choose a $\rho \in \mathcal{C}^1([0,T])$ with $\rho \geq 0$ and $\rho(T) = 0$ such that we still have $\langle \rho\tilde{y}, P_{\mathrm{rg}(A)^\perp} \tilde{y} \rangle_{\mathcal{V}_{p_1,\dots,p_d}} > 0$. But, $\rho\tilde{y}$ is also a valid test function for E, hence

$$\langle \rho\tilde{y}, P_{\mathrm{rg}(A)^\perp} \tilde{y} \rangle_{\mathcal{V}_{p_1,\dots,p_d}} = E(P_{\mathrm{rg}(A)^\perp} \tilde{y}, \rho\tilde{y}) = 0 \ ,$$

since $P_{\mathrm{rg}(A)^\perp} \tilde{y} \in \mathrm{rg}(A)^\perp$, again by Remark 3.21. This is a contradiction, so $P_{\mathrm{rg}(A)^\perp} \tilde{y} = 0$.

Now verify the uniqueness of the solution y in $W_{p_1,\dots,p_d}(0,T)$. For linear equations one only has to verify this for $(u, y_0) = (0,0)$ in the respective spaces. So let y be a solution of (3.10) with zero data. Choose a $\rho \in \mathcal{C}_0^\infty(]0,T[)$. It is easy to check that $\rho y \in W_{p_1,\dots,p_d}(0,T)$ is a valid test function according to Definition 3.17. With the product rule of Proposition 3.16 and "integration by parts" one obtains

$$- \Big\langle \frac{\partial}{\partial t}(\rho y), y \Big\rangle_{\mathcal{V}^* \times \mathcal{V}} = \int_0^T \Big\langle \frac{\partial y}{\partial t}(t), y(t) \Big\rangle_{V(t)^* \times V(t)} \rho(t) \ \mathrm{d}t$$

$$= \int_0^T \frac{1}{2} \Big(\frac{\partial}{\partial t} \|y(t)\|_2^2 \Big) \rho(t) \ \mathrm{d}t \ .$$

Hence, the weak formulation gives

$$\int_0^T \frac{1}{2} \Big(\frac{\partial}{\partial t} \|y(t)\|_2^2 \Big) \rho(t) \ \mathrm{d}t + \int_0^T \| D_{p_1(t),\dots,p_d(t)} \nabla y^{\mathsf{T}} \|_2^2 \rho(t) \ \mathrm{d}t = 0 \ .$$

In particular, this holds for each $\rho \in \mathcal{C}_0^\infty(]0,T[)$ with $\rho \geq 0$, implying that $\frac{\partial}{\partial t} \|y(t)\|_2^2 \leq 0$ almost everywhere in $]0,T[$. But this is only possible for $y = 0$. ◀

Collecting the previous results establishes the main existence and uniqueness theorem.

Theorem 3.25. *Let $\sigma, p_1, \ldots, p_d, u, y_0$ be given according to Definition 3.17. Then there exists a weak solution $y \in W_{p_1,\ldots,p_d}(0,T)$ of (3.1) in the sense of Definition 3.17. Moreover, y is unique in $W_{p_1,\ldots,p_d}(0,T)$ and depends linearly and continuously on the data, i.e. there exists a $C > 0$, only depending on T, such that*

$$\|y\|^2_{W_{p_1,\ldots,p_d}} \le C\left(\|u\|^2_{\mathcal{V}^*_{p_1,\ldots,p_d}} + \|y_0\|^2_2\right) .$$

Corollary 3.26. *The solution operator $S_{p_1,\ldots,p_d}(u, y_0) = y$ (where y is the weak solution obtained in Proposition 3.22) is a linear and continuous mapping*

$$S_{p_1,\ldots,p_d} : \mathcal{V}^*_{p_1,\ldots,p_d} \times L^2(\Omega) \to \mathcal{C}\left(0,T; L^2(\Omega)\right) .$$

Its norm is independent of $p_1, \ldots, p_d$.
The same holds true for

$$S_{p_1,\ldots,p_d} : L^2\left(0,T; L^2(\Omega)\right) \times L^2(\Omega) \to \mathcal{C}\left(0,T; L^2(\Omega)\right) .$$

Proof. These statements follow by combining Theorem 3.25 and Proposition 3.16 as well as noting that $\|u\|_{\mathcal{V}^*_{p_1,\ldots,p_d}} \le \|u\|_2$ (see Remark 3.13). ◀

Remark 3.27. Let us remark that a weak solution in $W_{p_1,\ldots,p_d}(0,T)$ attains $y(0) = y_0$ and that there are other weak formulations for (3.1) which are equivalent to the one given in Definition 3.17.

For a $y \in W_{p_1,\ldots,p_d}(0,T)$ we have for $z \in W_{p_1,\ldots,p_d}(0,T)$ with $z(T) = 0$ that

$$-\left\langle \frac{\partial z}{\partial t}, y\right\rangle_{V^* \times V} = \langle y(0), z(0)\rangle_{L^2} + \left\langle \frac{\partial y}{\partial t}, z\right\rangle_{V^* \times V} ,$$

see Proposition 3.16. Plugged into (3.10), this reads

$$\langle y(0), z(0)\rangle_{L^2} + \left\langle \frac{\partial y}{\partial t}, z\right\rangle_{V^* \times V} + \langle D_{p_1,\ldots,p_d}\nabla y^{\mathrm{T}}, D_{p_1,\ldots,p_d}\nabla z^{\mathrm{T}}\rangle_{L^2}$$
$$= \langle u, z\rangle_{V^* \times V} + \langle y_0, z(0)\rangle_{L^2} .$$

Now test with $z = \rho(st)z_0$ where $z_0 \in H^1(\Omega)$ and $\rho \in \mathcal{C}^\infty([0,\infty[)$ with $\rho(0) = 1$ as well as $\rho(t) = 0$ for $t \ge 1$. This makes sense for $s \ge T^{-1}$ and yields, as $s \to \infty$, that $\langle y(0), z_0\rangle_{L^2} = \langle y_0, z_0\rangle_{L^2}$ which in turn implies $y(0) = y_0$. Thus, the initial value is indeed attained.

Furthermore, by testing with $\rho(t)z(t)$ where $z \in W_{p_1,\ldots,p_d}(0,T)$ and $\rho \in \mathcal{C}_0^\infty(]0,T[)$, we get the identity

$$\langle \frac{\partial y}{\partial t}, \rho z\rangle_{\mathcal{V}^* \times \mathcal{V}} + \langle D_{p_1,\ldots,p_d}\nabla y^{\mathrm{T}}, D_{p_1,\ldots,p_d}\nabla(\rho z)^{\mathrm{T}}\rangle_{L^2} = \langle u, \rho z\rangle_{\mathcal{V}^* \times \mathcal{V}}$$

which can be extended to $z \in \mathcal{V}_{p_1,\ldots,p_d}$ by density. Regarding the definitions of the respective scalar products and taking into account that the above holds for each $\rho \in \mathcal{C}_0^\infty(]0,T[)$, it follows

$$\langle \frac{\partial y}{\partial t}(t), z(t)\rangle_{V(t)^* \times V(t)} + \langle D_{p_1,\ldots,p_d}\nabla y(t)^{\mathrm{T}}, D_{p_1,\ldots,p_d}\nabla z(t)^{\mathrm{T}}\rangle_{L^2}$$
$$= \langle u(t), z(t)\rangle_{V(t)^* \times V(t)} \quad (3.12)$$

for each $z \in \mathcal{V}_{p_1,\ldots,p_d}$ and almost every $t \in]0,T[$. Together with $y(0) = y_0$, this can be interpreted as an alternate weak formulation of (3.1).

Conversely, if one has a $y \in W_{p_1,\ldots,p_d}(0,T)$ with $y(0) = y_0$ which satisfies (3.12), then one can easily verify that y is a weak solution according to Definition 3.17.

In the remainder of this section, we will present a few remarks on the solution of the equation for the non-degenerate case $\|p\|_\infty < 1$, which can be treated with standard theory.

Remark 3.28.

(a) For $p_1,\ldots,p_d \in L^\infty(\Omega,\mathbb{R}^d)$ satisfying (3.2b) and $\|p_i\|_\infty < 1$ for $i = 1,\ldots,d$, the space $\mathcal{V}_{p_1,\ldots,p_d}$ obviously coincides with $H^1(\Omega)$. Analogously, for vector fields $p_1,\ldots,p_d$ satisfying Condition 3.2 as well as $\|p_i\|_\infty < 1$, the space $\mathcal{V}_{p_1,\ldots,p_d}$ coincides with $L^2(0,T;H^1(\Omega))$ and consequently,

$$W_{p_1,\ldots,p_d}(0,T) = W(0,T)$$

according to (3.6).

(b) In the latter case, the differential operator E can be regarded as a continuous linear mapping $W(0,T) \to L^2(\Omega) \times L^2(0,T;H^1(\Omega)^*)$ via

$$E(y) = \left(y(0), \frac{\partial y}{\partial t} - \mathrm{div}(D_{p(t)}^2\nabla y^{\mathrm{T}})\right)$$

in the weak sense. On the other hand, Theorem 3.25 also shows that for each $(y_0,u) \in L^2(\Omega) \times L^2(0,T;H^1(\Omega)^*)$ there is a unique solution and the a-priori estimates on the solution give the continuity of the inverse mapping E^{-1}. Hence, E is a linear isomorphism between $W(0,T)$ and $L^2(\Omega) \times L^2(0,T;H^1(\Omega)^*)$.

3.3 Weak compactness results

Since our goal is to solve the optimal control problem (2.8), where the parameters p vary over a certain set, it is also important to know how the solution operator of the underlying equation (2.4) or, more general, (3.1) acts in dependency of $p_1, \ldots, p_d$. In the previous section, we have already seen that the solution spaces associated with the variational formulation of the weak problem may vary with $p_1, \ldots, p_d$. Nevertheless, like in the non-degenerate setting, all solution spaces are continuously contained in $\mathcal{C}(0, T; L^2(\Omega))$.

Unfortunately, the important property of compactness fails for the degenerate equations, i.e. the of $W_{p_1, \ldots, p_d}(0, T)$ embedding in a space such as $L^2(0, T; L^2(\Omega))$ is in general not compact. Recall that in the usual parabolic setting with uniformly elliptic coefficients with respect to the space variable we have

$$W(0, T) \hookrightarrow L^2(0, T; L^2(\Omega)) .$$

Such a property cannot be established for all vector fields $p_1, \ldots, p_d$ since in general $V_{p_1(t), \ldots, p_d(t)} \hookrightarrow L^2(\Omega)$ does not hold. The uniform a-priori estimates of Theorem 3.25 and Corollary 3.26, however, suggest some kind of weak compactness in $\mathcal{C}(0, T; L^2(\Omega))$. In fact, the weak solutions of (3.1) associated with bounded sets of data (u, y_0) and all $p_1, \ldots, p_d$ fulfilling Condition 3.2 form a weakly precompact set in in the slightly weaker space $\mathcal{C}^*(0, T; L^2(\Omega))$, which will be introduced in the following. The aim of this section is to prove such a weak compactness result.

In the following, Banach spaces of bounded functions with values in a Hilbert space H will be introduced, which are weaker than $\mathcal{C}(0, T; H)$ but still allow for point evaluation. A characterization of the dual spaces as well as the weak sequential convergence is then given. Moreover, as a tool for proving the desired compactness, the compact metric spaces $\mathcal{L}_R^2(\Omega)$ are introduced, which are the closed L^2-balls $\{\|y\|_2 \leq R\}$ equipped with a special metric. The associated spaces $\mathcal{C}(0, T; \mathcal{L}_R^2(\Omega))$ can be related to $\mathcal{C}^*(0, T; L^2(\Omega))$. Such a connection is useful for applying the compactness theorem by Arzelà and Ascoli and for establishing the main result of this section.

3.3.1 Spaces of weakly continuous functions

For the further analysis, it is useful to introduce spaces of bounded functions with values in a Hilbert space H which are "weakly continuous" with respect to t. We will later utilize this construction with $H = L^2(\Omega)$.

Definition 3.29. Let H be a separable Hilbert space. Define the space

$$C^*(0, T; H) = \{y : [0, T] \to H \mid \sup_{t \in [0,T]} \|y(t)\| < \infty,$$

$$\langle y(\cdot), \varphi \rangle \in \mathcal{C}([0, T]) \text{ for all } \varphi \in H\}$$

which is equipped with the norm

$$\|y\|_{\mathcal{C}^*} = \sup_{t \in [0,T]} \|y(t)\| .$$

Such a construction yields Banach spaces which are intermediate between $\mathcal{C}(0, T; H)$ and $L^\infty(0, T; H)$ in the following sense.

Proposition 3.30. *Each $C^*(0, T; H)$ is a Banach space with continuous embeddings*

$$\mathcal{C}(0, T; H) \hookrightarrow C^*(0, T; H) \hookrightarrow L^\infty(0, T; H) .$$

Moreover, for each $y \in C^*(0, T; H)$, the function $\|y(\cdot)\|$ is lower semi-continuous.

Proof. It is easy to verify that $\|\cdot\|_{\mathcal{C}^*}$ is indeed a norm. It remains to show that $C^*(0, T; H)$ is complete with respect to this norm. Let a Cauchy sequence $\{y_k\}$ in $C^*(0, T; H)$ be given. The completeness of H immediately gives a bounded pointwise limit function $y(t) = \lim_{k \to \infty} y_k(t)$. With standard arguments one can then see that $\lim_{k \to \infty} \|y_k - y\|_{\mathcal{C}^*} = 0$. Moreover, for each $\varphi \in H$, the functions defined by $y_{k,\varphi}(t) = \langle y_k(t), \varphi \rangle$ are Cauchy sequences in $\mathcal{C}([0, T])$ and therefore admit limits $y_{\infty,\varphi} \in \mathcal{C}([0, T])$ which satisfy

$$y_{\infty,\varphi}(t) = \lim_{k \to \infty} \langle y_k(t), \varphi \rangle = \langle \lim_{k \to \infty} y_k(t), \varphi \rangle = \langle y(t), \varphi \rangle .$$

Hence, $\{y_k\}$ indeed converges to a $y \in C^*(0, T; H)$.

To establish the embeddings, first note that the norms coincide on all three spaces. The embedding $\mathcal{C}(0, T; H) \hookrightarrow C^*(0, T; H)$ is immediate while the embedding $C^*(0, T; H) \hookrightarrow L^\infty(0, T; H)$ follows by Pettis' theorem: each $y \in C^*(0, T; H)$ is weakly measurable by definition and of course separably-valued. Hence, y is measurable in the sense of Bochner and belongs to $L^\infty(0, T; H)$ by equality of the norms. Finally, let us see that the embedding is injective. Therefore, let $y \in C^*(0, T; H)$ such that $\|y\|_\infty = 0$ in $L^\infty(0, T; H)$. This means that $y(t) = 0$ almost everywhere and consequently $\langle y(\cdot), z \rangle = 0$ in $\mathcal{C}([0, T])$ for each $z \in H$ since there are no continuous functions which are zero except for a non-empty null-set. It follows that $y(t) = 0$ for every $t \in [0, T]$.

To prove the lower semi-continuity property, fix a $y \in \mathcal{C}^*(0, T; H)$ and an orthonormal basis $\{\varphi_k\}$ of H. Observe that for each $t \in [0, T]$,

$$u(t) = \|y(t)\| = \lim_{k \to \infty} u_k(t) \quad , \quad u_k(t) = \left(\sum_{l=1}^{k} |\langle y(t), \varphi_l \rangle|^2 \right)^{1/2}$$

where $u_k \in \mathcal{C}([0, T])$ by assumption. The sequence u_k is therefore a monotone increasing sequence of continuous functions, hence its pointwise limit u is lower semi-continuous. ◄

Remark 3.31. Note that $\mathcal{C}^*(0, T; H)$ still allows for point evaluation, i.e. for a given $t \in [0, T]$, the mapping $y \mapsto y(t)$ is well-defined, linear and continuous between $\mathcal{C}^*(0, T; H) \to H$ with norm 1.

In the following, we are interested in weak convergence in $\mathcal{C}^*(0, T; H)$, especially in a useful characterization. Hence, we have to take a closer look at the elements of the dual space $\mathcal{C}^*(0, T; H)^*$. This is closely related to generalizations of the Riesz representation theorem to linear and continuous mappings $\mathcal{C}(S, X) \to Y$ where S is a compact Hausdorff space and X, Y a locally convex topological vector spaces (see [Goo70], for example). Unfortunately, this general setting does not quite fit and also requires special integral notions which are not easy to handle.

In the particular setting of $S = [0, T]$ and X a Hilbert space however, the representation involving integrals simplifies significantly. Here, we develop a representation for elements in $\mathcal{C}^*(0, T; H)^*$ which only makes use of the Bochner integral notion.

Proposition 3.32. *Let H be a separable Hilbert space and $z \in \mathcal{C}^*(0, T; H)^*$ (with norm $\|z\|_*$) be given. Then there exists a regular Borel measure μ and a $\kappa \in L^\infty_\mu(0, T; H)$ with $\|\kappa\|_\infty \leq 1$, such that*

$$z(y) = \int_0^T \langle y(t), \kappa(t) \rangle \, \mathrm{d}\mu(t)$$

for every $y \in \mathcal{C}^*(0, T; H)$.

Proof. The main ideas can also be found in [EG92] where the proof is carried out for $H = \mathbb{R}^k$. The difficulty here lies in the adaptation for infinite-dimensional spaces which is fortunately a Hilbert space here.

Let $z \in \mathcal{C}^*(0, T; H)^*$ be given. To obtain the Borel measure μ, we will "reduce" the functional to act on real-valued functions. For a $u \in \mathcal{C}([0, T])$ with $u \geq 0$ define $\lambda(u)$ by

$$\lambda(u) = \sup \{z(y) \mid y \in \mathcal{C}^*(0, T; H) \text{ with } \|y(t)\| \leq u(t) \text{ on } [0, T]\} .$$

We will show that λ is linear. Trivially, $\lambda(0) = 0$. It is furthermore easy to see that for $c \geq 0$ we have $\lambda(cu) = c\lambda(u)$ as well as $\lambda(u_1) + \lambda(u_2) \leq \lambda(u_1 + u_2)$ for $u_1, u_2 \in \mathcal{C}([0,T])$ with $u_1, u_2 \geq 0$. On the other hand, for any $y \in \mathcal{C}^*(0,T;H)$ with $\|y(t)\| \leq u_1(t) + u_2(t)$ we can set $y_i = \frac{u_i}{u_1+u_2}y$ ($y_i = 0$ where $u_1 + u_2 = 0$) and produce functions in $\mathcal{C}^*(0,T;H)$ satisfying $\|y_i(t)\| \leq u_i(t)$ for $i = 1,2$ respectively. This implies that also $\lambda(u_1 + u_2) \leq \lambda(u_1) + \lambda(u_2)$.

Hence, we can extend λ to a positive linear functional on $\mathcal{C}([0,T])$ by $\lambda(u) = \lambda(u_+) - \lambda(-u_-)$. Furthermore, one easily gets $\lambda(u) \leq \|z\|_* \|u\|_\infty$, so this functional is also continuous. By Riesz's representation theorem, there exists a regular Borel measure μ such that

$$\lambda(u) = \int_0^T u(t) \, d\mu(t) .$$

Note that, since the function defined by $u(t) = \|y(t)\|$ is lower semi-continuous for each $y \in \mathcal{C}^*(0,T;H)$ (see Proposition 3.30), it can be plugged into the definition for λ as well as into the integral. Hence, one also has that $z(y) \leq \lambda(u) = \|y\|_{1,\mu}$, so z can be extended to an element of $L^1_\mu(0,T;H)^*$. Since H is a Hilbert space, the dual can be identified with $L^\infty_\mu(0,T;H)$, confer [DU77], thus there exists a $\kappa \in L^\infty_\mu(0,T;H)$ satisfying $\|\kappa\|_{\infty,\mu} \leq 1$ such that

$$z(y) = \int_0^T \langle y(t),\, \kappa(t)\rangle \, d\mu(t)$$

for each $y \in \mathcal{C}^*(0,T;H)$, which is the desired representation.   ◀

With the help of this representation result, we can characterize weak convergence in $\mathcal{C}^*(0,T;H)$ as follows.

Proposition 3.33. *Let H be a separable Hilbert space. Then, a sequence $\{y_k\}$ in $\mathcal{C}^*(0,T;H)$ converges weakly to a $y \in \mathcal{C}^*(0,T;H)$ if and only if $\{y_k\}$ is bounded and $y_k(t) \rightharpoonup y(t)$ for every $t \in [0,T]$.*

Proof. First, every weakly convergent sequence is bounded. For $(t,v) \in [0,T] \times H$ define the mapping $y \mapsto \delta_{t,v}(y) = \langle y(t), v \rangle$ which belongs to $\mathcal{C}^*(0,T;H)^*$ since

$$|\delta_{t,v}(y)| \leq \|v\|\,\|y\|_{\mathcal{C}^*} .$$

Thus, if $y_k \rightharpoonup y$ in $\mathcal{C}^*(0,T;H)$, then $y_k(t) \rightharpoonup y(t)$ for each $t \in [0,T]$.

To prove the converse direction, let $z \in \mathcal{C}^*(0,T;H)^*$ be given. Then by Proposition 3.32 there is a regular Borel measure μ and a $\kappa \in L^\infty_\mu(0,T;H)$ such that

$$z(y_k) = \int_0^T \langle y_k(t),\, \kappa(t)\rangle \, d\mu(t) .$$

Now, we have $\lim_{k\to\infty}\langle y_k(t),\,\kappa(t)\rangle = \langle y(t),\,\kappa(t)\rangle$ μ-almost everywhere as well as $|\langle y_k(t),\,\kappa(t)\rangle| \le C$ since $\{y_k\}$ is bounded. By the dominated-convergence theorem of Lebesgue,

$$\lim_{k\to\infty} z(y_k) = \int_0^T \langle y(t),\,\kappa(t)\rangle \; \mathrm{d}\mu(t) = z(y) \; ,$$

which implies the desired weak convergence. ◀

3.3.2 Relative weak compactness in $C^*\big(0,T;L^2(\Omega)\big)$

In the following, we will utilize the notion of $C^*\big(0,T;H\big)$ for the Hilbert space $H = L^2(\Omega)$ to prove relative weak compactness results for certain sets of weak solutions of (3.1). In order to obtain these results, a compact metric space associated with $C\big(0,T;L^2(\Omega)\big)$ is introduced, the space where all weak solutions are contained in.

Definition 3.34. Let $R > 0$ be given. Assume that Ω is a subset of the positive cone, i.e. there is an L such that $\Omega \subset [0,L]^d$. For a $k_i \in \mathbb{N}$ denote by

$$\varphi_{k_i}(x_i) = \begin{cases} \sqrt{\frac{2}{L}}\sin\left(\frac{k_i\pi x_i}{L}\right) & \text{if } k_i \ge 1 \\ \frac{1}{\sqrt{L}} & \text{if } k_i = 0 \; , \end{cases}$$

and set, for a multi-index $k \in \mathbb{N}^d$,

$$\varphi_k(x) = \prod_{i=1}^d \varphi_{k_i}(x_i) \; .$$

Then define the *weak L_R^2-space* as

$$\mathcal{L}_R^2(\Omega) = \{y \in L^2(\Omega) \mid \|y\|_2 \le R\}$$

equipped with the metric

$$\rho(y_1,y_2) = \sum_{k\in\mathbb{N}^d} 2^{-|k|} \frac{|\langle y_1 - y_2,\,\varphi_k\rangle_{L^2}|}{1 + |\langle y_1 - y_2,\,\varphi_k\rangle_{L^2}|} \; .$$

The construction of ρ is close to the usual metric used to prove that certain bounded sets in separable Banach spaces are metrizable in the weak topology, see [DS57], for instance. It differs, however, in the fact that a special basis is utilized instead of an arbitrary dense subset. We will point out the special properties of this basis in the following as well as present the characterization of the $\mathcal{L}^2_R(\Omega)$-topology for convenience.

First note that, possibly after translation, $\Omega \subset [0, L]^d$ can always be ensured for a suitable $L > 0$. Also, it is known that the $\{\varphi_{k_i}\}$, $k_i \in \mathbb{N}$ form an orthonormal basis of $L^2([0, L])$ (see [Can84], for example) and consequently $\{\varphi_k\}$, $k \in \mathbb{N}^d$ forms an orthonormal basis of $L^2([0, L]^d)$. This is summarized in the following lemma.

Lemma 3.35. *Let φ_k according to Definition 3.34 and $y \in L^2(\Omega)$ be given. Then*

$$y = \sum_{k \in \mathbb{N}^d} \langle y, \varphi_{k|\Omega} \rangle_{L^2} \varphi_{k|\Omega} \,,$$

as well as $\|\varphi_{k|\Omega}\|_2 \leq 1$ for all $k \in \mathbb{N}^d$.

Lemma 3.36. *For each $\alpha \geq 0$, the following summation is finite:*

$$\sum_{k \in \mathbb{N}^d} |k|^\alpha 2^{-|k|} < \infty \,.$$

Proof. For a given $l \in \mathbb{N}$, let us estimate the number of multi-indices k with $|k| = l$ roughly by $(l+1)^d$. Thus,

$$\sum_{k \in \mathbb{N}^d} |k|^\alpha 2^{-|k|} \leq 2 \sum_{l=1}^\infty l^{d+\alpha} 2^{-l} < \infty \,. \qquad \blacktriangleleft$$

Proposition 3.37. *The space $\mathcal{L}^2_R(\Omega)$ is a compact, complete metric space. Its topology coincides with the relative weak topology of $\{\|y\|_2 \leq R\}$ in $L^2(\Omega)$, i.e.*

$$\|y_l\|_2 \leq R \text{ and } y_l \rightharpoonup y \quad \Leftrightarrow \quad y_l \in \mathcal{L}^2_R(\Omega) \text{ and } y_l \to y \,.$$

Proof. This is in part an extension/application of Theorems V.5.1 and V.5.2 in [DS57]. The main ideas here are very similar to the ones used there.

We start with noting that ρ is indeed a metric. The non-trivial part, which is showing $\rho(y_1, y_2) = 0$ that implies $y_1 = y_2$, follows from the representation formula for $y_1 - y_2$ in Lemma 3.35.

The next step is to prove that $y_l \in \mathcal{L}^2_R(\Omega)$ is a Cauchy sequence if and only if $\|y_l\|_2 \leq R$ and $\langle y_l, \varphi_k \rangle_{L^2}$ is a Cauchy sequence for each $k \geq 0$. So if $\{y_l\}$ is a Cauchy sequence in $\mathcal{L}^2_R(\Omega)$ and $\rho(y_{l_1}, y_{l_2}) < \varepsilon$, then, by Lemma 3.35,

$$2^{-|k|} \frac{|\langle y_{l_1} - y_{l_2}, \varphi_k \rangle_{L^2}|}{1 + |\langle y_{l_1} - y_{l_2}, \varphi_k \rangle_{L^2}|} < \varepsilon$$

$$\Rightarrow \quad |\langle y_{l_1} - y_{l_2}, \varphi_k \rangle_{L^2}| < 2^{|k|} (1 + \|y_{l_1} - y_{l_2}\|_2 \|\varphi_k\|_2) \varepsilon \leq 2^{|k|+1}(R+1)\varepsilon \,.$$

On the other hand, if each $\langle y_l, \varphi_k \rangle_{L^2}$ is a Cauchy sequence, each summand
satisfies

$$\frac{|\langle y_{l_1} - y_{l_2}, \varphi_k \rangle_{L^2}|}{1 + |\langle y_{l_1} - y_{l_2}, \varphi_k \rangle_{L^2}|} \leq 1 \quad , \quad \frac{|\langle y_{l_1} - y_{l_2}, \varphi_k \rangle_{L^2}|}{1 + |\langle y_{l_1} - y_{l_2}, \varphi_k \rangle_{L^2}|} \leq |\langle y_{l_1} - y_{l_2}, \varphi_k \rangle_{L^2}|$$

and since $\sum_{k \in \mathbb{N}^d} 2^{-|k|} < \infty$ (Lemma 3.36), there is a $k_0 \in \mathbb{N}$ such that

$$\sum_{\substack{k \in \mathbb{N}^d \\ |k| \geq k_0}} 2^{-|k|} \frac{|\langle y_{l_1} - y_{l_2}, \varphi_k \rangle_{L^2}|}{1 + |\langle y_{l_1} - y_{l_2}, \varphi_k \rangle_{L^2}|} < \frac{\varepsilon}{2}$$

independent of l_1, l_2. Furthermore, for the finitely many $k \in \mathbb{N}^d$ with $|k| < k_0$
one can choose an l_0 such that for each $l_1, l_2 \geq l_0$ we have

$$\sum_{\substack{k \in \mathbb{N}^d \\ |k| < k_0}} 2^{-|k|} \frac{|\langle y_{l_1} - y_{l_2}, \varphi_k \rangle_{L^2}|}{1 + |\langle y_{l_1} - y_{l_2}, \varphi_k \rangle_{L^2}|} < \frac{\varepsilon}{2} \; .$$

Hence, $\{y_l\}$ is a Cauchy sequence in $\mathcal{L}_R^2(\Omega)$.

Now notice that for each sequence $\|y_l\|_2 \leq R$ the statement that $\langle y_l, \varphi_k \rangle_{L^2}$
is a Cauchy sequence for each $k \in \mathbb{N}^d$ is equivalent to $\langle y_l, z \rangle_{L^2}$ being a Cauchy
sequence for each $z \in L^2(\Omega)$. The non-trivial direction can be seen by representing z with the help of φ_k and cutting of at a $k_0 \geq 0$ such that

$$z_{k_0} = \sum_{\substack{k \in \mathbb{N}^d \\ |k| \geq k_0}} \langle z, \varphi_k \rangle \varphi_k \quad , \quad \|z - z_{k_0}\|_2 < \frac{\varepsilon}{4R} \; .$$

Hence, one can choose an $l_0 \geq 0$ such that for each $l_1, l_2 \geq l_0$ we have

$$\left| \sum_{\substack{k \in \mathbb{N}^d \\ |k| < k_0}} \langle y_{l_1} - y_{l_2}, \varphi_k \rangle_{L^2} \right| < \frac{\varepsilon}{2}$$

which implies

$$|\langle y_{l_1} - y_{l_2}, z \rangle_{L^2}| \leq |\langle y_{l_1} - y_{l_2}, z - z_{k_0} \rangle_{L^2}| + |\langle y_{l_1} - y_{l_2}, z_{k_0} \rangle_{L^2}|$$
$$< \frac{\varepsilon}{2} + 2R\|z - z_{k_0}\|_2 < \varepsilon \; .$$

The above gives that $\{y_l\}$ is a Cauchy sequence in $\mathcal{L}_R^2(\Omega)$ if and only if
$\|y_l\|_2 \leq R$ and $\{y_l\}$ is a weak Cauchy sequence in $L^2(\Omega)$. But $L^2(\Omega)$ is weakly

complete (since $L^2(\Omega)$ is reflexive, see Corollary II.3.29 in [DS57]) so there is a weak limit $y \in L^2(\Omega)$ with $\|y\|_2 \leq R$. This shows the completeness of $\mathcal{L}^2_R(\Omega)$.

Furthermore, since $L^2(\Omega)^* = L^2(\Omega)$ is separable, we know that the closed ball $\{\|y\|_2 \leq R\}$ with the weak topology is metrizable (confer Theorem V.5.2 in [DS57]), hence convergence can be characterized by weak sequential convergence. Since $\mathcal{L}^2_R(\Omega)$ is also a metric space which characterizes weak sequential convergence, the topologies must coincide.

Finally, $\{\|y\|_2 \leq R\}$ is weakly sequentially compact in $L^2(\Omega)$ and so is $\mathcal{L}^2_R(\Omega)$.   ◀

Next, we establish the relative weak compactness, in $\mathcal{C}(0, T; \mathcal{L}^2_R(\Omega))$, of the set of weak solutions of (3.1) associated with data (u_0, y_0) bounded in the space $L^2(0, T; L^2(\Omega)) \times L^2(\Omega)$ and all $p_1, \ldots, p_d$ satisfying Condition 3.2.

Lemma 3.38. *The norm of φ_k in $H^1(\Omega)$ satisfies*

$$\|\varphi_k\|_{H^1} \leq C(|k| + 1)$$

with a constant C independent of k.

Proof. The derivatives of φ_k with respect to x_i read as

$$\frac{\partial \varphi_k}{\partial x_i}(x) = \begin{cases} \sqrt{\frac{2}{L}} \frac{k_i \pi}{L} \cos\left(\frac{k_i \pi x_i}{L}\right) \prod_{j \neq i} \varphi_{k_j}(x_j) & \text{if } k_i \geq 1 \\ 0 & \text{if } k_i = 0 \end{cases}$$

and one easily sees that $\|\frac{\partial \varphi_k}{\partial x_i}\|_2 \leq k_i \pi / L$. Hence the estimate

$$\|y\|_{H^1} \leq 1 + \frac{\pi}{L}|k| \leq C(|k| + 1)$$

holds with $C = \max\{1, \pi/L\}$.   ◀

Proposition 3.39. *Let $\{(u_l, y_{l,0})\}$ in $L^2(0, T; L^2(\Omega)) \times L^2(\Omega)$ be a bounded sequence and $\{(p_{l,1}, \ldots, p_{l,d})\}$ a sequence of vector fields satisfying Condition 3.2.*

Then there is an $R > 0$ such that the weak solutions $y_l \in \mathcal{C}(0, T; L^2(\Omega))$ of the evolution equation (3.1) associated with u_l and p_l, respectively, admit a subsequence which converges in $\mathcal{C}(0, T; \mathcal{L}^2_R(\Omega))$.

Proof. By the uniform norm-estimate in Corollary 3.26, the sequence $\{y_l\}$ is bounded in $\mathcal{C}(0, T; L^2(\Omega))$, thus choose R as such a bound. Then $y_l \in \mathcal{C}(0, T; \mathcal{L}^2_R(\Omega))$ and the sequences $\{y_l(t)\}$ are precompact in $\mathcal{L}^2_R(\Omega)$ for each $t \in [0, T]$.

We will now show that $\{y_l\}$ is also equicontinuous. Let $t \in [0, T]$ and $\varepsilon > 0$. For every y_l and $k \in \mathbb{N}^d$ we have that

$$\frac{\partial}{\partial t}\langle y_l(t), \varphi_k\rangle_{L^2} = \Big\langle \frac{\partial y_l}{\partial t}(t), \varphi_k\Big\rangle_{V_l(t)^* \times V_l(t)}$$

by Proposition 3.16. Here $\langle \cdot, \cdot \rangle_{V_l(t)^* \times V_l(t)}$ denotes the dual pairing between the spaces $V^*_{p_{l,1}(t),\dots,p_{l,d}(t)}$ and $V_{p_{l,1}(t),\dots,p_{l,d}(t)}$. For every y_l as well as $0 \le s \le t \le T$ we can integrate and obtain, with the help of Lemma 3.38,

$$
\begin{aligned}
|\langle y_l(t) - y_l(s), \varphi_k\rangle_{L^2}| &= \Big|\int_s^t \Big\langle \frac{\partial y_l}{\partial t}(\tau), \varphi_k\Big\rangle_{V_l(\tau)^* \times V_l(\tau)}\, d\tau\Big| \\
&\le \Big\|\frac{\partial y_l}{\partial t}\Big\|_{V^*_{p_{l,1},\dots,p_{l,d}}} \|\chi_{[s,t]}\varphi_k\|_{V_{p_{l,1},\dots,p_{l,d}}} \\
&\le C_1\sqrt{t-s}\|\varphi_k\|_{H^1} \le C_2\sqrt{t-s}(|k| + 1)
\end{aligned}
$$

with C_1 being an $W_{p_{l,1},\dots,p_{l,d}}(0,T)$-estimate for all y_l and C_2 suitable constant. The distance of $y_l(s)$ to $y_l(t)$ in terms of the metric ρ can be uniformly estimated by

$$\rho\big(y_l(s), y_l(t)\big) \le C_2\sqrt{t-s} \sum_{k \in \mathbb{N}^d} 2^{-|k|}(|k| + 1) \le C_3\sqrt{t-s}$$

where C_3 can be suitably chosen, see Lemma 3.36.

Choosing $0 < \delta < \left(\frac{\varepsilon}{C_3}\right)^2$ for a given $\varepsilon > 0$ yields the equicontinuity. By the compactness theorem of Arzelà and Ascoli, $\{y_l\}$ is precompact in $\mathcal{C}\big(0, T; \mathcal{L}^2_R(\Omega)\big)$ which implies the assertion of the proposition.   ◀

The notion of convergence in $\mathcal{C}\big(0, T; \mathcal{L}^2_R(\Omega)\big)$ can be shifted back to the Banach-space setting by considering weak convergence in $\mathcal{C}^*\big(0, T; L^2(\Omega)\big)$. This will imply the main result of the this section.

Theorem 3.40. *Let $\{(u_k, y_{k,0})\}$ in $L^2\big(0, T; L^2(\Omega)\big) \times L^2(\Omega)$ be a bounded sequence and $\{(p_{k,1}, \dots, p_{k,d})\}$ be given according to Condition 3.2.*

Then the sequence of associated weak solutions $\{y_k\}$ of (3.1) possesses a subsequence which converges weakly in $\mathcal{C}^*\big(0, T; L^2(\Omega)\big)$.

Proof. By Proposition 3.39 we know that $\{y_k\}$ admits a subsequence which converges to a y in $\mathcal{C}\big(0, T; \mathcal{L}^2_R(\Omega)\big)$.

This implies pointwise convergence of $y_k(t)$ to $y(t)$ in $\mathcal{L}_R^2(\Omega)$ for every $t \in [0, T]$. Since the topologies coincide, see Proposition 3.37, this means that

$$y_k(t) \rightharpoonup y(t) \quad \text{in } L^2(\Omega) \text{ and for all } t \in [0, T] .$$

Now, y is contained in $\mathcal{C}^*\big(0, T; L^2(\Omega)\big)$: Of course, $\|y\|_{\mathcal{C}^*} \leq R$, so we only have to verify the weak continuity. Let $z \in L^2(\Omega)$, $t \in [0, T]$ and $\varepsilon > 0$ be given. Choosing L large enough, we can approximate z for each $\varepsilon_0 > 0$ such that

$$z_0 = \sum_{\substack{l \in \mathbb{N}^d \\ |l| \leq L}} \langle \varphi_l, z \rangle_{L^2} \varphi_l \quad , \quad \|z - z_0\|_2 < \varepsilon_0 .$$

Furthermore, note that for $|l| \leq L$, the scalar product $\langle y(t) - y(s), \varphi_l \rangle_{L^2}$ can be estimated by

$$|\langle y(t) - y(s), \varphi_l \rangle_{L^2}| \leq (1 + 2R) 2^L 2^{-|l|} \frac{|\langle y(t) - y(s), \varphi_l \rangle_{L^2}|}{1 + |\langle y(t) - y(s), \varphi_l \rangle_{L^2}|} ,$$

hence

$$|\langle y(t) - y(s), z \rangle_{L^2}| \leq \left| \sum_{\substack{l \in \mathbb{N}^d \\ |l| \leq L}} \langle z, \varphi_l \rangle_{L^2} \langle y(t) - y(s), \varphi_l \rangle_{L^2} \right|$$

$$+ |\langle y(t) - y(s), z - z_0 \rangle_{L^2}|$$

$$\leq \|z\|_2 (1 + 2R) 2^L \rho\big(y(t), y(s)\big) + 2R\varepsilon_0 < \varepsilon .$$

if $\varepsilon_0 < (4R)^{-1}\varepsilon$ and $\delta > 0$ is chosen such that $|s - t| < \delta$ always implies $\rho\big(y(t), y(s)\big) < \|z\|_2^{-1}(2 + 4R)^{-1} 2^{-L}\varepsilon$. Thus, $\langle y(\cdot), z \rangle_{L^2} \in \mathcal{C}([0, T])$ and consequently $y \in \mathcal{C}^*\big(0, T; L^2(\Omega)\big)$.

Applying Proposition 3.33 finally yields that we have indeed weak convergence in $\mathcal{C}^*\big(0, T; L^2(\Omega)\big)$. ◀

Chapter 4

Weighted and directional Sobolev spaces

In this chapter, the analysis is carried out for special spaces of weak weighted and directional derivatives which are analog to the usual Sobolev spaces. The motivation for the consideration of these kinds of spaces comes, on the one hand, from finding an optimal solution for the control problem (2.8) and on the other hand, from finding a handy device which allows a characterization of the spaces V_p and $\mathcal{V}_p$, respectively, which contain, by definition, classes of Cauchy sequences. Additionally, the characterization in terms of weak derivatives is useful for obtaining additional properties of the set of spaces V_p when p is varied, since these vector fields enter the defining variational equations of the weak weighted and directional derivative in a way which can be handled analytically. Especially, the closedness of the weak weighted and directional derivative with respect to both y and $p_1, \ldots, p_d$ can be proven. This is crucial for the existence of solutions for the optimal control problem (2.8).

In the following, we first introduce appropriate vector spaces of admissible weights and directions for the respective definitions of weak derivatives. These will be spaces which are closely related to spaces of Lipschitz functions considered as dual spaces equipped with a weak*-convergence. Then, the weak weighted and directional derivative with respect to these admissible weights and directions are introduced and analyzed. Furthermore, spaces of weak weighted and directional derivatives analog to the Sobolev spaces are investigated and dense sets are derived. It will turn out that $\mathcal{C}^\infty(\Omega)$ (with finite norm) is dense leading to a statement which is analog to the famous $H = W$ by Meyers and Serrin [MS64], especially for V_p. Additionally, the density of $\mathcal{C}^\infty(\overline{\Omega})$ will be proven under the prerequisite that Ω is a Lipschitz domain. Finally, in Chapter 5, this setting is transferred to the situation of time-dependent spaces, which gives a characterization of $\mathcal{V}_p$.

4.1 The space $H^{1,\infty}$ as a dual space

To introduce a variational definition of weak weighted and directional derivatives, it turns out to be necessary that for the weights w and directions q the
derivatives ∇w and $\operatorname{div} q$ exist in the weak sense and are bounded almost everywhere. Of course, this can also be described in terms of Sobolev spaces, one
would e.g. require

$$w \in H^{1,\infty}(\Omega) \quad , \quad \|w\|_\infty \leq 1 \quad , \quad \|\nabla w\|_\infty \leq M_x^{-1} \tag{4.1}$$

for a "small" $M_x > 0$, which can interpreted as a "minimal feature size" (since,
roughly speaking, oscillations of amplitude 1 of size less that M_x cannot occur). However, it is known that $H^{1,\infty}(\Omega)$ lacks important properties, such as
reflexivity. Hence, a sequence w_k satisfying the above assumption does not necessarily have a weakly convergent subsequence. In this situation, remedy can be
found by interpreting $H^{1,\infty}(\Omega)$ as the dual space of a separable normed space
and to consider weak*-convergence. We want to emphasize that this notion of
weak*-convergence is sometimes mentioned in the literature, however only as
the analog to weak convergence in $H^{1,r}(\Omega, \mathbb{R}^k)$ for some $r < \infty$. In the following, we will prove some extensions and generalizations which are not common
in the literature.

Before turning towards the definition of a predual space of $H^{1,\infty}(\Omega, \mathbb{R}^k)$, note
that an element $v \in H^{1,\infty}(\Omega, \mathbb{R}^k)$ is locally Lipschitz in Ω (in a certain sense, see
[EG92]). Thus one can say that functions in $H^{1,\infty}(\Omega, \mathbb{R}^k)$ are "almost Lipschitz
functions". We will give a characterization in Proposition 4.5. Moreover, it is
important to mention that spaces whose duals are spaces of Lipschitz functions
have already been constructed in general vector space theory. For instance, the
Arens-Eells space or Lipschitz free space (see [AE56]) associated with a metric
space X containing a special point x^* is the predual of the space of Lipschitz
functions vanishing at that special point. Our construction is more specialized
since we only consider the metric space Ω with the Euclidean distance and there
is no special point given.

Let us introduce the following normed space, which depends on the constant
M_x.

Definition 4.1. Let Ω a bounded domain, $M_x > 0$ and $k \geq 1$ be given. Define
the space $Y_k = L^1(\Omega, \mathbb{R}^k)$ equipped with the norm

$$\|v\|_{Y_k} = \inf \left\{ \|v_1\|_1 + \|v_2\|_1 \mid (v_1, v_2) \in \Sigma(v) \right\} \tag{4.2}$$

where

$$\Sigma(v) = \{(v_1, v_2) \in L^1(\Omega, \mathbb{R}^k) \times H_0^{1,1}(\Omega, \mathbb{R}^{k \times d}) \mid v = v_1 - M_x \operatorname{div} v_2\} \ .$$

Remark 4.2. Obviously, the definition of Y_k depends on the domain Ω and the parameter M_x. Throughout the following, it is assumed that M_x is a fixed constant. Therefore, it will be omitted in the notation of Y_k. The dependence of Y_k on Ω, however, will sometimes be denoted by $Y_k(\Omega)$, especially when there is a chance of ambiguity.

Proposition 4.3. *The space Y_k according to Definition 4.1 is a separable normed space. Its dual is given by*

$$Y_k^* = H^{1,\infty}(\Omega, \mathbb{R}^k) \quad , \quad \|w\|_{Y_k^*} = \max\{\|w\|_\infty, M_x\|\nabla w\|_\infty\} \ .$$

Proof. First, $\Sigma(v) \neq \emptyset$ for each $v \in L^1(\Omega, \mathbb{R}^k)$, hence the norm is well-defined as a function on $L^1(\Omega, \mathbb{R}^k)$. The positive homogeneity and the triangle inequality can be directly obtained from the definition. The positive definiteness follows by a closedness argument. Let $v \in Y_k$ such that $\|v\|_{Y_k} = 0$. This means there are sequences $v_{l,1} \in L^1(\Omega, \mathbb{R}^k)$ and $v_{l,2} \in H_0^{1,1}(\Omega, \mathbb{R}^{k \times d})$ such that

$$v = v_{l,1} - M_x \operatorname{div} v_{l,2} \quad , \quad \|v_{l,1}\|_1 + \|v_{l,2}\|_1 \to 0 \ .$$

Consequently, $v_{l,2} \to 0$ in $L^1(\Omega, \mathbb{R}^{k \times d})$ as well as $\operatorname{div} v_{l,2} = M_x^{-1}(v_{l,1} - v) \to -M_x^{-1}v$ in $L^1(\Omega, \mathbb{R}^k)$. By the closedness of the weak divergence mapping between $H_0^{1,1}(\Omega, \mathbb{R}^{k \times d}) \to L^1(\Omega, \mathbb{R}^k)$, it follows that $-M_x^{-1}v = \operatorname{div} 0 = 0$.

Next, by definition, the Y_k-norm is weaker than the L^1-norm,

$$\|v\|_{Y_k} \leq \|v\|_1 \quad \text{or} \quad L^1(\Omega, \mathbb{R}^k) \hookrightarrow Y_k \ .$$

We know that $L^1(\Omega, \mathbb{R}^k)$ is separable, and so is Y_k by density of the embedding. However, the space is not complete.

The above dense embedding also implies that $Y_k^* \hookrightarrow L^\infty(\Omega, \mathbb{R}^k)$. Hence, an element $w \in Y_k^*$ can be interpreted as a L^∞-function which acts on $v \in L^1(\Omega, \mathbb{R}^k)$ via

$$v \mapsto \int_\Omega w \cdot v \, \mathrm{d}x \ .$$

Additionally, for each $v \in \mathcal{C}_0^\infty(\Omega, \mathbb{R}^{k \times d})$,

$$\left| \int_\Omega w \cdot \operatorname{div} v \, \mathrm{d}x \right| \leq M_x^{-1}\|w\|_{Y_k^*}\|v\|_1$$

holds, meaning that the distributional derivative of w is in $L^\infty(\Omega, \mathbb{R}^{k\times d})$. Consequently, $w \in H^{1,\infty}(\Omega, \mathbb{R}^k)$. Furthermore, for each $(v_1, v_2) \in L^1(\Omega, \mathbb{R}^k) \times H_0^{1,1}(\Omega, \mathbb{R}^{k\times d})$,

$$\left| \int_\Omega w \cdot v_1 - M_x w \cdot \operatorname{div} v_2 \, \mathrm{d}x \right| \leq \left| \int_\Omega w \cdot v_1 + M_x \nabla w \cdot v_2 \, \mathrm{d}x \right| \leq$$

$$\leq \max\left\{ \|w\|_\infty, M_x \|\nabla w\|_\infty \right\} (\|v_1\|_1 + \|v_2\|_1),$$

implying $\max\left\{ \|w\|_\infty, M_x \|\nabla w\|_\infty \right\} \leq \|w\|_{Y_k^*}$. Conversely, the above also holds for each $w \in H^{1,\infty}(\Omega, \mathbb{R}^k)$. This shows $H^{1,\infty}(\Omega, \mathbb{R}^k) \hookrightarrow Y_k^*$ with $\|w\|_{Y_k^*} \leq \max\left\{ \|w\|_\infty, M_x \|\nabla w\|_\infty \right\}$.

Taking both implications together yields $Y_k^* = H^{1,\infty}(\Omega, \mathbb{R}^k)$ with the above stated norm in the sense of Banach-space isometry. ◀

Remark 4.4. Note that the definition of Y_k (and consequently the definition of Y_k^*) depends on the constant M_x. In the following, whenever we write Y_k, we implicitly assume that this M_x is already given.

The following connection between locally Lipschitz continuous functions and $H^{1,\infty}(\Omega, \mathbb{R}^k)$ holds for general bounded domains Ω.

Proposition 4.5. *A function $w : \Omega \to \mathbb{R}^k$ belongs to $H^{1,\infty}(\Omega, \mathbb{R}^k) = Y_k^*$ if and only if w is bounded and locally Lipschitz continuous with Lipschitz constant C not depending on Ω' for each convex $\Omega' \subset\subset \Omega$.*

In particular, each $w \in Y_k^*$ is differentiable almost everywhere and the gradient satisfies $\|\nabla w\|_\infty \leq C'C$ with some $C' > 0$ and C the above Lipschitz constant.

Proof. This is an extension of Theorem 4.2.5 in [EG92] which states that w being locally Lipschitz is equivalent to $w \in H_{\mathrm{loc}}^{1,\infty}(\Omega, \mathbb{R}^k)$.

To prove the only if part, it remains to show the independence of the Lipschitz constant on convex sets $\Omega' \subset\subset \Omega$. With the standard mollifier G and its dilated versions G_ε, we define $w_\varepsilon = w * G_\varepsilon$ on Ω' for sufficiently small $\varepsilon > 0$. We have

$$w_\varepsilon(x) - w_\varepsilon(\xi) = \int_0^1 \nabla w_\varepsilon(\xi + s(x - \xi))(x - \xi) \, \mathrm{d}x$$

since Ω' is convex and w_ε is smooth, hence

$$|w(x) - w(\xi)| = \lim_{\varepsilon \to 0} |w_\varepsilon(x) - w_\varepsilon(\xi)| \leq \|\nabla w_\varepsilon\|_\infty |x - \xi| \leq \|\nabla w\|_\infty |x - \xi|,$$

using that w_ε converges uniformly in Ω' as well as $\nabla w_\varepsilon = \nabla w * G_\varepsilon$ (see Theorem 4.2.1 in [EG92]).

Conversely, we only have to show that the weak gradient ∇w belongs to $L^\infty(\Omega, \mathbb{R}^{k \times d})$. It is known that the L^∞-norm of the gradient in a $\Omega' \subset \Omega$ can be estimated by the Lipschitz constant in Ω'. In particular, denoting by C' the constant which arises from the change from operator matrix norm to the Frobenius norm (i.e. $|A| \leq C' \sup_{|u| \leq 1} |Au|$),

$$\operatorname*{ess\,sup}_{\xi \in B_\varepsilon(x)} |\nabla w(\xi)| \leq C'C \quad \text{for all} \quad \overline{B_\varepsilon(x)} \subset \Omega$$

by assumption, hence $\nabla w \in L^\infty(\Omega, \mathbb{R}^{k \times d})$ since Ω is open.

The last statement as well as the norm estimate follow from the fact that one can choose a sequence Ω'_k of open convex sets satisfying

$$\bigcup_k \Omega'_k = \Omega \quad , \quad \overline{\Omega'_k} \subset\subset \Omega \text{ for } k = 1, 2, \ldots$$

and applying Rademacher's theorem to each $\overline{\Omega'_k}$ thus reducing the set where w is not differentiable to a countable union of null-sets which is also a null-set. ◀

Remark 4.6. Local Lipschitz continuity of $w \in Y_k^*$ in the above sense means the existence of a neighborhood (of each $x \in \Omega$) in which the function is Lipschitz continuous. This can be extended to Lipschitz continuity in $\Omega' \subset\subset \Omega$ as follows.

First suppose that Ω' is connected. This is not less general since each $\Omega' \subset\subset \Omega$ is subset of a connected compact set in Ω. Choose an $\varepsilon > 0$ such that $\Omega_\varepsilon = \Omega' + B_\varepsilon(0)$ satisfies $\overline{\Omega_\varepsilon} \subset\subset \Omega$ and define for each $x, \xi \in \Omega'$ the usual distance

$$\rho(x, \xi) = \inf \left\{ \int_0^1 |\gamma'(s)| \, ds \ \Big| \ \gamma \in C^1([0,1], \overline{\Omega_\varepsilon}), \gamma(0) = x, \gamma(1) = \xi \right\}.$$

One can easily deduce the triangle inequality for ρ as well as $\rho(x, x) = 0$ for $x \in \Omega'$. It moreover follows that $\rho(x, \xi) = |x - \xi|$ if $|x - \xi| < \varepsilon$, hence

$$|\rho(\tilde{x}, \tilde{\xi}) - \rho(x, \xi)| \leq |\tilde{x} - x| + |\tilde{\xi} - \xi| < \delta$$

if $|\tilde{x} - x| < \delta/2 < \varepsilon, |\tilde{\xi} - \xi| < \delta/2 < \varepsilon$, yielding the continuity of ρ on $\Omega' \times \Omega'$. In particular, $\rho(x, \xi)/|x - \xi|$ is continuous on $(\Omega' \times \Omega') \backslash \{|x - \xi| < \varepsilon\}$, admitting a maximum $C_{\Omega'} \geq 1$. On $(\Omega' \times \Omega') \cap \{0 < |x - \xi| < \varepsilon\}$ we have $\rho(x, \xi)/|x - \xi| = 1$, hence the function is bounded on $\Omega' \times \Omega'$.

Recalling the above proof we deduce for each $w \in Y_k^*$ that

$$|w(x) - w(\xi)| \leq \rho(x, \xi) \|\nabla w\|_\infty \leq \sup_{\tilde{x}, \tilde{\xi} \in \Omega', \tilde{x} \neq \tilde{\xi}} \frac{\rho(\tilde{x}, \tilde{\xi})}{|\tilde{x} - \tilde{\xi}|} \|\nabla w\|_\infty |x - \xi|$$

$$\leq C_{\Omega'} \|\nabla w\|_\infty |x - \xi|.$$

It follows that the Lipschitz constant of w on the compact subset Ω' can be estimated by $C_{\Omega'}\|\nabla w\|_\infty$ where $C_{\Omega'}$ only depends on Ω'. In particular, $C_{\Omega'} = 1$ if Ω' is convex.

4.1.1   Sequential weak*-convergence and compactness

Remark 4.7. The requirement (4.1) can now be expressed equivalently by $w \in Y_1^*$ with $\|w\|_{Y_1^*} \leq 1$. It is moreover possible to speak of weak*-convergence of a sequence of functions in Y_k^*, i.e. $w_l \overset{*}{\rightharpoonup} w$ in Y_k^* if and only if

$$\lim_{l\to\infty} \int_\Omega w_l \cdot v \, \mathrm{d}x = \int_\Omega w \cdot v \, \mathrm{d}x \quad \text{for all } v \in L^1(\Omega, \mathbb{R}^k) \,.$$

Proposition 4.8. *Let $\{w_l\}$ in Y_k^* be a bounded sequence. Then w_l possesses a weakly* convergent subsequence in the sense of Remark 4.7.*

Proof. By Proposition 4.3, we know that the predual space Y_k is separable, hence bounded sequences in Y_k^* admit weakly* convergent subsequences (confer [Yos80], for example). ◀

The notion of weak*-convergence can be characterized by means of strong convergence in $\mathcal{C}(\Omega)$ and weak* convergence of the derivatives.

Lemma 4.9. *Let $w_l, w \in Y_k^*$, $l \in \mathbb{N}$ be given such that*

$$w_l \overset{*}{\rightharpoonup} w \text{ in } L^\infty(\Omega, \mathbb{R}^k) \quad , \quad \nabla w_l \overset{*}{\rightharpoonup} \nabla w \text{ in } L^\infty(\Omega, \mathbb{R}^{k\times d}) \,.$$

Then all w_l are continuous and the sequence also converges in $\mathcal{C}(\Omega, \mathbb{R}^k)$.

Proof. We first recall that each w_l is Lipschitz continuous in $\Omega' \subset\subset \Omega$ with a constant $C_{\Omega'}$ which is only dependent on Ω' and on the bound for $\|\nabla w_l\|_\infty$, see Proposition 4.5 and Remark 4.6. Since this holds for all such Ω', $w_l \in \mathcal{C}(\Omega, \mathbb{R}^k)$.

Now pick a $\Omega' \subset\subset \Omega$ with $\Omega' = \overline{\mathrm{int}(\Omega')}$. We want to show that $\{w_l\}$ converges in $\mathcal{C}(\Omega', \mathbb{R}^k)$ by applying the theorem of Arzelà and Ascoli. We have to verify the equicontinuity in Ω'. For each $x \in \Omega'$ and $\varepsilon > 0$ we can choose $0 < \delta < C_{\Omega'}^{-1}\varepsilon$ which gives

$$|w_l(x) - w_l(\xi)| \leq C_{\Omega'}|x - \xi| < \varepsilon$$

for all l and $|x - \xi| < \delta$. Now, there exists a subsequence of $\{w_l\}$ which converges in $\mathcal{C}(\Omega', \mathbb{R}^k)$ to a $\tilde{w}$. For the subsequence there also holds $\nabla w_l \overset{*}{\rightharpoonup} \nabla w$ in $L^\infty(\Omega', \mathbb{R}^{k\times d})$, hence the closedness of the gradient yields that $\nabla \tilde{w} = \nabla w$. But this means that on each connected component $\Omega'' \subset \Omega'$ with non-empty interior we have $\tilde{w} = w + c$ with a $c \in \mathbb{R}^k$. But testing $w_l \in L^\infty(\Omega, \mathbb{R}^k)$ with $\chi_{\Omega''} c'$, $c' \in \mathbb{R}^k$ leads to

$$\int_{\Omega''} \tilde{w} \cdot c' \, \mathrm{d}x = \lim_{l \to \infty} \int_{\Omega''} w_l \cdot c' \, \mathrm{d}x = \int_{\Omega''} w \cdot c' \, \mathrm{d}x$$

for all $c' \in \mathbb{R}^k$ which means that actually $c = 0$. Since this holds for each connected component of Ω' we conclude $\tilde{w} = w$. Additionally, this is true for each subsequence, so the whole sequence converges to w in $\mathcal{C}(\Omega', \mathbb{R}^k)$.

Now choose for $\varepsilon > 0$ the compact set $\Omega'_\varepsilon = \{x \in \Omega \mid \mathrm{dist}(x, \partial\Omega) \geq \varepsilon\}$ which satisfies $\Omega'_\varepsilon = \overline{\mathrm{int}(\Omega'_\varepsilon)}$ for ε small enough. By the above arguments, $w_l \to w$ in $\mathcal{C}(\Omega'_\varepsilon, \mathbb{R}^k)$, hence for $l \geq l_0$ we have

$$|w_l(x) - w(x)| < \varepsilon$$

for each $x \in \Omega'_\varepsilon$. Moreover, for each $\xi \in \Omega$ an $x \in \Omega'_\varepsilon$ with $|x - \xi| < \varepsilon_0 < \varepsilon$ can be found and by local Lipschitz continuity on $\overline{B_{\varepsilon_0}(x)} \subset\subset \Omega$ we have

$$|w_l(\xi) - w(\xi)|$$
$$\leq |w_l(\xi) - w_l(x)| + |w_l(x) - w(x)| + |w(x) - w(\xi)| \leq (2C + 1)\varepsilon$$

where C is a bound on $\|\nabla w_l\|_\infty$. Consequently, $\|w_l - w\|_\infty \to 0$ on Ω which proves the desired statement. ◄

Corollary 4.10. *In particular, each $w \in Y_k^*$ belongs to $\mathcal{C}(\Omega, \mathbb{R}^k)$.*

The previous lemma enables us to show the following characterization:

Proposition 4.11. *Let $\{w_l\}$ in Y_k^*. Then $w_l \overset{*}{\rightharpoonup} w$ is equivalent to $w_l \to w$ in $\mathcal{C}(\Omega, \mathbb{R}^k)$ and $\nabla w_l \overset{*}{\rightharpoonup} \nabla w$ in $L^\infty(\Omega, \mathbb{R}^{k \times d})$.*

Proof. We will first assume that $w_l \overset{*}{\rightharpoonup} w$ in Y_k^*. It immediately follows that $w_l \overset{*}{\rightharpoonup} w$ in $L^\infty(\Omega, \mathbb{R}^k)$ and $\nabla w_l \overset{*}{\rightharpoonup} \nabla w$ in $L^\infty(\Omega, \mathbb{R}^{k \times d})$. Thus we get the convergence $w_l \to w$ in $\mathcal{C}(\Omega, \mathbb{R}^k)$ by applying Lemma 4.9 to the sequence w_l.

On the other hand, suppose $w_l \to w$ in $\mathcal{C}(\Omega, \mathbb{R}^k)$ as well as $\nabla w_l \overset{*}{\rightharpoonup} \nabla w$ in $L^\infty(\Omega, \mathbb{R}^{k \times d})$. With the usual dual pairing we get that $w_l \in Y_k^*$ as well as the convergence

$$\int_\Omega w_l \cdot v \, \mathrm{d}x \to \int_\Omega w \cdot v \, \mathrm{d}x$$

for each $v \in L^1(\Omega, \mathbb{R}^k)$. Hence $w_l \overset{*}{\rightharpoonup} w$ in Y_k^* (see Remark 4.7). ◄

In addition to spaces of continuous functions with bounded weak derivatives we also need a weaker space of vectors which only requires the existence of a bounded divergence.

Definition 4.12. In analogy to Definition 4.1, let

$$Y_{\mathrm{div}} = L^1(\Omega, \mathbb{R}^d) \qquad \|v\|_{Y_{\mathrm{div}}} = \inf \left\{ \|v_1\|_1 + \|v_2\|_1 \mid (v_1, v_2) \in \Sigma^{\mathrm{div}}(v) \right\}$$

with

$$\Sigma^{\mathrm{div}}(v) = \left\{ (v_1, v_2) \in L^1(\Omega, \mathbb{R}^d) \times H_0^{1,1}(\Omega) \mid v = v_1 - M_x \nabla v_2 \right\}.$$

Remark 4.13.

(a) In complete analogy to the proof of Proposition 4.3, it is possible to obtain that Y_{div} is a separable normed space with dual space

$$Y_{\mathrm{div}}^* = \left\{ q \in L^\infty(\Omega, \mathbb{R}^d) \mid \mathrm{div}\, q \in L^\infty(\Omega) \right\}$$
$$\|q\|_{Y_{\mathrm{div}}^*} = \max \left\{ \|q\|_\infty, M_x \|\mathrm{div}\, q\|_\infty \right\}.$$

Here, the divergence is taken in the weak sense which arises from the duality given by integration. Also observe the weakly* sequentially continuous embedding $Y_d^* \hookrightarrow Y_{\mathrm{div}}^*$.

(b) A statement analog to Proposition 4.11, however, is not true. The reason for this is that the existence of a bounded weak divergence does not imply continuity. But one still has that bounded sequences of Y_{div}^* admit weakly* convergent subsequences, again by the separability of Y_{div} (see Proposition 4.8).

4.1.2   Superposition operators in Y_k^*

In the remainder of this section, we focus our interest on the mapping properties of superposition operators on Y_k^*. Note that the Sobolev spaces with exponent ∞ play a special role here, the cases with smaller exponents are completely characterized [MM79]. It will turn out that superposition is even weakly* sequentially continuous in many cases. Moreover, we show under which circumstances Fréchet-differentiability can be established. Here, it is important into which spaces the superposition operator maps.

Proposition 4.14. *Let $\varphi : \mathbb{R}^k \to \mathbb{R}^l$ be a locally Lipschitz continuous mapping. Then*

$$T_\varphi : Y_k^* \to Y_l^* \quad \text{defined by} \quad (T_\varphi w)(x) = \varphi\big(w(x)\big)$$

maps bounded sets of Y_k^* to bounded sets in Y_l^* and is weakly sequentially continuous.*

Proof. Let $w \in Y_k^*$ be given. Choose an arbitrary convex $\Omega' \subset\subset \Omega$. Since $|w(x)| \leq \|w\|_\infty$, the superposed function $\tilde{w} = T_\varphi(w)$ satisfies

$$|\tilde{w}(x) - \tilde{w}(\xi)| \leq C_1|w(x) - w(\xi)| \leq C_1 C|x - \xi|$$

for all $x, \xi \in \Omega'$, where C_1 and C are the Lipschitz constants of φ in $B_{\|w\|_\infty}(0)$ and w in Ω' (which is independent of Ω'), respectively. By Remark 4.6, there moreover holds $C \leq \|\nabla w\|_\infty$. It follows that $\tilde{w}$ is bounded as well as locally Lipschitz continuous and consequently in the space $H^{1,\infty}(\Omega, \mathbb{R}^l) = Y_l^*$ (confer Propositions 4.3 and 4.5). Moreover, the norm can be estimated by

$$\|\tilde{w}\|_{Y_l^*} = \max\{\|\tilde{w}\|_\infty, M_x\|\nabla\tilde{w}\|_\infty\}$$
$$\leq \max\{\varphi(0) + C_1\|w\|_\infty, M_x C_1 C'\|\nabla w\|_\infty\},$$

where C' is the constant from Proposition 4.5, establishing the stated boundedness.

Let $w_m \overset{*}{\rightharpoonup} w$ in Y_k^* be given. Then, according to Proposition 4.11, $w_m \to w$ in $\mathcal{C}(\Omega, \mathbb{R}^k)$. By the above arguments as well as Proposition 4.8, the $\tilde{w}_m = T_\varphi(w_m)$ are bounded in Y_l^* and admit a weakly*-convergent subsequence, not relabeled, with limit $\tilde{w}$. This also means that $\tilde{w}_m \to \tilde{w}$ in $\mathcal{C}(\Omega, \mathbb{R}^l)$. Now, T_φ maps $\mathcal{C}(\Omega, \mathbb{R}^k) \to \mathcal{C}(\Omega, \mathbb{R}^l)$ continuously, thus $\tilde{w} = T_\varphi(w)$. The usual subsequence argument then yields the stated weak* sequential continuity. ◀

Proposition 4.15. *Let $\varphi : \mathbb{R}^k \to \mathbb{R}^l$ and consider the superposition operator* $T_\varphi(w)(x) = \varphi(w(x))$.

1. *If φ is continuously differentiable, then T_φ is continuously differentiable from $Y_k^* \to L^\infty(\Omega, \mathbb{R}^l)$.*

2. *Moreover, if φ is twice continuously differentiable, then T_φ is continuously differentiable from $Y_k^* \to Y_l^*$.*

In both cases, the derivative is given by $DT_\varphi(w)h = \nabla\varphi(w)h$ and is uniformly continuous on bounded sets in Y_k^*.

Proof. We will verify that $h \mapsto \nabla\varphi(w)h$ is the desired derivative in the respective spaces and is continuous with respect to w.

Consider the first case where φ is continuously differentiable and the superposition operator maps into the space $L^\infty(\Omega, \mathbb{R}^l)$. Let $w \in Y_k^*$ be given. First, we will show that $h \mapsto \nabla\varphi(w)h$ is a linear and continuous mapping from $Y_k^* \to L^\infty(\Omega, \mathbb{R}^l)$. The linearity is immediate, and the continuity can easily be verified by estimating

$$\|\nabla\varphi(w)h\|_\infty \leq \|\nabla\varphi(w)\|_\infty\|h\|_\infty$$

and noting that, due to the continuity of $\nabla\varphi$ as well as the boundedness of w, the respective norms are finite.

Since φ is differentiable, we have for each $h \in Y_k^*$ and $x \in \Omega$ that

$$\varphi\big(w(x) + h(x)\big) - \varphi\big(w(x)\big) - \nabla\varphi\big(w(x)\big)h(x)$$
$$= \int_0^1 \Big(\nabla\varphi\big(w(x) + sh(x)\big) - \nabla\varphi\big(w(x)\big)\Big)h(x) \, ds \ . \quad (*)$$

Now $\nabla\varphi$ is uniformly continuous on the compact set $\overline{B_{2\|w\|_\infty}(0)}$, hence for each $\varepsilon > 0$ one can choose a $\delta \in \left]0, \|w\|_\infty\right[$ such that for $|h(x)| < \delta$ we have

$$\big|\nabla\varphi\big(w(x) + h(x)\big) - \nabla\varphi\big(w(x)\big)\big| < \varepsilon$$

independently of x and implying

$$\|\varphi(w + h) - \varphi(w) - \nabla\varphi(w)h\|_\infty \leq \varepsilon\|h\|_\infty \leq \varepsilon\|h\|_{Y_k^*} \quad (**)$$

whenever $\|h\|_\infty < \delta$, establishing the differentiability.

Next, we show that $w \mapsto \mathrm{D}T_\varphi(w)$ is mapping $Y_k^* \to \mathcal{L}\big(Y_k^*, L^\infty(\Omega, \mathbb{R}^l)\big)$ in a uniformly continuous way on bounded sets in Y_k^*. With a choice of constants analog to the above, we have for each $w \in Y_k^*$ with $\|w\|_\infty \leq C$ and for each $\bar{w}$ with $\|\bar{w} - w\|_\infty < \delta$ that

$$\|\mathrm{D}T_\varphi(\bar{w}) - \mathrm{D}T_\varphi(w)\| \leq \sup_{\|h\|_\infty \leq 1} \|\nabla\varphi(\bar{w})h - \nabla\varphi(w)h\|_\infty < \varepsilon$$

which yields the claimed uniform continuity since δ only depends on C and ε.

To prove the respective statements with Y_l^* instead of $L^\infty(\Omega, \mathbb{R}^l)$, we additionally have to verify that analogous properties hold for the L^∞-norm of the gradient. That T_φ is indeed mapping between $Y_k^* \to Y_l^*$ was already subject of Proposition 4.14. The derivative $\mathrm{D}\varphi(w)h = \nabla\varphi(w)h$ is in $\mathcal{L}\big(Y_k^*, Y_l^*\big)$ since

$$\|\nabla\big(\nabla\varphi(w)h\big)\|_\infty \leq \|\nabla^2\varphi(w)\|_\infty\|\nabla w\|_\infty\|h\|_\infty + \|\nabla\varphi(w)\|_\infty\|\nabla h\|_\infty$$

in addition to the L^∞-norm estimate above.

Regarding the differentiability, we already established

$$\|\varphi(w + h) - \varphi(w) - \nabla\varphi(w)h\|_\infty < \varepsilon\|h\|_{Y_k^*} \ .$$

By proceeding analogously to the above, we can moreover choose $\delta > 0$, only depending on $\|w\|_{Y_k^*}$, such that simultaneously

$$\begin{aligned}\big|\nabla^2\varphi\big(w(x) + h(x)\big) - \nabla^2\varphi\big(w(x)\big)\big| &< \varepsilon \ , \\ \big|\nabla\varphi\big(w(x) + h(x)\big) - \nabla\varphi\big(w(x)\big)\big| &< \varepsilon\end{aligned} \qquad (***)$$

for all $x \in \Omega$ if $\|h\|_\infty < \delta$. Fix such an h and take a point $x \in \Omega$ at which w and h are differentiable. The derivative of the remainder term $(*)$ at x then satisfies

$$
\begin{aligned}
\nabla\big(\varphi(w+h) &- \varphi(w) - \nabla\varphi(w)h\big)(x) \\
&= \nabla\varphi\big(w(x)+h(x)\big)\big(\nabla w(x)+\nabla h(x)\big) - \nabla\varphi\big(w(x)\big)\big(\nabla w(x)+\nabla h(x)\big) \\
&\quad - \nabla^2\varphi\big(w(x)\big)\big(\nabla w(x), h(x)\big) \\
&= \int_0^1 \Big(\nabla^2\varphi\big(w(x)+sh(x)\big) - \nabla^2\varphi\big(w(x)\big)\Big)\big(\nabla(w+h)(x), h(x)\big)\,\mathrm{d}s \\
&\quad + \nabla^2\varphi\big(w(x)\big)\big(\nabla h(x), h(x)\big) .
\end{aligned}
$$

Now for fixed w, h and each $x \in \Omega$ one can find a neighborhood in which w and h are Lipschitz continuous and, due to Rademacher's theorem, differentiable almost everywhere. Consequently, the above identity holds a.e. in Ω leading to

$$
\begin{aligned}
\|\nabla\big(\varphi(w+h) &- \varphi(w) - \nabla\varphi(w)h\big)\|_\infty \\
&\leq \varepsilon\|\nabla(w+h)\|_\infty\|h\|_\infty + \|\nabla^2\varphi(w)\|_\infty\|\nabla h\|_\infty\|h\|_\infty .
\end{aligned}
$$

Together with $(**)$, this yields for each $\|h\|_{Y_k^*} < \delta$,

$$
\begin{aligned}
\|\varphi(w+h) &- \varphi(w) - \nabla\varphi(w)h\|_{Y_l^*} \\
&\leq \varepsilon(1 + M_x\|\nabla w\|_\infty + \delta)\|h\|_{Y_k^*} + \|\nabla^2\varphi(w)\|_\infty\|h\|_{Y_k^*}^2 ,
\end{aligned}
$$

establishing the differentiability.

Finally, $w \mapsto DT_\varphi(w)$ is also continuous as a mapping between the spaces $Y_k^* \to \mathcal{L}(Y_k^*, Y_l^*)$. Choose a $w \in Y_k^*$ with $\|w\|_{Y_k^*} \leq C$ and $\bar{w} \in Y_k^*$ with $\|w - \bar{w}\|_{Y_k^*} < \delta$ where $\delta > 0$ is such that the inequalities $(***)$ hold for $h = \bar{w} - w$. Choose a $h_1 \in Y_k^*$ and consider the derivative of $\nabla\varphi(w)h_1$ applied to a $h_2 \in \mathbb{R}^k$ at some point $x \in \Omega$ where $w, \bar{w}$ and h_1 are differentiable:

$$
\nabla\big(\nabla\varphi(w)h_1\big)(x)h_2 = \nabla^2\varphi\big(w(x)\big)\big(h_1(x), \nabla w(x)h_2\big) + \nabla\varphi\big(w(x)\big)\nabla h_1(x)h_2
$$

and the analog statement for $\bar{w}$, so

$$
\begin{aligned}
\big|\nabla\big(\nabla\varphi(\bar{w})h_1 &- \nabla\varphi(w)h_1\big)(x)h_2\big| \\
&\leq \big|\big(\nabla^2\varphi(w) - \nabla^2\varphi(\bar{w})\big)\big(h_1, \nabla w h_2\big)(x)\big| \\
&\quad + \big|\nabla^2\varphi(\bar{w})\big(h_1, (\nabla w - \nabla\bar{w})h_2\big)(x)\big| \\
&\quad + \big|\big(\big(\nabla\varphi(w) - \nabla\varphi(\bar{w})\big)\nabla h_1\big)(x)h_2\big| \\
&< \varepsilon|h_1(x)||\nabla w(x)||h_2| + \delta M_x^{-1}|\nabla^2\varphi(\bar{w})(x)||h_1(x)||h_2| \\
&\quad + \varepsilon|\nabla h_1(x)||h_2| .
\end{aligned}
$$

Taking the supremum over $|h_2| \leq 1$ and over a.e. $x \in \Omega$ yields

$$\left\|\nabla\big(\nabla\varphi(\bar{w})h_1 - \nabla\varphi(w)h_1\big)(x)\right\|_\infty$$
$$\leq \varepsilon C_1\big(\|\nabla w\|_\infty\|h_1\|_\infty + \|\nabla h_1\|_\infty\big) + \delta C_1 M_x^{-1}\|\nabla^2\varphi(\bar{w})\|_\infty\|h_1\|_\infty$$
$$\leq \max\{\varepsilon,\delta\}C_2\|h_1\|_{Y_k^*} ,$$

where C_1 collects the change from the operator matrix-norm to the Frobenius matrix-norm and $C_2 = C_1\big(\|\nabla w\|_\infty + M_x^{-1}(1 + \|\nabla^2\varphi(\bar{w})\|_\infty)\big)$. This implies the desired continuity estimate if the supremum over all $\|h_1\|_{Y_k^*} \leq 1$ is taken and $\delta < \varepsilon$. Analog to the above, the δ only depends on $\|w\|_{Y_k^*}$ and ε, giving the uniform continuity on bounded sets in Y_k^*.   ◀

4.2   Weak weighted and directional derivatives

This section is devoted to the analysis of the properties of weak weighted and directional derivatives. Weighted Sobolev spaces with weight $w \geq 0$, which are usually defined as the set of locally integrable functions y for which

$$\int_\Omega \big(|y|^r + |\nabla y|^r\big)w \,\mathrm{d}x < \infty ,$$

holds, are well-analyzed and widely known in the literature (see [Kuf80] and [Tri95], for example). In particular, it is known that for general measurable weights, approximation with smooth functions fails ($H \neq W$, see [Zhi98]). But this is an important property since it is closely connected with the closedness of the gradients in the weighted L^r-space and the existence of Poincaré-Wirtinger inequalities, see [Kil97]. The latter is mainly important for degenerate elliptic equations and the failure is not severe if degenerate parabolic equations like (3.1) are considered. The former however is important for optimal control. Often, one is able to choose minimizing sequences of controls and states whose limits should be the optimal state associated with the optimal control. This can be ensured if certain closedness properties hold. There are sufficient conditions for weights such that the associated weighted Sobolev space satisfies $H = W$, see [Kil94]. Roughly speaking, for the exponent r, it is sufficient that w is in the A_r-class of Muckenhoupt weights which amounts to $w^{1/(r-1)} \in L_{\mathrm{loc}}^1(\Omega)$ as well as

$$\sup_{Q \subset \Omega} \Big(\frac{1}{|Q|}\int_Q w \,\mathrm{d}x\Big)\Big(\frac{1}{|Q|}\int_\Omega w^{1/(1-r)} \,\mathrm{d}x\Big)^{r-1} < \infty$$

where the supremum over all cubes $Q \subset \Omega$ is taken.

However, it is hard to find an analytic structure which reflects these conditions properly and is simultaneously feasible with respect to optimal control.

Moreover, the local integrability of $w^{1/(1-r)}$ may fail if w vanishes "too fast". The reason for this is that the classical weak derivative is not able to capture arbitrarily rapid decay. It will turn out that with the notion of the weak weighted derivative, it is possible to overcome these difficulties. Moreover, it possesses convenient properties with respect to the analysis of optimal control problems such as (2.8).

4.2.1 The weak weighted derivative

First, we introduce the weak weighted derivative with respect to a weight $w \in Y_1^*$ and examine its properties.

Definition 4.16. Let $y : \Omega \to \mathbb{R}$ be locally integrable and $w \in Y_1^*$. Then the *weak weighted gradient* $v = w\nabla y$ is defined as the locally integrable function $v : \Omega \to \mathbb{R}^d$ fulfilling

$$\int_\Omega v \cdot z \, \mathrm{d}x = -\int_\Omega y(w \operatorname{div} z + \nabla w \cdot z) \, \mathrm{d}x \quad \text{for all } z \in \mathcal{C}_0^\infty(\Omega, \mathbb{R}^d) . \tag{4.3}$$

Remark 4.17.

(a) Note that this definition allows for functions which are of lower regularity than weakly differentiable functions: Since w can be arbitrarily close to zero on non-null sets in Ω, the weak weighted gradient may exist in $L^r(\Omega)$ even if y is not weakly differentiable as long as w gets small in the respective regions.

(b) It is immediate that $w\nabla y$ has to be unique, since it is tested against all functions $z \in \mathcal{C}_0^\infty(\Omega, \mathbb{R}^d)$.

(c) For $y \in H^1(\Omega)$, it is easily verified that $(w\nabla y) = w(\nabla y)$, i.e. the weak weighted derivative is the pointwise product of w and ∇y. The operation $(y, w) \to w\nabla y$ mapping $H^1(\Omega) \times L^\infty(\Omega) \to L^2(\Omega, \mathbb{R}^d)$ is of course bilinear and continuous since

$$\|w\nabla y\|_2 \le \|w\|_\infty (\|y\|_2^2 + \|\nabla y\|_2^2)^{1/2} .$$

This also applies if the space Y_1^* is chosen for w.

In the following, basic properties of the weak weighted gradient are derived. It is pointed out what the connection to the classical weak gradient is. This also turns out to be useful when proving closedness properties. A further discussion will be carried out in the section about Sobolev spaces associated with the weak weighted gradient.

Lemma 4.18. *If $w\nabla y$ exists for a locally integrable y, then (4.3) also holds for vector fields $z \in H^{1,\infty}(\Omega, \mathbb{R}^d)$ with compact support in Ω.*

Proof. This can be seen by using the standard mollifier G and its dilated versions G_ε. Denoting $z_\varepsilon = z * G_\varepsilon$ with $\varepsilon > 0$ small enough such that $z_\varepsilon \in \mathcal{C}_0^\infty(\Omega, \mathbb{R}^d)$, it is clear that the left-hand side of

$$\int_\Omega w\nabla y \cdot z_\varepsilon \, \mathrm{d}x = -\int_\Omega y(w \operatorname{div} z_\varepsilon + \nabla w \cdot z_\varepsilon) \, \mathrm{d}x$$

converges to $\int_\Omega w\nabla y \cdot z \, \mathrm{d}x$ for $\varepsilon \to 0$ since z is Lipschitz continuous in $\operatorname{supp} z$ (Remark 4.6) and hence the convergence is uniform. Moreover, the right-hand side converges pointwise a.e. and can be bounded by an integrable function, thus it also converges by virtue of Lebesgue's dominated-convergence theorem. This can be seen by

$$\int_\Omega \left| y(w \operatorname{div} z_\varepsilon + \nabla w \cdot z_\varepsilon) \right| \, \mathrm{d}x \leq (\|w\|_\infty \|\operatorname{div} z\|_\infty + \|\nabla w\|_\infty \|z\|_\infty) \int_{\Omega'} |y| \, \mathrm{d}x$$

where Ω' denotes a compact subset of Ω in which the supports of z_ε are contained. ◀

Proposition 4.19. *Let $w \in Y_1^*$ and y be locally integrable on Ω such that the weak weighted gradient $w\nabla y \in L^1_{\mathrm{loc}}(\Omega)$ exists. Then*

1. *∇y exists where $w \neq 0$ and there we have $\nabla y = w^{-1}(w\nabla y)$ almost everywhere and*

2. *$w\nabla y = 0$ almost everywhere where $w = 0$.*

Proof. Due to Corollary 4.10, w is continuous in Ω; hence the subset $\Omega_+ = \{x \in \Omega \mid w(x) \neq 0\}$ is relatively open. Moreover, for each $z \in \mathcal{C}_0^\infty(\Omega_+)$ there exists a $\delta > 0$ such that $|w(x)| \geq \delta$ for each $x \in \operatorname{supp} z$. Hence $\bar{z} = w^{-1}z$ is still in $H^{1,\infty}(\Omega_+)$ with compact support in Ω_+ and can be used as a test function (see Lemma 4.18). This yields

$$-\int_{\Omega_+} y \operatorname{div} z \, \mathrm{d}x = -\int_{\Omega_+} y(w \operatorname{div} \bar{z} + \nabla w \cdot \bar{z}) \, \mathrm{d}x = \int_{\Omega_+} \frac{w\nabla y}{w} \cdot z \, \mathrm{d}x$$

so $\nabla y = w^{-1}(w\nabla y)$ exists in the weak sense in Ω_+. Especially, the asserted identity holds for almost every $x \in \Omega_+$.

Now fix an $1 \leq i \leq d$ and a compact $\Omega' \subset \Omega_0 = \{x \in \Omega \mid w(x) = 0\}$ and let $z_\varepsilon = \chi_{\Omega'} * G_\varepsilon$ where G denotes the standard mollifier with $\operatorname{supp} G \subset B_1(0)$ and G_ε its dilated versions. Also note that we choose $\varepsilon > 0$ sufficiently small, i.e. such that $z_\varepsilon \in C_0^\infty(\Omega)$. It is clear that $|\frac{\partial z_\varepsilon}{\partial x_i}(x)| \leq \|\chi_{\Omega'}\|_\infty \|\frac{\partial G_\varepsilon}{\partial x_i}\|_1 \leq \varepsilon^{-1}C_1$ with a suitable $C_1 > 0$. Since $w \in Y_1^*$, we moreover know that $\max\{\|w\|_\infty, \|\nabla w\|_\infty\} \leq C_2$ for some $C_2 > 0$. Hence, if $x^* \in \Omega'$ then for each $x \in \Omega$ with $|x - x^*| \leq \varepsilon$ follows $|w(x)| \leq C_2 C_{\Omega'}\varepsilon$ (with a $C_{\Omega'} > 0$ given by Remark 4.6). Furthermore, $w = 0$ implies $\nabla w = 0$ almost everywhere on the respective subset. Now consider the estimate

$$\left| \int_\Omega y(w\frac{\partial z_\varepsilon}{\partial x_i} + \frac{\partial w}{\partial x_i} z_\varepsilon) \, \mathrm{d}x \right| \leq C_2(C_1 C_{\Omega'} + 1) \int_{\mathrm{supp}\, z_\varepsilon \setminus \Omega'} |y| \, \mathrm{d}x$$

which tends to zero as $\varepsilon \to 0$ since from the compactness of Ω' follows $|\mathrm{supp}\, z_\varepsilon \setminus \Omega'| \to 0$. The weak weighted derivative exists, hence we have, by Lebesgue's dominated-convergence theorem that $\int_{\Omega'} (w\nabla y)_i \, \mathrm{d}x = 0$ for all compact $\Omega' \subset \Omega_0$ and $1 \leq i \leq d$. This yields $w\nabla y = 0$ almost everywhere in Ω_0, the desired result. ◄

4.2.2 The weak directional derivative

The remainder of this section is devoted to the notion of the weak directional derivative. Such a construction can be regarded as a generalization of the weak partial derivative $\frac{\partial y}{\partial x_i}$ to differentiation with respect to vector fields q indicating the direction of derivation. If the classical weak gradient exists (i.e. is a locally integrable function), such a directional derivative is given by $\partial_q y = q \cdot \nabla y$. In order to describe such terms for less regular functions y, the weak directional derivative is introduced.

For the following definition, vector fields $q \in Y_{\mathrm{div}}^*$ are sufficient while later, we require that $q \in Y_d^*$.

Definition 4.20. Let $q \in Y_{\mathrm{div}}^*$ be given and $y \in L_{\mathrm{loc}}^1(\Omega)$. The *weak directional derivative* of y with respect to q is defined as the function $\partial_q y = v \in L_{\mathrm{loc}}^1(\Omega)$ satisfying

$$\int_\Omega y(z \, \mathrm{div}\, q + \nabla z \cdot q) \, \mathrm{d}x = -\int_\Omega vz \, \mathrm{d}x \quad \text{for every } z \in C_0^\infty(\Omega) . \tag{4.4}$$

Remark 4.21.

(a) Again, the weak directional derivative is unique.

(b) Also, $\partial_q y = q \cdot \nabla y$ for $y \in H^1(\Omega)$. Furthermore, the operation $(y, q) \mapsto \partial_q y$ mapping $H^1(\Omega) \times L^\infty(\Omega, \mathbb{R}^d) \to L^2(\Omega)$ is bilinear and continuous (compare with Remark 4.17). Again, this also applies, if the space for q is Y_d^*.

In analogy to the weak weighted derivative, there are also connections to usual weak derivative in one direction. Indeed, one can say that functions, whose weak directional derivative with respect to a q exists locally, are the images of functions weakly differentiable in one direction under a certain coordinate transform induced by q. To establish such connection, however, some work is necessary. In particular, one requires some facts about solving autonomous ordinary differential equations.

Lemma 4.22. *Let $q \in Y_d^*$ and a point $x \in \Omega$ be given such that $q(x) \neq 0$. Then there exists an open neighborhood U of 0 in $\mathbb{R}^d$ and a Lipschitz mapping j on $U \to \Omega' \subset \Omega$ such that Ω' is an open neighborhood of x and j has a Lipschitz-continuous inverse. Moreover, j is partially differentiable in U with respect to the d-th component (denoted by $\xi = (\bar{\xi}, \xi_d)$) with derivative*

$$\frac{\partial j}{\partial \xi_d}(\bar{\xi}, \xi_d) = q\big(j(\bar{\xi}, \xi_d)\big) . \tag{4.5}$$

Proof. Denote by $q_1, \ldots, q_{d-1}$ a set of orthonormal vectors perpendicular to $q(x)$ such that $\{q_1, \ldots q_{d-1}, q(x)\}$ becomes a positive orthogonal basis. By local Lipschitz continuity (with Lipschitz constant C_1), there exists an open neighborhood $\Omega_0' \subset \Omega$ of x such that for each $x' \in \Omega_0'$, the vectors $\{q_1, \ldots, q_{d-1}, q(x')\}$ are still linearly independent and fulfilling

$$q(x) \cdot q(x') \geq \tfrac{1}{2}|q(x)||q(x')| . \tag{4.6}$$

Let j be defined as the solution of the ordinary differential equation

$$\frac{\partial j}{\partial \xi_d}(\bar{\xi}, \xi_d) = q\big(j(\bar{\xi}, \xi_d)\big) \quad , \quad j(\bar{\xi}, 0) = \sum_{i=1}^{d-1} \bar{\xi}_i q_i + x ,$$

which makes sense on the product $U_0 \times]{-\varepsilon}, \varepsilon[$ where $U_0 \subset \mathbb{R}^{d-1}$ is a neighborhood of 0 and ε a sufficiently small constant which arises in the existence theorem of Picard and Lindelöf (and does not depend on $\bar{\xi}$, see [Ama90]).

Let $\xi_1, \xi_2 \in U_0 \times]{-\varepsilon}, \varepsilon[$ with $\xi_{1,d} \leq \xi_{2,d}$ and estimate

$$|j(\xi_1) - j(\xi_2)| \leq |j(\bar{\xi}_1, \xi_{1,d}) - j(\bar{\xi}_2, \xi_{1,d})| + |j(\bar{\xi}_2, \xi_{1,d}) - j(\bar{\xi}_2, \xi_{2,d})|$$

$$\leq |j(\bar{\xi}_1, 0) - j(\bar{\xi}_2, 0)| + \left| \int_0^{\xi_{1,d}} \left| q\big(j(\bar{\xi}_1, s)\big) - q\big(j(\bar{\xi}_2, s)\big) \right| ds \right|$$

$$+ \int_{\xi_{1,d}}^{\xi_{2,d}} \left| q\big(j(\bar{\xi}_2, s)\big) \right| ds$$

$$\leq |\bar{\xi}_1 - \bar{\xi}_2| + C_1 \left| \int_0^{\xi_{1,d}} |j(\bar{\xi}_1, s) - j(\bar{\xi}_2, s)| \, ds \right|$$
$$+ \|q\|_\infty |\xi_{2,d} - \xi_{1,d}|$$
$$\leq e^{\varepsilon C_1} \left(|\bar{\xi}_1 - \bar{\xi}_2| + \|q\|_\infty |\xi_{2,d} - \xi_{1,d}| \right)$$
$$\leq C_2 |\xi_1 - \xi_2|$$

with the help of Gronwall's inequality and a suitable $C_2 > 0$. This yields the Lipschitz continuity of j in $U_0 \times\ ]{-\varepsilon}, \varepsilon[$. Note that in this domain, j is also partially differentiable with respect to ξ_d with derivative $\frac{\partial j}{\partial \xi_d}(\bar{\xi}, \xi_d) = q(j(\bar{\xi}, \xi_d))$ satisfying

$$\left| \frac{\partial j}{\partial \xi_d}(\xi_1) - \frac{\partial j}{\partial \xi_d}(\xi_2) \right| = \left| q(j(\xi_1)) - q(j(\xi_2)) \right|$$
$$\leq C_1 |j(\xi_1) - j(\xi_2)| \leq C_1 C_2 |\xi_1 - \xi_2| \ ,$$

meaning that $\frac{\partial j}{\partial \xi_d}$ is Lipschitz continuous. The partial derivatives of j with respect to $\bar{\xi}$ in 0 trivially exist, leading to

$$\nabla j(0) = \begin{bmatrix} q_1 & \cdots & q_{d-1} & q(x) \end{bmatrix}$$

if j is also (totally) differentiable in 0. But this follows from

$$|j(\xi) - j(0) - \nabla j(0)\xi| = |j(\bar{\xi}, \xi_d) - j(\bar{\xi}, 0) - \xi_d q(x)|$$
$$= \left| \int_0^1 \Big(q(j(\bar{\xi}, s)) - q(j(0, 0)) \Big) \xi_d \, ds \right| \leq C_1 C_2 |\xi| |\xi_d| \ .$$

In particular, j is Hadamard differentiable in 0 meaning that the generalized Jacobian is a singleton and moreover non-singular by the above. This allows us to use the Lipschitz inverse function theorem (see Theorem 3.12 and Chapter 2 in [CLSW98] for details) to deduce the existence of a neighborhood $U = \tilde{U}_0 \times\ ]{-\tilde{\varepsilon}}, \tilde{\varepsilon}[$ of 0 such that the restriction $j : U \to j(U) = \Omega' \subset \Omega_0'$ is invertible with Lipschitz-continuous inverse. ◀

An illustration of the mapping j is given in Figure 4.1. The technical device of coordinate transformations can now be used to pull a function y locally back to a straight coordinate system. I.e. one can also interpret y locally as the image of a function on V under the "curvilinear flow" given by q. Now if y has a weak directional derivative with respect to q, this can be described by the behavior of the pullback in the direction of the d-th component.

Before we turn towards proving a mathematically precise version of this statement, let us note how the determinant of the coordinate transformation behaves.

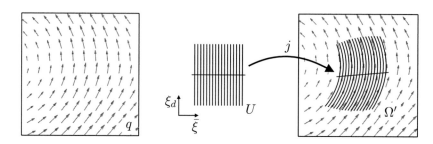

Figure 4.1: Illustration of a local coordinate transformation j associated with a vector field q. On the left-hand side, a sample vector field q is depicted, while in the middle and the right, the coordinate transform j defined on V is illustrated. Note that the mapping takes lines along ξ_d to integral lines with respect to q.

Lemma 4.23. *Let the situation of the last lemma be given. Then, the function* $J(\bar{\xi}, \xi_d) = \det \nabla j(\bar{\xi}, \xi_d)$ *fulfills*

$$\frac{\partial J}{\partial \xi_d}(\bar{\xi}, \xi_d) = \operatorname{div} q\big(j(\bar{\xi}, \xi_d)\big) J(\bar{\xi}, \xi_d) \quad , \quad J(\bar{\xi}, 0) = \frac{q(x)}{|q(x)|} \cdot q\big(j(\bar{\xi}, 0)\big)$$

for almost every $(\bar{\xi}, \xi_d) \in U$. *In particular, J is given a.e. by*

$$J(\bar{\xi}, \xi_d) = \frac{q(x)}{|q(x)|} \cdot q\big(j(\bar{\xi}, 0)\big) e^{\int_0^{\xi_d} \operatorname{div} q\big(j(\bar{\xi}, s)\big) \, ds} \tag{4.7}$$

Proof. We prove that $\frac{\partial}{\partial \xi_d} \nabla j(\bar{\xi}, \xi_d) = \nabla q\big(j(\bar{\xi}, \xi_d)\big) \nabla j(\bar{\xi}, \xi_d)$ with certain initial conditions almost everywhere in U so the desired identity follows from Liouville's formula for Wronski determinants [Ama90].

First recall from the proof in Lemma 4.22 that $\frac{\partial j}{\partial \xi_d} = q\big(j(\bar{\xi}, \xi_d)\big)$ is Lipschitz continuous. Thus, one can differentiate almost everywhere with respect to ξ_d and gets

$$\frac{\partial^2 j}{\partial \xi_d^2}(\bar{\xi}, \xi_d) = \nabla q\big(j(\bar{\xi}, \xi_d)\big) \frac{\partial j}{\partial \xi_d}(\bar{\xi}, \xi_d) \quad , \quad \frac{\partial j}{\partial \xi_d}(\bar{\xi}, 0) = q\bigg(x + \sum_{i=1}^{d-1} \xi_i q_i\bigg)$$

with the q_i already chosen in the proof of Lemma 4.22. Now consider the initial value problems

$$\frac{\partial \varphi_i}{\partial \xi_d}(\bar{\xi}, \xi_d) = \nabla q\big(j(\bar{\xi}, \xi_d)\big) \varphi_i(\bar{\xi}, \xi_d) \quad , \quad \varphi_i(\bar{\xi}, 0) = q_i \quad , \quad i = 1, \dots, d-1 \ .$$

These are time-variant linear equations. The set of $\bar{\xi}$ where $\nabla q\big(j(\bar{\xi}, \xi_d)\big)$ does not exist on a non-null set cannot have positive measure since q would not be almost everywhere differentiable anymore. Thus, for almost every $\bar{\xi}$ the derivative ∇q along the trajectory $j(\bar{\xi}, \xi_d)$ exists almost everywhere with respect to ξ_d. Hence, the above initial value problems admit solutions in the sense of Carathéodory almost everywhere in U [Wal96]. Moreover, due to the boundedness of ∇q, the solutions have to be Lipschitz continuous with respect to ξ_d with the same constant for almost every $\bar{\xi}$.

Pick a $\bar{\xi}$ such that the trajectory $j(\bar{\xi}, \cdot)$ meets almost everywhere points where q is differentiable. By the same arguments as above, we can also suppose that $\frac{\partial j}{\partial \xi_i}(\bar{\xi}, \cdot)$ with $i = 1, \ldots, d-1$ exists almost everywhere since this has to be the case for almost all $\bar{\xi}$. Now

$$\frac{1}{\varepsilon}\left(j(\bar{\xi} + \varepsilon e_i, \xi_d) - j(\bar{\xi}, \xi_d) \right) - \varphi_i(\bar{\xi}, \xi_d)$$
$$= \int_0^{\xi_d} \frac{1}{\varepsilon}\left(q(j(\bar{\xi} + \varepsilon e_i, s)) - q(j(\bar{\xi}, s)) \right) - \nabla q(j(\bar{\xi}, s)) \varphi_i(\bar{\xi}, s) \, \mathrm{d}s$$

which becomes

$$\frac{\partial j}{\partial \xi_i}(\bar{\xi}, \xi_d) - \varphi_i(\bar{\xi}, \xi_d) = \int_0^{\xi_d} \nabla q(j(\bar{\xi}, s)) \left(\frac{\partial j}{\partial \xi_i}(\bar{\xi}, s) - \varphi_i(\bar{\xi}, s) \right) \, \mathrm{d}s$$

in the limit due to a.e. pointwise convergence and the estimate

$$\left| \frac{1}{\varepsilon}\left(q(j(\bar{\xi} + \varepsilon e_i, \xi_d)) - q(j(\bar{\xi}, \xi_d)) \right) \right| \leq C_1 C_2$$

which follows easily from the Lipschitz estimate in the proof of Lemma 4.22 and allows to use Lebesgue's dominated-convergence theorem. By Gronwall's inequality, it follows that $\varphi_i(\bar{\xi}, \xi_d) = \frac{\partial j}{\partial \xi_i}(\bar{\xi}, \xi_d)$ almost everywhere in U.

Finally, denoting

$$q_d = \frac{q(x)}{|q(x)|} \quad , \quad \bar{q}(\xi) = q(j(\bar{\xi}, 0)) \quad , \quad (\tilde{q}(\xi)) = q_i \cdot \bar{q}(\xi)$$

leads to, remembering that $q_1, \ldots, q_d$ are orthonormal vectors,

$$\nabla j(\bar{\xi}, 0) = \begin{bmatrix} q_1 & \cdots & q_{d-1} & \bar{q}(\xi) \end{bmatrix} = \begin{bmatrix} q_1 & \cdots & q_d \end{bmatrix} \begin{bmatrix} e_1 & \cdots & e_{d-1} & \tilde{q}(\xi) \end{bmatrix}$$
$$\Rightarrow \quad \det \nabla j(\bar{\xi}, 0) = q_d \cdot \bar{q}(\xi) = \frac{q(x)}{|q(x)|} \cdot q(j(\bar{\xi}, 0))$$

by the determinant product formula, implying the desired initial-value problem as well as (4.7) by the solution formula for scalar time-variant ordinary differential equations. ◀

Lemma 4.24. *Let U a bounded domain and $y, \frac{\partial y}{\partial \xi_i} \in L^1_{\text{loc}}(U)$. Then*

$$\int_U y \frac{\partial z}{\partial \xi_i} \, d\xi = - \int_U \frac{\partial y}{\partial \xi_i} z \, d\xi$$

for each $z \in L^\infty(U)$ which has compact support in U and $\frac{\partial z}{\partial \xi_i} \in L^\infty(U)$.

Proof. The proof is analog to the proof presented for Lemma 4.18. Again, take the dilated versions G_ε of the standard mollifier G and $\varepsilon > 0$ small enough such that $z_\varepsilon = z * G_\varepsilon$ has its support inside of U. Then

$$\int_U y \frac{\partial z}{\partial \xi_i} \, d\xi = \lim_{\varepsilon \to 0} \int_U y \frac{\partial z_\varepsilon}{\partial \xi_i} \, d\xi = - \lim_{\varepsilon \to 0} \int_U \frac{\partial y}{\partial \xi_i} z_\varepsilon \, d\xi = - \int_U \frac{\partial y}{\partial \xi_i} z \, d\xi$$

where in both limits we have pointwise a.e. convergence of $z_\varepsilon \to z$ and $\frac{\partial z_\varepsilon}{\partial \xi_i} \to \frac{\partial z}{\partial \xi_0}$ with $|z_\varepsilon| \leq |z|$ and $\left|\frac{\partial z_\varepsilon}{\partial \xi_i}\right| \leq \left|\frac{\partial z}{\partial \xi_i}\right|$ respectively, allowing the interchange of integration and passage to the limit.   ◀

Proposition 4.25. *Let $q \in Y_d^*$ be given. Then there exists an open locally finite covering Ω_k, $k = 1, 2, \ldots$ of $\Omega_+ = \{x \in \Omega \mid q(x) \neq 0\}$, Lipeomorphisms $j_k : U_k \to \Omega_k$ (with associated $J_k = \det \nabla j_k$) and a partition of unity ζ_k subordinate to this covering such that the following characterization holds:*

A locally integrable y possesses a weak directional derivative $\partial_q y$ with respect to q if and only if each $\bar{y}_k = y \circ j_k$ has a weak derivative with respect to ξ_d in U_k with

$$\sum_{\{k' \mid \Omega_{k'} \cap \Omega' \neq \emptyset\}} \int_{U_{k'}} J_{k'}(\zeta_{k'} \circ j_{k'}) \left|\frac{\partial \bar{y}_{k'}}{\partial \xi_d}\right| \, d\xi < \infty \tag{4.8}$$

for each $\Omega' \subset\subset \Omega$. In the case $\partial_q y$ exists, the identity

$$\partial_q y = \sum_{k=1}^{\infty} \zeta_k \left(\frac{\partial \bar{y}_k}{\partial \xi_d} \circ j_k^{-1}\right) \tag{4.9}$$

holds almost everywhere in Ω.

In particular, $\partial_q y = 0$ almost everywhere where $q = 0$.

Proof. We first construct the open covering. For each $k = 1, 2, \ldots$ set $\Omega'_k = \{x \in \Omega \mid |q(x)| \geq \frac{1}{k} \wedge \text{dist}(x, \partial\Omega) \geq \frac{1}{k}\}$ which is compact in Ω. For $k < 1$, let $\Omega'_k = \emptyset$. Now, cover the compact sets $\Omega'_k \setminus \text{int}(\Omega'_{k-1})$ with finitely many open $\Omega'_{k,l}$ associated with the anchor points $x_{k,l} \in \Omega'_k$ and Lipeomorphisms $j_{k,l} : U_{k,l} \to \Omega'_{k,l}$. Note that due to the construction in Lemma 4.22, each $\Omega'_{k,l}$ has to

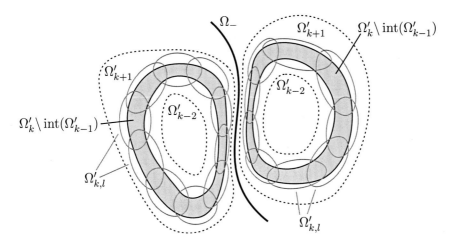

Figure 4.2: Illustration of the covering constructed in the proof of Proposition 4.25. For a fixed k, the nested sets $\Omega'_{k-2}, \dots, \Omega'_{k+1}$ are depicted as well as the open $\Omega'_{k,l}$ (indicated by ellipses) which cover $\Omega'_k \setminus \mathrm{int}(\Omega'_{k-1})$ (shaded regions). Note that all $\Omega'_{k,l}$ are contained in $\mathrm{int}(\Omega'_{k+1}) \setminus \Omega'_{k-2}$ (the regions within the dashed lines). Additionally, the set $\Omega_- = \{x \in \Omega \mid q(x) = 0\}$ is shown.

be contained in $\Omega_+ = \{x \in \Omega \mid q(x) \neq 0\}$. Without loss of generality one can also assume that

$$\Omega'_{k,l} \subset \mathrm{int}(\Omega'_{k+1}) \setminus \Omega'_{k-2} .$$

for each l. This covering has to be locally finite: Each $x \in \Omega_+$ has to be contained in some open $\Omega'_{k,l}$ which has empty intersection with each $\Omega'_{k',l'}$ for $k' > k + 2$, leaving only finitely many possibilities for non-empty intersection. See Figure 4.2 for an illustration of this construction.

Now choose for each k a smooth partition of unity $\eta_{k,l}$ of $\Omega'_k \setminus \mathrm{int}(\Omega'_{k-1})$ subordinate to $\Omega'_{k,l}$. Summing all up $\eta = \sum_{k,l} \eta_{k,l}$ yields a function which is finite on Ω_+ and not less than 1. Thus one can set $\zeta_{k,l} = \eta_{k,l}/\eta$ and gets a partition of unity with $\zeta_{k,l} \in \mathcal{C}_0^\infty(\Omega_+)$. By renumbering the sets $U_{k,l}$, $\Omega'_{k,l}$ as well as the mappings $j_{k,l}$ and functions $\zeta_{k,l}$, we can drop the index l and change the notation to U_k, Ω_k, j_k and ζ_k, respectively.

Now let y be locally integrable with weak directional derivative $\partial_q y$. With $\bar{y}_k = y \circ j_k$, the function y admits the representation $y = \sum_{k=1}^\infty \zeta_k \bar{y}_k \circ j_k^{-1}$ in $L^1_{\mathrm{loc}}(\Omega_+)$. Let us check that for each k, the weak derivative $\frac{\partial \bar{y}_k}{\partial \xi_d}$ exists. Choose a $z \in \mathcal{C}_0^\infty(U_k)$ which can also be written as $z = \tilde{z} J_k$ where $\tilde{z} = J_k^{-1} z$. Note that J_k is weakly differentiable with respect to ξ_d since it satisfies (4.7) and in particular

$$c_k e^{-d\|\nabla q\|_\infty |\xi_d|} \leq |J(\bar{\xi}, \xi_d)| \leq \|q\|_\infty e^{d\|\nabla q\|_\infty |\xi_d|}$$

on U_k with sufficiently small $c_k > 0$ which can be chosen due to the above construction as well as (4.6). Hence, $\tilde{z}$ can be used as a test function (see Lemma 4.24). Transforming the integral with the help of the Lipschitz coordinate transform theorem [EG92], applying the definition of J_k and Equation (4.5) gives

$$-\int_{U_k} \bar{y}_k \frac{\partial z}{\partial \xi_d} \, d\xi = -\int_{U_k} \bar{y}_k \frac{\partial}{\partial \xi_d}(\tilde{z} J_k) \, d\xi$$

$$= -\int_{U_k} \bar{y}_k \left(\tilde{z} \frac{\partial J_k}{\partial \xi_d} + \frac{\partial \tilde{z}}{\partial \xi_d} J_k \right) \, d\xi$$

$$= -\int_{U_k} \bar{y}_k \left(\tilde{z}((\operatorname{div} q) \circ j_k) + \frac{\partial \tilde{z}}{\partial \xi_d} \right) J_k \, d\xi$$

$$= -\int_{\Omega_k} y \left((\tilde{z} \circ j_k^{-1}) \operatorname{div} q + \nabla(\tilde{z} \circ j_k^{-1}) \cdot q \right) \, dx$$

which becomes, using the defining property of the weak directional derivative,

$$-\int_{\Omega_k} y \left((\tilde{z} \circ j_k^{-1}) \operatorname{div} q + \nabla(\tilde{z} \circ j_k^{-1}) \cdot q \right) \, dx = \int_{\Omega_k} \partial_q y (\tilde{z} \circ j_k^{-1}) \, dx$$

$$= \int_{U_k} (\partial_q y \circ j_k) \tilde{z} J_k \, d\xi = \int_{U_k} (\partial_q y \circ j_k) z \, d\xi \, .$$

Hence, $\bar{y}_k$ is weakly differentiable with respect to ξ_d with derivative $\frac{\partial \bar{y}_k}{\partial \xi_d} = \partial_q y \circ j_k$. Moreover, for $\Omega' \subset\subset \Omega$

$$\sum_{\{k'|\Omega_{k'} \cap \Omega' \neq \emptyset\}} \int_{U_{k'}} J_{k'}(\zeta_{k'} \circ j_{k'}) \left| \frac{\partial \bar{y}_{k'}}{\partial \xi_d} \right| \, d\xi$$

$$= \sum_{\{k'|\Omega_{k'} \cap \Omega' \neq \emptyset\}} \int_{\Omega_{k'}} \zeta_{k'} |\partial_q y| \, dx = \int_{\Omega''} |\partial_q y| \, dx < \infty$$

where $\Omega'' = \bigcup_{\{k'|\Omega_{k'} \cap \Omega' \neq \emptyset\}} \Omega_{k'}$ whose closure is also compact in Ω due to construction. This establishes (4.8) and therefore the "only if" part.

Before proceeding to the "if part", let us note that, using the partition of unity, one obtains (4.9) in $L^1_{\text{loc}}(\Omega_+)$. Moreover, we first verify that $\partial_q y = 0$ almost everywhere on the set $\Omega_- = \{x \in \Omega \mid q(x) = 0\}$. Choose a $\Omega' \subset \Omega_-$ which is compact in Ω and enlarge it to $\Omega'_{-,\varepsilon} = \overline{\Omega'_- + B_\varepsilon(0)}$ for small $\varepsilon > 0$. If ε is chosen small enough, the test functions $z_\varepsilon = \chi_{\Omega'} * G_\varepsilon$, where G_ε denote the dilated versions of the standard mollifier G, are still in $\mathcal{C}_0^\infty(\Omega)$. Let C be the Lipschitz constant of q which is common to all considered $\operatorname{supp} z_\varepsilon \subset\subset \Omega$.

Hence, we can estimate $|q(x')| \leq \varepsilon C$ whenever $x' \in B_\varepsilon(x)$ with $x \in \Omega'_-$. Moreover $|\operatorname{div} q| \leq dC$ as well as $\operatorname{div} q = 0$ almost everywhere in Ω_-, so the defining integral can be estimated by

$$\left| \int_\Omega \partial_q y z_\varepsilon \, \mathrm{d}x \right| = \left| \int_\Omega y(z_\varepsilon \operatorname{div} q + \nabla z_\varepsilon \cdot q) \, \mathrm{d}x \right|$$

$$\leq \int_{\operatorname{supp} z_\varepsilon \backslash \Omega_-} |y| \big(dC\|G\|_1 + C\|\nabla G\|_1\big) \, \mathrm{d}x \, , \qquad (*)$$

which vanishes as $\varepsilon \to 0$ since, due to the compactness of Ω'_- as well as $\Omega'_- \subset \Omega_-$, the Lebesgue measure of $\operatorname{supp} z_\varepsilon \backslash \Omega_-$ tends to zero. But since $\partial_q y$ is locally integrable, we can conclude $\int_{\Omega'} \partial_q y \, \mathrm{d}x = \lim_{\varepsilon \to 0} \int_\Omega \partial_q y z_\varepsilon \, \mathrm{d}x = 0$. Consequently, $\partial_q y = 0$ almost everywhere in Ω_-.

It remains to prove the converse direction of the equivalence statement. Suppose that each $\bar{y}_k$ is weakly differentiable in U_k with respect to ξ_d and (4.8) holds. Our aim is to show that $\partial_q y$ exists and is given by (4.9). First, take a test function $z \in \mathcal{C}_0^\infty(\Omega_+)$. Hence, testing y with $z \operatorname{div} q + \nabla z \cdot q$ leads to

$$\int_\Omega \sum_{k=1}^\infty \zeta_k y(z \operatorname{div} q + \nabla z \cdot q) \, \mathrm{d}x = \sum_{k=1}^K \int_{\Omega_k} \zeta_k y(z \operatorname{div} q + \nabla z \cdot q) \, \mathrm{d}x \, ,$$

where K is large enough such that $\Omega_{k'} \cap \operatorname{supp} z = \emptyset$ for $k' > K$. Each summand can be transformed to U_k which gives

$$\int_{\Omega_k} \zeta_k y(z \operatorname{div} q + \nabla z \cdot q) \, \mathrm{d}x = \int_{U_k} \bar{\zeta}_k \bar{y}_k \left(\frac{\partial J_k}{\partial \xi_d} \bar{z}_k + J_k \frac{\partial \bar{z}_k}{\partial \xi_d} \right) \, \mathrm{d}\xi$$

$$= - \int_{U_k} \frac{\partial}{\partial \xi_d} (\bar{\zeta}_k \bar{y}_k) J_k \bar{z}_k \, \mathrm{d}\xi$$

denoting $\bar{\zeta}_k = \zeta_k \circ j_k$, $\bar{z}_k = z \circ j_k$ and using (4.5) with the chain rule as well as (4.7). Furthermore,

$$-\int_{U_k} \frac{\partial}{\partial \xi_d} (\bar{\zeta}_k \bar{y}_k) J_k \bar{z}_k \, \mathrm{d}\xi = -\int_{U_k} \left(\frac{\partial \bar{\zeta}_k}{\partial \xi_d} \bar{y}_k + \bar{\zeta}_k \frac{\partial \bar{y}_k}{\partial \xi_d} \right) J_k \bar{z}_k \, \mathrm{d}\xi$$

$$= - \int_{\Omega_k} \left((\nabla \zeta_k \cdot q) y + \zeta_k \left(\frac{\partial \bar{y}_k}{\partial \xi_d} \circ j_k^{-1} \right) \right) z \, \mathrm{d}x$$

leading to, summed up,

$$\int_\Omega \left(\sum_{k=1}^\infty \zeta_k (\bar{y}_k \circ j_k^{-1}) \right) (z \operatorname{div} q + \nabla z \cdot q) \, \mathrm{d}x = - \int_\Omega \left(\sum_{k=1}^\infty \zeta_k \left(\frac{\partial \bar{y}_k}{\partial \xi_d} \circ j_k^{-1} \right) \right) z \, \mathrm{d}x$$

since $\sum_{k=1}^{K} \nabla \zeta_k \cdot q = \nabla \left(\sum_{k=1}^{K} \zeta_k \right) \cdot q = 0$ on supp z. This proves the weak directional differentiability on Ω_+.

Finally, for the case where $z \in C_0^\infty(\Omega)$ we split z into a part with compact support in Ω_+ and a part for which the associated integral vanishes. This can be achieved by $z = \tilde{z}_\varepsilon + z_\varepsilon$ where z_ε is constructed analog to the above with $\Omega'_- = \text{supp } z \cap \Omega_-$ and using $z\chi_{\Omega'_{-,\varepsilon}}$ instead of $\chi_{\Omega'_-}$. Set $\tilde{z}_\varepsilon = z - z_\varepsilon$. Testing then gives

$$\int_\Omega y(z \operatorname{div} q + \nabla z \cdot q) \, \mathrm{d}x = \int_\Omega y\big((\tilde{z}_\varepsilon + z_\varepsilon) \operatorname{div} q + (\nabla \tilde{z}_\varepsilon + \nabla z_\varepsilon) \cdot q \big) \, \mathrm{d}x$$

$$= -\int_\Omega \sum_{k=1}^\infty \zeta_k \left(\frac{\partial \bar{y}_k}{\partial \xi_d} \circ j_k^{-1} \right) \tilde{z}_\varepsilon \, \mathrm{d}x + \int_\Omega y(z_\varepsilon \operatorname{div} q + \nabla z_\varepsilon \cdot q) \, \mathrm{d}x \ .$$

Consider the limits of both terms on the right-hand side as $\varepsilon \to 0$. The functions in the first term converge pointwise almost everywhere and can be estimated by $\|z\|_\infty \sum_{k=1}^\infty \zeta_k \left| \frac{\partial \bar{y}_k}{\partial \xi_d} \right|$ which is integrable on $\overline{\text{supp } z \backslash \Omega_-}$ due to

$$\int_{\text{supp } z} \sum_{k=1}^\infty \zeta_k \left| \frac{\partial \bar{y}_k}{\partial \xi_d} \circ j_k^{-1} \right| \, \mathrm{d}x \le \sum_{\{k' | \Omega_{k'} \cap \text{supp } z \ne \emptyset\}} \int_{\Omega_{k'}} \zeta_{k'} \left| \frac{\partial \bar{y}_{k'}}{\partial \xi_d} \circ j_{k'}^{-1} \right| \, \mathrm{d}x$$

$$= \sum_{\{k' | \Omega_{k'} \cap \text{supp } z \ne \emptyset\}} \int_{U_{k'}} J_{k'}(\zeta_{k'} \circ j_{k'}) \left| \frac{\partial \bar{y}_{k'}}{\partial \xi_d} \right| \, \mathrm{d}\xi < \infty \ .$$

By Lebesgue's dominated-convergence theorem, the integral converges to $-\int_{\text{supp } z \cap \Omega_+} \sum_{k=1}^\infty \zeta_k \left(\frac{\partial \bar{y}_k}{\partial \xi_d} \circ j_k^{-1} \right) z \, \mathrm{d}x$. The second term tends to zero by virtue of $(*)$. This establishes the weak directional differentiability on Ω. ◀

4.3   Weighted and directional Sobolev spaces

In the following, we study spaces of Lebesgue-functions whose weak weighted gradients and directional derivatives are also in some Lebesgue space. Since this leads to spaces which are defined in analogy to the classical Sobolev spaces, we will call them weighted and directional Sobolev spaces.

To obtain a Banach space structure, closedness properties of the weighted and directional derivative with respect to L^r are important. Moreover, weak closedness with respect to the defining weights w and directions q with weak*-convergence in Y_1 and Y_d plays a central role for the solution of associated optimal control problems. Since the variational definitions (4.3) and (4.4) of these new notions of differentiability exhibit a non-linear structure with respect to both y and w, q respectively, we have to obtain certain compactness results for the underlying Sobolev space. It will turn out that this is the reason that the latter closedness can only be established for the weak weighted gradient and fails for the weak directional derivative.

But first, we give the definition of the weighted and directional Sobolev space associated with one weight w and a finite collection of directions denoted by $q_1, \ldots, q_K$.

Definition 4.26. Let $1 \leq r \leq \infty$, $w \in Y_1$ and $q_1, \ldots, q_K \in Y_{\mathrm{div}}^*$ for some $K \geq 0$. Then we define the *weighted and directional Sobolev space* $H_{w,\partial q_1,\ldots,\partial q_K}^r$ as the vector space of functions satisfying

$$y \in L^r(\Omega) \quad , \quad w\nabla y \in L^r(\Omega, \mathbb{R}^d) \quad , \quad \partial_{q_1} y, \ldots, \partial_{q_K} y \in L^r(\Omega)$$

with norm

$$\|y\|_{H_{w,\partial q_1,\ldots,\partial q_K}^r} = \left(\int_\Omega |y|^r + |w\nabla y|^r + \sum_{k=1}^K |\partial_{q_k} y|^r \, \mathrm{d}x \right)^{1/r} \quad \text{if } 1 \leq r < \infty$$

$$\|y\|_{H_{w,\partial q_1,\ldots,\partial q_K}^\infty} = \max \left\{ \|y\|_\infty, \|w\nabla y\|_\infty, \|\partial_{q_1} y\|_\infty, \ldots, \|\partial_{q_K} y\|_\infty \right\}.$$

$$(4.10)$$

Remark 4.27. For notational ease, we write $H_{\partial q_1,\ldots,\partial q_K}^r = H_{w,\partial q_1,\ldots,\partial q_K}^r$ when $w = 0$.

The first important question is whether $H_{w,\partial q_1,\ldots,\partial q_K}^r$ are Banach spaces. The affirmative answer will be given at the end of this section. Moreover, reflexivity for $1 < r < \infty$ as well as a compactness result is derived. As already mentioned, the essential ingredient for this is the (strong) closedness of the involved weak differential operators. The main part of this section is devoted to proving variants of weak closedness which is a stronger property and will be utilized later on.

4.3.1 Closedness of the weak weighted gradient

Generally, weak closedness of an operation ensures that if a weak-convergent sequence is given and the images under this operation also converge weakly, then the weak limit of the preimages is mapped to the weak limit of the images. This immediately implies the weaker property of strong closedness, a notion similar to weak closedness, where the weak convergence is replaced by strong convergence. The main purpose of this subsection is to examine the weak weighted derivative with respect to weak closedness in the setting that $w_k \overset{*}{\rightharpoonup} w$ in Y_1^*, $y_k \rightharpoonup y$ in some $L^r(\Omega)$ and $w_k \nabla y_k \rightharpoonup \theta$ in $L^r(\Omega, \mathbb{R}^d)$.

The following two lemmas are preliminary results for the closedness property we want to obtain for the weak weighted derivative. The first deals with weak* sequential continuity while the second can be regarded as some compactness result.

Lemma 4.28. *Let $\{w_k\}$ in Y_1^*, $w \in Y_1^*$ as well as $z \in H^{1,\infty}(\Omega, \mathbb{R}^d)$ be given. If $w_k \overset{*}{\rightharpoonup} w$ in Y_1^* then $-(w_k \operatorname{div} z + \nabla w_k \cdot z) \overset{*}{\rightharpoonup} -(w \operatorname{div} z + \nabla w \cdot z)$ in $L^\infty(\Omega)$.*

Proof. We want to show that from the weak*-convergence of $\{w_k\}$ in Y_1^* follows the weak*-convergence of $\{w_k \operatorname{div} z + \nabla w_k \cdot z\}$ in $L^\infty(\Omega)$. For this purpose, we study the mapping $J : v \mapsto \nabla v \cdot z$ from $L^1(\Omega) \to Y_1$. Our aim is to prove that this mapping is continuous on a dense subset. Thus, let $v \in \mathcal{C}^\infty(\overline{\Omega})$ and rewrite $\nabla v \cdot z = \operatorname{div}(zv) - v \operatorname{div} z$. With this identity one can estimate

$$\|\nabla v \cdot z\|_{Y_1} \leq (M_x^{-1}\|z\|_\infty + \|\operatorname{div} z\|_\infty)\|v\|_1 ,$$

remembering (4.2), where the norm in Y_1 is defined. This yields the continuity of the operation.

Since $\mathcal{C}^\infty(\overline{\Omega})$ is dense in $L^1(\Omega)$, we are able to conclude that the adjoint of J maps weak*-convergent sequences in Y_1^* to weak*-convergent ones in $L^1(\Omega)^* = L^\infty(\Omega)$. Assume that $w_k \overset{*}{\rightharpoonup} w$ in Y_1^*. Calculations yield that $J^*w = -(w \operatorname{div} z + \nabla w \cdot z)$ whenever $w \in Y_1^*$, hence $-(w_k \operatorname{div} z + \nabla w_k \cdot z) \overset{*}{\rightharpoonup} -(w \operatorname{div} z + \nabla w \cdot z)$ in $L^\infty(\Omega)$. ◄

Lemma 4.29. *Let Ω fulfill the cone condition, $\{w_k\} \subset Y_1^*$, $w \in Y_1^*$ be given with $w_k \overset{*}{\rightharpoonup} w$. Moreover, let $1 < r < \infty$, $\{y_k\} \subset L^r(\Omega)$, $y \in L^r(\Omega)$ with $y_k \rightharpoonup y$ as well as $\{w_k \nabla y_k\} \subset L^r(\Omega, \mathbb{R}^d)$, $\theta \in L^r(\Omega, \mathbb{R}^d)$ with $w_k \nabla y_k \rightharpoonup \theta$. For $\varepsilon > 0$ denote by $\Omega_\varepsilon = \{x \in \Omega \mid |w(x)| > \varepsilon\}$.*

Then, for each $\varepsilon > 0$, we have $y_k \to y$ in $L^r(\Omega_\varepsilon)$ and $\theta = w\nabla y$ in Ω_ε.

Proof. First note that $w_k \to w$ in $\mathcal{C}(\Omega)$ (see Proposition 4.11), so for sufficiently large k we have $|w_k| > \varepsilon/2$ on Ω_ε and thus ∇y_k as well as ∇y exist there, confer Proposition 4.19. Moreover, we know that $\nabla y_k = w_k^{-1}(w_k \nabla y_k)$ on Ω_ε, so

$$\|\nabla y_k\|_r \leq 2\varepsilon^{-1}\|w_k \nabla y_k\|_r \leq C$$

by assumption. Hence, we can extract a subsequence, also denoted by $\{\nabla y_k\}$, which converges weakly in $L^r(\Omega_\varepsilon, \mathbb{R}^d)$ with limit ∇y since taking the gradient is closed. Now $H^{1,r}(\Omega_\varepsilon)$ is compactly embedded in $L^r(\Omega_\varepsilon)$ (since Ω_ε also satisfies the cone condition, see [AF03]) so actually $y_k \to y$ in $L^r(\Omega_\varepsilon)$. Finally, we know from Lemma 4.28 that $w_k \operatorname{div} z + \nabla w_k \cdot z \overset{*}{\rightharpoonup} w \operatorname{div} z + \nabla w \cdot z$ which implies

$$\int_\Omega y_k(w_k \operatorname{div} z + \nabla w_k \cdot z) \, \mathrm{d}x = - \int_\Omega w_k \nabla y_k \cdot z \, \mathrm{d}x$$

for each $z \in \mathcal{C}_0^\infty(\Omega_\varepsilon, \mathbb{R}^d)$ so taking limits yields

$$\int_\Omega y(w \operatorname{div} z + \nabla w \cdot z) \, \mathrm{d}x = - \int_\Omega \theta \cdot z \, \mathrm{d}x$$

since $y_k \to y$ in $L^r(\Omega_\varepsilon)$ implies the weak convergence to the product of the limits in the left-hand side of the integral. Thus, $\theta = w \nabla y$ in Ω_ε. ◀

The following proposition states the closedness of the weak weighted deriva-tive considered in $L^r(\Omega) \times Y_1^* \to L^r(\Omega, \mathbb{R}^d)$ with weak convergence. It will turn out that the sequence of weak weighted gradients converges weakly to the weak weighted gradient of the limit only on the set where the limit weight does not vanish.

Proposition 4.30. *Suppose that Ω satisfies the cone condition, $1 < r < \infty$ and let the sequences $\{y_k\}$ in $L^r(\Omega)$ and $\{w_k\}$ in Y_1^* with limits $y_k \rightharpoonup y$ and $w_k \overset{*}{\rightharpoonup} w$ be given. If $w_k \nabla y_k \rightharpoonup \theta$ in $L^r(\Omega, \mathbb{R}^d)$, then $w \nabla y = \chi_{\{w \neq 0\}} \theta$.*

Proof. Choose an arbitrary $z \in \mathcal{C}_0^\infty(\Omega, \mathbb{R}^d)$ and $C_1 > 0$ such that C_1 is an estimate of the Lipschitz constant for all w_k on $\operatorname{supp} z$ (see also Remark 4.6). For each $0 < \varepsilon < \operatorname{dist}(\operatorname{supp} z, \partial\Omega)$, we define $\Omega_\varepsilon = \{|w| > \varepsilon\}$ and observe that for the dilated standard mollifier G_ε the function $z_\varepsilon^1 = (G_{\varepsilon/(2C_1)} * \chi_{\Omega_\varepsilon}) z$ belongs to $\mathcal{C}_0^\infty(\Omega_{\varepsilon/2}, \mathbb{R}^d)$ since

$$|w(\xi)| \geq |w(x)| - |w(\xi) - w(x)| > \varepsilon - C_1|\xi - x| > \varepsilon/2$$

whenever $x \in \Omega_\varepsilon$ and $\xi \in B_{\varepsilon/(2C_1)}(x)$. As we have proven in Lemma 4.29, $y_k \to y$ in $L^r(\Omega_{\varepsilon/2})$ which implies

$$\int_\Omega y(w \operatorname{div} z_\varepsilon^1 + \nabla w \cdot z_\varepsilon^1) \, \mathrm{d}x = \lim_{k \to \infty} \int_\Omega y_k(w_k \operatorname{div} z_\varepsilon^1 + \nabla w_k \cdot z_\varepsilon^1) \, \mathrm{d}x \,.$$

Setting $z_\varepsilon^2 = z - z_\varepsilon^1$, we moreover observe that $\operatorname{supp} z_\varepsilon^2 \subset \{|w| \leq 3\varepsilon/2\}$ since

$$|w(\xi)| \geq |w(x)| - |w(\xi) - w(x)| > 3\varepsilon/2 - C_1|\xi - x| \geq \varepsilon$$

whenever $|w(x)| > 3\varepsilon/2$ and $|\xi - x| \leq \varepsilon/(2C_1)$. Hence, we deduce

$$\left| \int_\Omega y(w \operatorname{div} z_\varepsilon^2 + \nabla w \cdot z_\varepsilon^2) \, dx \right|$$

$$\leq \max_{\{|w| \leq 3\varepsilon/2\}} |w \operatorname{div} z_\varepsilon^2 + \nabla w \cdot z_\varepsilon^2| \int_{\{0 < |w| \leq 3\varepsilon/2\}} |y| \, dx$$

and estimate

$$|w \operatorname{div} z_\varepsilon^2| \leq 3\varepsilon/2 \|(1 - G_{\varepsilon/(2C_1)} * \chi_{\Omega_\varepsilon}) \operatorname{div} z - (\nabla G_{\varepsilon/(2C_1)} * \chi_{\Omega_\varepsilon}) \cdot z\|_\infty$$
$$\leq 3\varepsilon/2 (\|\operatorname{div} z\|_\infty + 2C_1/\varepsilon \|\nabla G\|_1 \|z\|_\infty) \leq C_2$$

as well as

$$\|\nabla w \cdot z_\varepsilon^2\|_\infty \leq \|\nabla w\|_\infty \|z\|_\infty \leq C_3$$

with suitable $C_2, C_3 > 0$ which can be chosen to be independent of ε. Additionally, $\int_{\{0 < |w| \leq 3\varepsilon/2\}} |y| \, dx$ tends to zero as $\varepsilon \to 0$, which allows to choose $\varepsilon > 0$ such that

$$\left| \int_\Omega y(w \operatorname{div} z_\varepsilon^2 + \nabla w \cdot z_\varepsilon^2) \, dx \right| < \delta$$

if $\delta > 0$ is arbitrarily small. Together with the above, we have

$$\int_\Omega y(w \operatorname{div} z + \nabla w \cdot z) \, dx = \lim_{k \to \infty} \int_\Omega y_k(w_k \operatorname{div} z_\varepsilon^1 + \nabla w_k \cdot z_\varepsilon^1) \, dx$$
$$+ \int_\Omega y(w \operatorname{div} z_\varepsilon^2 + \nabla w \cdot z_\varepsilon^2) \, dx$$
$$= - \lim_{k \to \infty} \int_\Omega w_k \nabla y_k \cdot z_\varepsilon^1 \, dx$$
$$+ \int_\Omega y(w \operatorname{div} z_\varepsilon^2 + \nabla w \cdot z_\varepsilon^2) \, dx$$

such that

$$\left| \int_\Omega y(w \operatorname{div} z + \nabla w \cdot z) \, dx + \int_\Omega \theta \cdot z_\varepsilon^1 \, dx \right| < \delta$$

for each $\delta > 0$. Finally, letting $\delta \to 0$ and $\varepsilon \to 0$ yields

$$\int_\Omega y(w \operatorname{div} z + \nabla w \cdot z) \, dx = - \int_\Omega \chi_{\{w \neq 0\}} \theta \cdot z \, dx$$

which implies the assertion.   ◀

Remark 4.31. In the case where $y_k \rightarrow y$ in $L^r(\Omega)$, the statement remains true for $1 \leq r \leq \infty$ (with weak*-convergence of $w_k \nabla y_k$ for $r = \infty$) and arbitrary bounded domains. Observe that for each $z \in \mathcal{C}_0^\infty(\Omega, \mathbb{R}^d)$

$$\lim_{k \rightarrow \infty} \int_\Omega y_k(w_k \operatorname{div} z + \nabla w_k \cdot z) \, dx = \int_\Omega y(w \operatorname{div} z + \nabla w \cdot z) \, dx$$

holds since it is a dual pairing between strongly and weakly (weakly*) convergent sequences. It follows

$$\int_\Omega \theta \cdot z \, dx = - \int_\Omega y(w \operatorname{div} z + \nabla w \cdot z) \, dx$$

meaning that $\theta = w \nabla y$.

In particular, the weak limit of $\{w_k \nabla y_k\}$ coincides with $w \nabla y$ a.e. on the whole domain Ω, not just on $\{w \neq 0\}$.

4.3.2 Closedness of the weak directional derivative

An analog result can also be obtained for the weak directional derivative. However, in contrast to the weak weighted gradient, bounded sequences in $L^r(\Omega)$ for which $\{\partial_q y_k\}$ is also bounded do not necessarily admit (strongly) convergent subsequences in $L^r(\Omega)$. To compensate this missing compactness, one has to employ an additional condition on the weakly and weakly*-convergent sequences $\{y_k\}$ and $\{q_k\}$, respectively.

Proposition 4.32. *Let* $1 < r < \infty$ *and the sequences* $\{y_k\}$ *in* $L^r(\Omega)$ *as well as* $\{q_k\}$ *in* Y_{div}^* *be given such that*

$$y_k \rightharpoonup y \text{ in } L^r(\Omega) \quad , \quad q_k \overset{*}{\rightharpoonup} q \text{ in } Y_{\mathrm{div}}^* \quad \text{and} \quad \partial_{q_k} y_k \rightharpoonup v \text{ in } L^r(\Omega) .$$

If, additionally, $y_k \rightarrow y$ *or* $q_k \rightarrow q$ *as well as* $\operatorname{div} q_k \rightarrow \operatorname{div} q$ *pointwise almost everywhere in* Ω, *then* $v = \partial_q y$.

Proof. This can be done in analogy to the proof of Proposition 4.30.

Note that for a fixed $z \in \mathcal{C}_0^\infty(\Omega)$ one can define $J : v \mapsto z \nabla v$ from $L^1(\Omega) \rightarrow Y_{\mathrm{div}}$ which is continuous on a dense subset since $z \nabla v = \nabla(zv) - v \nabla z$ leads to the estimate

$$\|z \nabla v\|_{Y_{\mathrm{div}}} \leq (M_x^{-1} \|z\|_\infty + \|\nabla z\|_\infty) \|v\|_1$$

if, say, $v \in \mathcal{C}^\infty(\overline{\Omega})$. This makes J^* weak*-convergence preserving, hence $J^* q_k = -(z \operatorname{div} q_k + \nabla z \cdot q_k) \overset{*}{\rightharpoonup} J^* q = -(z \operatorname{div} q + \nabla z \cdot q)$ in $L^\infty(\Omega)$.

Our aim now is to show convergence of the defining integrals in (4.4). First suppose that $y_k \rightarrow y$ pointwise a.e. By Egorov's Theorem one can choose for each $\varepsilon > 0$ a measurable $\Omega' \subset \Omega$ with $|\Omega \backslash \Omega'| < \varepsilon$ such that $y_k \to y$ uniformly on Ω'. In particular, $y_k \to y$ in $L^r(\Omega')$. Hence, splitting the defining integral as follows

$$\int_\Omega y_k(z \operatorname{div} q_k + \nabla z \cdot q_k) \, \mathrm{d}x$$
$$= \int_{\Omega'} y_k(z \operatorname{div} q_k + \nabla z \cdot q_k) \, \mathrm{d}x + \int_{\Omega \backslash \Omega'} y_k(z \operatorname{div} q_k + \nabla z \cdot q_k) \, \mathrm{d}x$$

leads to

$$\lim_{k \to \infty} \int_{\Omega'} y_k(z \operatorname{div} q_k + \nabla z \cdot q_k) \, \mathrm{d}x = \int_{\Omega'} y(z \operatorname{div} q + \nabla z \cdot q) \, \mathrm{d}x$$

as well as

$$\left| \int_{\Omega \backslash \Omega'} y_k(z \operatorname{div} q_k + \nabla z \cdot q_k) \, \mathrm{d}x \right|$$
$$\leq (M_x^{-1} \|z\|_\infty + \|\nabla z\|_\infty) \|q_k\|_{Y_{\operatorname{div}}^*} \int_{\Omega \backslash \Omega'} |y_k| \, \mathrm{d}x$$
$$\leq C_1 |\Omega \backslash \Omega'|^{1/r'} \|y_k\|_r \leq C_2 \varepsilon^{1/r'}$$

where $\frac{1}{r} + \frac{1}{r'} = 1$ and $C_1, C_2 > 0$ are suitable constants which can be chosen independently of k since y_k and q_k are bounded sequences. Letting $\varepsilon \to 0$ then implies that

$$\lim_{k \to \infty} \int_\Omega y_k(z \operatorname{div} q_k + \nabla z \cdot q_k) \, \mathrm{d}x = \int_\Omega y(z \operatorname{div} q + \nabla z \cdot q) \, \mathrm{d}x . \qquad (*)$$

Now suppose that $q_k \to q$, $\operatorname{div} q_k \to q$ pointwise almost everywhere in Ω. From the above considerations we have that $\|z \operatorname{div} q_k + \nabla z \cdot q_k\|_\infty$ is bounded, giving a pointwise a.e. as well as integrable majorant. By Lebesgue's convergence theorem, $z \operatorname{div} q_k + \nabla z \cdot q_k \to z \operatorname{div} q + \nabla z \cdot q$ in $L^{r'}(\Omega)$ from which it immediately follows that $(*)$ is also satisfied for this case.

So, convergence of both sides of the defining integral is guaranteed and one has

$$\int_\Omega vz \, \mathrm{d}x = - \int_\Omega y(z \operatorname{div} q + \nabla z \cdot q) \, \mathrm{d}x$$

consequently leading to $v = \partial_q y$.   ◀

Lebesgue space	Convergence of functions and parameters	Convergence of derivatives	Conclusion
$1 \le r \le \infty$	$y_k \to y$ $w_k \xrightarrow{*} w$	$w_k \nabla y_k \xrightarrow{(*)} \theta$	$w \nabla y = \theta$
$1 \le r \le \infty$	$y_k \to y$ $q_k \xrightarrow{*} q$	$\partial_{q_k} y_k \xrightarrow{(*)} v$	$\partial_q y = v$
$1 < r < \infty$	$y_k \rightharpoonup y$ $w_k \xrightarrow{*} w$	$w_k \nabla y_k \rightharpoonup \theta$	$w \nabla y = \chi_{\{w \ne 0\}} \theta$
$1 < r < \infty$	$y_k \rightharpoonup y$ + p.w. $q_k \xrightarrow{*} q$	$\partial_{q_k} y_k \rightharpoonup v$	$\partial_q y = v$
$1 < r < \infty$	$y_k \rightharpoonup y$ $q_k \xrightarrow{*} q$ + p.w. $\text{div } q_k \to \text{div } q$ p.w.	$\partial_{q_k} y_k \rightharpoonup v$	$\partial_q y = v$

Table 4.1: Overview of the strong and weak closedness of the weak weighted and directional derivatives in $L^r(\Omega)$. The table can be read as follows: In the Lebesgue space indicated by the leftmost column, if one has a sequence of functions and weights/directions whose weak weighted/directional derivatives exist and converge in the sense indicated by the middle columns, then the conclusion on the rightmost column can be drawn.

Remark 4.33.

(a) It can easily be seen that in the case of strong convergence $y_k \to y$ in $L^r(\Omega)$, the statement is also valid for $1 \le r \le \infty$ (with weak*-convergence of $\partial_{q_k} y_k$ for $r = \infty$).

(b) If each $q_k \in Y_d^*$ and $q_k \xrightarrow{*} q$ there, then it suffices that $\text{div } q_k \to \text{div } q$ pointwise almost everywhere since $q_k \to q$ uniformly in Ω, due to Proposition 4.11.

An overview of the closedness properties can be found in Table 4.1. We are now able to prove that the weighted and directional Sobolev spaces are Banach spaces.

Proposition 4.34. *Let* $H^r_{w, \partial q_1, \dots, \partial q_K}$ *be given according to Definition 4.26. Then* $H^r_{w, \partial q_1, \dots, \partial q_K}$ *is a Banach space which is reflexive if* $1 < r < \infty$.

Proof. The normed space $H^r_{w, \partial q_1, \dots, \partial q_K}$ can be interpreted as a subspace, denoted by $\tilde{H}^r_{w,\partial q_1,\dots,\partial q_K}$, of a Lebesgue space via

$$(y,\theta,v_1,\dots,v_K) \in \tilde{H}^r_{w,\partial q_1,\dots,\partial q_K} \subset L^r(\Omega) \times L^r(\Omega,\mathbb{R}^d) \times L^r(\Omega)^K$$
$$\Leftrightarrow \quad \theta = w\nabla y \quad, \quad v_k = \partial_{q_k} y \quad \text{for } k = 1,\dots,K \;.$$

All statements follow from the closedness of $\tilde{H}^r_{w,\partial q_1,\dots,\partial q_K}$ since closed sub-spaces of (reflexive) Banach spaces are (reflexive) Banach spaces as well. But this corresponds to the closedness of the differential operations $y \mapsto (w\nabla y, \partial_{q_1} y, \dots, \partial_{q_K} y)$ which has been established component-wise in Remarks 4.31 and 4.33. ◄

Remark 4.35. It is easy to check that $H^2_{w,\partial q_1,\dots,\partial q_K}$ is also a Hilbert space.

4.4   Density of smooth functions

In the following, we establish that $C^\infty(\Omega)$ is dense in $H^r_{w,\partial q_1,\dots,\partial q_K}$ for any choice of w and $q_1,\dots,q_K$ in Y_1^* and $(Y_d^*)^K$, respectively, and that $C^\infty(\overline{\Omega})$ is moreover dense whenever Ω is a Lipschitz domain. These are stronger assumptions for $q_1,\dots,q_K$ than the assumption originally made in Definition 4.26. As can be seen in the proofs, the local Lipschitz continuity of w as well as q_k is necessary for the argumentation. Such restrictions can be regarded as the price one has to pay for allowing general weights (and directions) which are able to vanish arbitrarily fast.

We start with two abstract approximation results providing the existence of smooth functions which approximate any function as well as the image under a certain closed linear (differential) operator in the respective Lebesgue spaces. The first deals with $C^\infty(\Omega)$-approximation while the second establishes density of $C^\infty(\overline{\Omega})$ which turns out to be crucial for the characterization of $H^r_{w,\partial q_1,\dots,\partial q_K}$ as the closure of $H^{1,r}(\Omega)$ under a certain norm. We will then apply these approximation results to the operators $y \mapsto w\nabla y$ and $y \mapsto \partial_{q_k} y$. The main tool for approximating a function with smooth functions is the well-known standard mollification. It will turn out that the boundedness of the image of the respective (differential) operator for the sequence of mollified functions is the essential ingredient for proving density.

The results then are that $C^\infty(\Omega)$ is always dense in $H^r_{w,\partial q_1,\dots,\partial q_K}$ while the density of $C^\infty(\overline{\Omega})$ requires that the domain Ω is a Lipschitz domain. The difficulty here is that, close to the boundary, the L^2-norm for the gradient may be stronger than the weighted norm, possibly causing a gap. Similar issues are known for the classical Sobolev spaces: For example, while $C^\infty(\Omega)$ is always dense in $H^{1,r}(\Omega)$, the density of the smaller set $C^\infty(\overline{\Omega})$ in $H^{1,r}(\Omega)$ requires the segment condition and fails for general bounded domains (see [AF03]).

Proposition 4.36. *Let $l \geq 1$, $1 < r < \infty$ and $\Lambda : \mathcal{D}(\Lambda) \subset L^r_{\mathrm{loc}}(\Omega) \to L^r_{\mathrm{loc}}(\Omega, \mathbb{R}^l)$ with $C^\infty_0(\Omega) \subset \mathcal{D}(\Lambda)$ a closed linear operator with the property that*

1. *$y \in \mathcal{D}(\Lambda)$ if and only if for each $\zeta \in C^\infty_0(\Omega)$ follows $\zeta y \in \mathcal{D}(\Lambda)$,*

2. *for each $y \in \mathcal{D}(\Lambda)$ and $\zeta \in C^\infty_0(\Omega)$ follows $\operatorname{supp} \Lambda(\zeta y) \subset \operatorname{supp} \zeta$,*

3. *for each $y \in \mathcal{D}(\Lambda)$ with compact support in Ω, there exists a $\delta_0 > 0$ such that the sequence of mollified $y_\delta = y * G_\delta$ satisfies $\|\Lambda y_\delta\|_r \leq C$ for every $0 < \delta < \delta_0$ and some $C > 0$.*

Then, for each $\varepsilon > 0$ and $y \in L^r(\Omega)$ with $\Lambda y \in L^r(\Omega, \mathbb{R}^l)$, there exists a $\bar{y} \in C^\infty(\Omega)$ such that

$$\|y - \bar{y}\|_r + \|\Lambda y - \Lambda \bar{y}\|_r < \varepsilon .$$

Proof. As in the proof of the density of $C^\infty(\Omega)$ in the usual Sobolev spaces, we introduce $\Omega_k = \{x \in \Omega \mid \operatorname{dist}(x, \partial\Omega) > \frac{1}{k}\}$, $\Omega_0 = \emptyset$ and $\Omega'_k = \Omega_{k+1} \backslash \overline{\Omega_{k-1}}$ for $k = 1, 2, \ldots$. Let $\varepsilon > 0$ be given and choose $\zeta_k \in C^\infty_0(\Omega'_k)$ as the usual partition of unity, i.e.

$$0 \leq \zeta_k \leq 1 \quad , \quad \zeta_k = 1 \text{ in } \Omega_k \quad , \quad \operatorname{supp} \zeta_k \subset\subset \Omega'_k .$$

For each $\zeta_k y$, the mollified versions satisfy $\|\Lambda((\zeta_k y) * G_\delta)\|_r \leq C_k$ for every $0 < \delta < \min\{\frac{1}{(k+1)^2}, \delta_{0,k}\}$, yielding a weakly convergent subsequence $\Lambda((\zeta_k y) * G_{\delta_{k'}})$ with limit θ_k as $\delta_{k'} \to 0$. By the theorem of Banach-Saks, there is a subsequence (not relabeled), such that

$$\lim_{m \to \infty} \frac{1}{m} \sum_{k'=1}^m \Lambda((\zeta_k y) * G_{\delta_{k'}}) = \theta_k \quad , \quad \lim_{m \to \infty} \frac{1}{m} \sum_{k'=1}^m (\zeta_k y) * G_{\delta_{k'}} = \zeta_k y$$

the latter being a consequence of $(\zeta_k y) * G_{\delta_{k'}} \to \zeta_k y$ as $k' \to \infty$. Due to the closedness of Λ, there has to be $\theta_k = \Lambda(\zeta_k y)$. Choosing m_k large enough and writing

$$y_k = \frac{1}{m_k} \sum_{k'=1}^{m_k} (\zeta_k y) * G_{\delta_{k'}}$$

we can achieve that, exploiting the linearity of Λ,

$$\|y_k - \zeta_k y\|_r + \|\Lambda y_k - \Lambda(\zeta_k y)\|_r < \frac{\varepsilon}{2^k} .$$

Note that by construction $y_k \in \mathcal{C}_0^\infty(\Omega_k')$ since δ is chosen small enough. Moreover, the sets $\Omega_1', \Omega_2', \ldots$ are locally finite, hence the summation $\bar{y} = \sum_{k'=1}^\infty y_{k'}$ is finite in a neighborhood of each $x \in \Omega$, yielding that $\bar{y} \in \mathcal{C}^\infty(\Omega)$. Furthermore, there has to be $\bar{y} \in \mathcal{D}(\Lambda)$ meaning that $\Lambda\bar{y} \in L_{\mathrm{loc}}^r(\Omega, \mathbb{R}^l)$: This follows from the first assumption on Λ since for each $\zeta \in \mathcal{C}_0^\infty(\Omega)$ there has to be a k such that $\mathrm{supp}\,\zeta \subset \Omega_{k-1}$ and consequently

$$\zeta\bar{y} = \sum_{k'=1}^k \zeta y_{k'} \quad \text{as well as} \quad \Lambda(\zeta\bar{y}) = \sum_{k'=1}^k \Lambda(\zeta y_{k'}) \,,$$

taking the second assumption on Λ into account. In particular, $\zeta y \in \mathcal{D}(\Lambda)$.

Also, observe that one can deduce similarly that

$$\chi_{\Omega_k}(\bar{y} - y) = \chi_{\Omega_k} \sum_{k'=1}^{k+1} (y_{k'} - \zeta_{k'} y) \,, \quad \chi_{\Omega_k}(\Lambda\bar{y} - \Lambda y) = \chi_{\Omega_k} \sum_{k'=1}^{k+1} \left(\Lambda y_{k'} - \Lambda(\zeta_{k'} y) \right)$$

leading to the estimate

$$\|\bar{y} - y\|_{r,\Omega_k} + \|\Lambda\bar{y} - \Lambda y\|_{r,\Omega_k} \leq \sum_{k'=1}^{k+1} \left(\|y_{k'} - \zeta_{k'} y\|_r + \|\Lambda y_{k'} - \Lambda(\zeta_{k'} y)\|_r \right) < \varepsilon \,.$$

The fact that $\Lambda\bar{y} \in L^r(\Omega, \mathbb{R}^l)$ and consequently the result then follows from the application of Levi's monotone-convergence theorem. ◀

Remark 4.37. The way $\bar{y}$ is constructed in the proof immediately gives that for each $\Omega' \subset\subset \Omega$, the restriction of the approximation $\bar{y}$ to Ω' fulfills $\bar{y}|_{\Omega'} \in \mathcal{C}^\infty(\overline{\Omega'})$, since $\bar{y}|_{\Omega'}$ can be written as the sum of finitely many y_k which are in turn a linear combination of the $\mathcal{C}^\infty(\overline{\Omega'})$-functions $(\zeta_k y) * G_{\delta_k'}$.

The function $\bar{y}$ as well as its derivatives may become unbounded when Ω' "approaches" Ω, which can be interpreted as the reason for $\mathcal{C}^\infty(\overline{\Omega})$ not always being dense in weighted and directional Sobolev spaces associated with general bounded domains.

To obtain the density of $\mathcal{C}^\infty(\overline{\Omega})$, we establish an analog approximation result which is much easier to prove.

Proposition 4.38. *Let $l \geq 1$, $1 < r < \infty$ and $\Lambda : \mathcal{D}(\Lambda) \subset L^r(\Omega) \to L^r(\Omega, \mathbb{R}^l)$ with $\mathcal{C}^\infty(\overline{\Omega}) \subset \mathcal{D}(\Lambda)$ a closed linear operator with the property that for each $y \in \mathcal{D}(\Lambda)$, there exists a sequence $\{y_\delta\}$ in $\mathcal{C}^\infty(\overline{\Omega})$ for $0 < \delta < \delta_0$ such that $y_\delta \to y$ in $L^r(\Omega)$ as $\delta \to 0$ and $\|\Lambda y_\delta\|_r \leq C$ with some $C > 0$.*

Then, for each $\varepsilon > 0$ and $y \in \mathcal{D}(\Lambda)$, there exists a $\bar{y} \in \mathcal{C}^\infty(\overline{\Omega})$, which is a finite linear combination of some y_δ, such that

$$\|y - \bar{y}\|_r + \|\Lambda y - \Lambda\bar{y}\|_r < \varepsilon \,.$$

Proof. Choose a $y \in \mathcal{D}(\Lambda)$ and δ_0 according to the assumptions. Then, the sequence $\{\Lambda y_{\delta_k}\}$ remains bounded in $L^r(\Omega, \mathbb{R}^l)$ as $\delta_k \to 0$, hence there exists a weakly convergent subsequence (still labeled with k) such that $y_{\delta_k} \rightharpoonup \theta$. Again, by the theorem of Banach-Saks,

$$\lim_{m \to \infty} \frac{1}{m} \sum_{k=1}^{m} \Lambda y_{\delta_k} = \theta \quad , \quad \lim_{m \to \infty} \frac{1}{m} \sum_{k=1}^{m} y_{\delta_k} = y$$

where the latter follows from $y_{\delta_k} \to y$ in $L^r(\Omega)$ as $k \to \infty$. Moreover, $\theta = \Lambda y$ since the operator is closed. Choosing m large enough and exploiting the linearity of Λ finally gives a $\bar{y}$ according to

$$\bar{y} = \frac{1}{m} \sum_{k=1}^{m} y_{\delta_k}$$

such that

$$\|y - \bar{y}\|_r + \|\Lambda y - \Lambda \bar{y}\|_r < \varepsilon$$

holds. From the construction then follows $\bar{y} \in \mathcal{C}^\infty(\overline{\Omega})$. ◄

The following two lemmas are necessary to obtain the prerequisites for Proposition 4.36 for the operator $\Lambda y = w\nabla y$. In the first, it is shown that the weak weighted derivative is stable under the multiplication with a sufficiently smooth ζ, while the second lemma deals with the boundedness of $\{w\nabla y_\varepsilon\}$ for the sequence of mollified $\{y_\varepsilon\}$ approximating y.

Lemma 4.39. *Let $1 \leq r \leq \infty$ and $w \in Y_1^*$. If $y \in H_w^r$ and $\zeta \in H^{1,\infty}(\Omega)$, then the product satisfies $\zeta y \in H_w^r$ with identity*

$$w\nabla(\zeta y) = \zeta(w\nabla y) + y(w\nabla\zeta) .$$

Moreover, building the product maps $H_w^r \to H_w^r$ continuously.

Proof. We first show that ζy has a weak weighted gradient. Note that for a $z \in \mathcal{C}_0^\infty(\Omega, \mathbb{R}^d)$ almost everywhere in Ω, the identity

$$\zeta(w \operatorname{div} z + \nabla w \cdot z) = \zeta \operatorname{div}(wz) = \operatorname{div}(w\zeta z) - \nabla\zeta \cdot (wz)$$

holds. The product ζz can be used as a test function (cf. Lemma 4.18), hence

$$\int_\Omega \zeta y(w \operatorname{div} z + \nabla w \cdot z) \, \mathrm{d}x = \int_\Omega y(w \operatorname{div}(\zeta z) + \nabla w \cdot (\zeta z)) \, \mathrm{d}x$$
$$- \int_\Omega y\nabla\zeta \cdot (wz) \, \mathrm{d}x$$

$$= -\int_\Omega w\nabla y \cdot \zeta z \, \mathrm{d}x - \int_\Omega wy\nabla\zeta \cdot z \, \mathrm{d}x$$

$$= -\int_\Omega (\zeta w\nabla y + yw\nabla\zeta) \cdot z \, \mathrm{d}x$$

which means that the weak weighted derivative satisfies $w\nabla(\zeta y) = \zeta w\nabla y + yw\nabla\zeta$. The norm can further be estimated by

$$\|w\nabla(\zeta y)\|_r \le \|\zeta\|_\infty \|w\nabla y\|_r + \|\nabla\zeta\|_\infty \|w\|_\infty \|y\|_r$$

which shows that $w\nabla(\zeta y) \in L^r(\Omega)$ and the continuity of the operation.   ◀

Lemma 4.40. *Let $1 \le r < \infty$ and $w \in Y_1^*$. For each $y \in H_w^r$ with $\operatorname{supp} y \subset \Omega' \subset\subset \Omega$ there exists an $\varepsilon_0 > 0$ and a $C > 0$ which only depends on $\|w\|_{Y_1^*}$ and Ω' such that for each $0 < \varepsilon < \varepsilon_0 < \operatorname{dist}(\operatorname{supp} y, \partial\Omega)$ the estimate*

$$\|y_\varepsilon\|_{H_w^r} \le C\|y\|_{H_w^r} \qquad \text{where} \qquad y_\varepsilon = y * G_\varepsilon$$

holds with G_ε being the dilated versions of the standard mollifier G with $\operatorname{supp} G \subset B_1(0)$.

Proof. Choose $0 < \varepsilon_0 < \operatorname{dist}(\operatorname{supp} y, \partial\Omega)$ such that the mollified versions y_ε belong to $\mathcal{C}_0^\infty(\Omega)$. Also, using the local Lipschitz continuity (see Proposition 4.5 and Remark 4.6),

$$|w(x) - w(\xi)| \le \|\nabla w\|_\infty \varepsilon \quad \text{for all} \quad \xi \in \operatorname{supp} y \, , \, |x - \xi| \le \varepsilon \, .$$

Note that this estimate only holds in the convex neighborhood $\overline{B_\varepsilon(x)}$ of each $x \in \operatorname{supp} y$, since in this situation, the Lipschitz constant does not depend on the domain.

By Young's inequality we get that $\|y_\varepsilon\|_r \le \|G\|_1 \|y\|_r$ so let us concentrate on the norm of the weak weighted gradient. Since y_ε is smooth, we can write $(w\nabla y_\varepsilon)(x) = w(x)\nabla y_\varepsilon(x)$ so

$$w(x)\nabla y_\varepsilon(x) = \int_\Omega w(x)y(\xi)\nabla G_\varepsilon(x - \xi) \, \mathrm{d}\xi$$

$$= \int_\Omega y(\xi)\big(w(x) - w(\xi)\big)\nabla G_\varepsilon(x - \xi) \, \mathrm{d}\xi$$

$$+ \int_\Omega y(\xi)w(\xi)\nabla G_\varepsilon(x - \xi) \, \mathrm{d}\xi \, . \qquad (*)$$

Note that $\nabla G_\varepsilon(x - \xi) = -\nabla G_{\varepsilon,x}(\xi)$ with $G_{\varepsilon,x}(\xi) = G_\varepsilon(x - \xi)$ thus the latter term also reads as

$$
\int_\Omega y(\xi) w(\xi) \nabla G_\varepsilon(x - \xi) \, d\xi = -\int_\Omega y \left(w \nabla G_{\varepsilon,x} + G_{\varepsilon,x} \nabla w \right) d\xi
$$
$$
+ \int_\Omega y G_{\varepsilon,x} \nabla w \, d\xi
$$
$$
= \int_\Omega \left((w \nabla y)(\xi) + y(\xi) \nabla w(\xi) \right) G_\varepsilon(x - \xi) \, d\xi \quad (**)
$$

using the defining property of $w \nabla y$ componentwise (note that $G_{\varepsilon,x} e_i$ is in $\mathcal{C}_0^\infty(\Omega, \mathbb{R}^d)$). We now estimate the norm for each of the terms in $(*)$ separately. The first term obeys, since the integrand is only non-zero if $\xi \in \operatorname{supp} y$ and $|x - \xi| \leq \varepsilon$,

$$
\left(\int_\Omega \left| \int_\Omega y(\xi) (w(x) - w(\xi)) \nabla G_\varepsilon(x - \xi) \, d\xi \right|^r dx \right)^{1/r}
$$
$$
\leq \|\nabla w\|_\infty \left(\int_\Omega \left| \int_\Omega \varepsilon^{-d} |y(\xi)| \left| \nabla G\left(\frac{x - \xi}{\varepsilon} \right) \right| d\xi \right|^r dx \right)^{1/r}
$$
$$
\leq \|\nabla w\|_\infty \|\nabla G\|_1 \|y\|_r \ ,
$$

by Young's inequality while the second can be estimated by

$$
\|(w \nabla y + y \nabla w) * G_\varepsilon\|_r \leq \|G\|_1 (\|w \nabla y\|_r + \|\nabla w\|_\infty \|y\|_r) \ ,
$$

taking $(**)$ into account. Both estimates together yield

$$
\|w \nabla y_\varepsilon\|_r \leq (\|\nabla G\|_1 + \|G\|_1) \|\nabla w\|_\infty \|y\|_r + \|G\|_1 \|w \nabla y\|_r
$$
$$
\leq \left(\|\nabla w\|_\infty^{r'} (\|\nabla G\|_1 + \|G\|_1)^{r'} + \|G\|_1^{r'} \right)^{1/r'} \|y\|_{H_w^r}
$$

with dual exponent r' satisfying $\frac{1}{r} + \frac{1}{r'} = 1$. This immediately implies the desired result $\|y_\varepsilon\|_{H_w^r} \leq C\|y\|_{H_w^r}$ by choosing

$$
C = \left(\left(\|\nabla w\|_\infty^{r'} (\|\nabla G\|_1 + \|G\|_1)^{r'} + \|G\|_1^{r'} \right)^{r-1} + \|G\|_1^r \right)^{1/r} . \quad \blacktriangleleft
$$

The following two lemmas provide the analog statements for the weak directional derivative $y \mapsto \partial_{q_k} y$, only differing in the fact that for the boundedness of $\{\partial_q y_\varepsilon\}$ we assume $q \in Y_d^*$. While the proofs are widely analog and do not reveal any essentially new ideas, we carry them out for the sake of completeness.

Lemma 4.41. *Let $1 \leq r \leq \infty$, $q \in Y_d^*$ and $\zeta \in H^{1,\infty}(\Omega)$. Then, from $y \in H_{\partial q}^r$ follows $\zeta y \in H_{\partial q}^r$ with the identity*

$$
\partial_q(\zeta y) = y \partial_q \zeta + \zeta \partial_q y \ .
$$

Moreover, the operation maps $H_{\partial q}^r \to H_{\partial q}^r$ continuously.

Proof. It is clear that for each $z \in C_0^\infty(\Omega)$ we have

$$\zeta(z \operatorname{div} q + \nabla z \cdot q) = \zeta \operatorname{div}(qz) = \operatorname{div}(q\zeta z) - z\nabla \zeta \cdot q$$

almost everywhere in Ω and since ζz can be used as a test function (what can be seen in complete analogy to Lemma 4.18), it follows

$$-\int_\Omega \zeta y(z \operatorname{div} q + \nabla z \cdot q) \, \mathrm{d}x = -\int_\Omega y(\operatorname{div}(q\zeta z) - z\nabla \zeta \cdot q) \, \mathrm{d}x$$
$$= \int_\Omega (\partial_q y \zeta + y \partial_q \zeta) z \, \mathrm{d}x \ .$$

Moreover, the norm can be estimated by

$$\|\partial_q(\zeta y)\|_r \leq \|\zeta\|_\infty \|\partial_q y\|_r + \|q\|_\infty \|\nabla \zeta\|_\infty \|y\|_r \ ,$$

yielding the continuity.   ◄

Lemma 4.42. *Let* $1 \leq r < \infty$ *and* $q \in Y_d^*$. *For each* $y \in H_{\partial q}^r(\Omega)$ *with* $\operatorname{supp} y \subset \Omega' \subset\subset \Omega$ *there exists an* $\varepsilon_0 > 0$ *and a* C *which only depends on* $\|q\|_{Y_d^*}$ *and* Ω' *such that for each* $0 < \varepsilon < \varepsilon_0$

$$\|y_\varepsilon\|_{H_{\partial q}^r} \leq C\|y\|_{H_{\partial q}^r} \qquad where \qquad y_\varepsilon = y * G_\varepsilon$$

holds with G_ε *being the dilated versions of the standard mollifier* G *with* $\operatorname{supp} G \subset B_1(0)$.

Proof. Choose $0 < \varepsilon_0 < \operatorname{dist}(\operatorname{supp} y, \partial\Omega)$, so all $y_\varepsilon \in C_0^\infty(\Omega)$. Note that $q \in Y_d^*$ means that q is bounded and Lipschitz continuous in the sense that

$$|q(x) - q(\xi)| \leq \|\nabla q\|_\infty \varepsilon \quad \text{for all } \xi \in \operatorname{supp} y \ , |x - \xi| \leq \varepsilon \ ,$$

again by Proposition 4.5 and Remark 4.6. It is moreover clear that $\|y_\varepsilon\|_r \leq \|G\|_1 \|y\|_r$. To estimate the approximation of the gradient consider $\partial_q y_\varepsilon$ which can, because of the smoothness of y_ε, be written as

$$\partial_q y_\varepsilon(x) = \int_\Omega y(\xi) q(x) \cdot \nabla G_\varepsilon(x - \xi) \, \mathrm{d}\xi$$
$$= \int_\Omega y(\xi)(q(x) - q(\xi)) \cdot \nabla G_\varepsilon(x - \xi) \, \mathrm{d}\xi$$
$$+ \int_\Omega y(\xi) q(\xi) \cdot \nabla G_\varepsilon(x - \xi) \, \mathrm{d}\xi$$

but $q(\xi) \cdot \nabla G_\varepsilon(x - \xi) = -q(\xi) \cdot \nabla G_{\varepsilon,x}(\xi)$ where $G_{\varepsilon,x}(\xi) = G_\varepsilon(x - \xi)$, so one is able to reformulate the latter term to

$$-\int_\Omega yq \cdot \nabla G_{\varepsilon,x} \, \mathrm{d}\xi = -\int_\Omega y\big(G_{\varepsilon,x} \operatorname{div} q + q \cdot \nabla G_{\varepsilon,x}\big) \, \mathrm{d}\xi + \int_\Omega yG_{\varepsilon,x} \operatorname{div} q \, \mathrm{d}\xi$$

$$= \int_\Omega \big(\partial_q y(\xi) + y(\xi) \operatorname{div} q(\xi)\big) G_\varepsilon(x - \xi) \, \mathrm{d}\xi$$

with the help of the definition of the weak directional derivative. This yields

$$\|(\partial_q y + y \operatorname{div} q) * G_\varepsilon\|_r \le \|G\|_1 (\|\partial_q y\|_r + d\|\nabla q\|_\infty \|y\|_r)$$

so let us estimate the other part according to

$$\left(\int_\Omega \left|\int_\Omega y(\xi)(q(x) - q(\xi)) \cdot \nabla G_\varepsilon(x - \xi) \, \mathrm{d}\xi\right|^r \mathrm{d}x\right)^{1/r}$$

$$\le \|\nabla q\|_\infty \left(\int_\Omega \left|\int_\Omega \varepsilon^{-d} |y(\xi)| \left|\nabla G\left(\frac{x - \xi}{\varepsilon}\right)\right| \mathrm{d}\xi\right|^r \mathrm{d}x\right)^{1/r}$$

$$\le \|\nabla q\|_\infty \|\nabla G\|_1 \|y\|_r .$$

Altogether, one gets

$$\|\partial_q y_\varepsilon\|_r \le (\|\nabla G\|_1 + d\|G\|_1)\|\nabla q\|_\infty \|y\|_r + \|G\|_1 \|\partial_q y\|_r$$

$$\le \big(\|\nabla q\|_\infty^{r'}(\|\nabla G\|_1 + d\|G\|_1)^{r'} + \|G\|_1^{r'}\big)^{1/r'} \|y\|_{H^r_{\partial q}}$$

again, with the dual exponent r' such that $\frac{1}{r} + \frac{1}{r'} = 1$. This leads to the desired result if C is chosen as follows

$$C = \left(\big(\|\nabla q\|_\infty^{r'}(\|\nabla G\|_1 + d\|G\|_1)^{r'} + \|G\|_1^{r'}\big)^{r-1} + \|G\|_1^r\right)^{1/r} . \qquad \blacktriangleleft$$

Collecting the results, the first main result of this section can be proven.

Theorem 4.43. *Let $1 < r < \infty$, $w \in Y_1^*$ and $q_1, \dots, q_K \in Y_d^*$. Then $C^\infty(\Omega)$ is dense in $H^r_{w, \partial q_1, \dots, \partial q_K}$. In particular, $H^r_{w, \partial q_1, \dots, \partial q_K}$ can be regarded as the closure of $C^\infty(\Omega)$ under the norm (4.10).*

Proof. To show the density, we want to apply Proposition 4.36 to the closed differential operator

$$\Lambda : y \mapsto (w\nabla y, \partial_{q_1} y, \dots, \partial_{q_K} y)$$

whose domain contains $\mathcal{C}_0^\infty(\Omega)$ and which can be interpreted in $\mathcal{D}(\Lambda) \subset L_{\mathrm{loc}}^r(\Omega) \to L_{\mathrm{loc}}^r(\Omega, \mathbb{R}^l)$. By virtue of Lemmas 4.39 and 4.41, Λ meets the first requirement of the proposition (the statement that $\zeta y \in \mathcal{D}(\Lambda)$ for each $\zeta \in \mathcal{C}_0^\infty(\Omega)$ implies $y \in \mathcal{D}(\Lambda)$ follows directly from the definitions of the weak weighted and directional derivative). The second is easily obtained from the product formulas for the respective weak derivatives, while the third is an direct consequence of the Lemmas 4.40 and 4.42. Hence, the asserted statement follows from the application of Proposition 4.36. ◀

Note that we used $\mathcal{C}^\infty(\Omega) \subset H_{w,\partial q_1,\ldots,\partial q_K}^r$ as a short-hand notation for $\mathcal{C}^\infty(\Omega) \cap H_{w,\partial q_1,\ldots,\partial q_K}^r$ being a subset of $H_{w,\partial q_1,\ldots,\partial q_K}^r$, a slight abuse of notation which seems to be common in the literature.

The next step is to focus on showing the density of the set $\mathcal{C}^\infty(\overline{\Omega})$, which is in general a proper subset of $\mathcal{C}^\infty(\Omega) \cap H_{w,\partial q_1,\ldots,\partial q_K}^r$. For this purpose, we have to impose an additional condition on the domain Ω, namely that Ω is a Lipschitz domain. Such a condition allows us to translate the standard mollifier G such that locally, the convolution of the corresponding dilated version with y only involves integration in Ω. This yields smooth functions in $\mathcal{C}^\infty(\overline{\Omega})$ whose $H_{w,\partial q_1,\ldots,\partial q_K}^r$-norm is still bounded, allowing the application of Proposition 4.38. Such a procedure, with all its technical details, is carried out in the following lemma.

Lemma 4.44. *Let Ω be a bounded Lipschitz domain, $1 < r < \infty$ and $w \in Y_1^*$ as well as $q_1, \ldots, q_K \in Y_d^*$. Then, there exists family of linear operators $\mathcal{M}_\varepsilon$ for $0 < \varepsilon < \tilde{\varepsilon}$ which only depends on Ω, with the following properties:*

1. *For each $y \in H_{w,\partial q_1,\ldots,\partial q_K}^r$, the functions $y_\varepsilon = \mathcal{M}_\varepsilon y$ are in $\mathcal{C}^\infty(\overline{\Omega})$, $y_\varepsilon \to y$ in $L^r(\Omega)$ as $\varepsilon \to 0$ and*

$$\left(\|w\nabla y_\varepsilon\|_r^r + \sum_{k=1}^K \|\partial_{q_k} y_\varepsilon\|_r^r \right)^{1/r} \leq C\|y\|_{H_{w,\partial q_1,\ldots,\partial q_K}^r}$$

   for some $C > 0$.

2. *There moreover exists a $C' > 0$ such that*

$$\|y_\varepsilon\|_{H^{1,r}} \leq \varepsilon^{-1} C'\|y\|_r .$$

Proof. We first construct the operators $\mathcal{M}_\varepsilon$. By assumption, Ω is a bounded Lipschitz domain, so according to that, there exist finitely many $U_l \subset \mathbb{R}^{d-1}$

which are neighborhoods of zero (for $1 \le l \le L$), Lipschitz continuous mappings $\varphi_l : U_l \to \mathbb{R}$, linear isometries $S_l : \mathbb{R}^d \to \mathbb{R}^d$ and $\bar{x}_l \in \partial\Omega$ such that

$$
\begin{aligned}
S_l\big(\bar{\xi}, \varphi_l(\bar{\xi})\big) + \bar{x}_l \in \partial\Omega && \text{for } \bar{\xi} \in U_l \\
S_l\big(\bar{\xi}, \varphi_l(\bar{\xi}) + \xi_d\big) + \bar{x}_l \in \Omega && \text{for } \bar{\xi} \in U_l , \ \xi_d \in \,]0, \varepsilon_l[\\
S_l\big(\bar{\xi}, \varphi_l(\bar{\xi}) + \xi_d\big) + \bar{x}_l \in \mathbb{R}^d \backslash \overline{\Omega} && \text{for } \bar{\xi} \in U_l , \ \xi_d \in \,]-\varepsilon_l, 0[
\end{aligned}
$$

with a suitable $\varepsilon_l > 0$. One can moreover assume that the

$$
\tilde{\Omega}_l = \{x \in \mathbb{R}^d \mid x = S_l\big(\bar{\xi}, \varphi_l(\bar{\xi}) + \xi_d\big) + \bar{x}_l , \bar{\xi} \in U_l , \ |\xi_d| < \varepsilon_l\}
$$

are open and that $\partial\Omega \subset \bigcup_{1 \le l \le L} \tilde{\Omega}_l$. Moreover, there has to be an $\varepsilon_0 > 0$ such that the sets

$$
\Omega_{l,\varepsilon_0} = \{x \in \tilde{\Omega}_l \mid \text{dist}(x, \partial\tilde{\Omega}_l) > \varepsilon_0\}
$$

still cover $\partial\Omega$: Otherwise, there would be a sequence $\{x_\varepsilon\}$ in $\partial\Omega$ for $\varepsilon > 0$ such that $x_\varepsilon \notin \Omega_{l,\varepsilon}$ for each $1 \le l \le L$. But $\partial\Omega$ is compact, so there exists a subsequence which converges to some $x \in \partial\Omega$. Now there is an l such that $x \in \tilde{\Omega}_l$ and since $\tilde{\Omega}_l$ is open, we have $\varepsilon_0 = \text{dist}(x, \partial\tilde{\Omega}_l)/2 > 0$. Consequently, x is in Ω_{l,ε_0} as well as almost every member of the above subsequence, which results in a contradiction. Thus, there is an $\varepsilon_0 > 0$ such that $\Omega_l = \Omega_{l,\varepsilon_0}$ cover $\partial\Omega$. Moreover, by standard arguments, an open Ω_0 with $\overline{\Omega_0} \subset\subset \Omega$ can be found such that $\overline{\Omega} \subset \bigcup_{0 \le l \le L} \Omega_l$.

Denote by η_l the image of e_d under S_l, i.e. the unit direction which locally points from $\partial\Omega$ into Ω. Now consider the mollifiers $G_{l,\varepsilon,x}(\xi) = G_\varepsilon(x + \varepsilon t_l \eta_l - \xi)$ for $x \in \overline{\Omega} \cap \Omega_l$, $\xi \in \mathbb{R}^d$ and $t_l = 2(C_l + 1)$ with C_l being the Lipschitz constant for the boundary mapping φ_l. For suitably small $\varepsilon > 0$ we want to obtain that $\text{supp}\, G_{l,\varepsilon,x}$ is compact in Ω. First, by choosing $0 < \varepsilon < (1 + t_l)^{-1}\varepsilon_0$ we can ensure that $\text{supp}\, G_{l,\varepsilon,x} \subset\subset \tilde{\Omega}_l$: If $\xi \in \text{supp}\, G_{l,\varepsilon,x}$, then

$$
|x + \varepsilon t_l \eta_l - \xi| \le \varepsilon \quad \Rightarrow \quad |x - \xi| \le |x + \varepsilon t_l \eta_l - \xi| + \varepsilon t_l \le (1 + t_l)\varepsilon < \varepsilon_0
$$

meaning that $\xi \in \Omega_l + B_{\varepsilon_0}(0) \subset \tilde{\Omega}_l$. Moreover, for $\xi \in \text{supp}\, G_{l,\varepsilon,x}$ we can estimate the distance to the boundary part $\partial\Omega \cap \tilde{\Omega}_l$ from below as follows. Since $x \in \tilde{\Omega}_l$, there has to be a $\bar{\xi}_1 \in U_l$ and a $\xi_d \in \,]0, \varepsilon_l[$ such that $x = S_l(\bar{\xi}_1, \varphi_l(\bar{\xi}_1) + \xi_d) + \bar{x}_l$. Now any point $\bar{x} \in \partial\Omega \cap \tilde{\Omega}_l$ can be written as $\bar{x} = S_l\big(\bar{\xi}_2, \varphi(\bar{\xi}_2)\big) + \bar{x}_l$ for $\bar{\xi}_2 \in U_l$. Thus, we know that

$$
\begin{aligned}
|x + \varepsilon t_l \eta_l - \bar{x}| &= \big|S_l\big(\bar{\xi}_1, \varphi_l(\bar{\xi}_1) + \xi_d + \varepsilon t_l\big) - S_l\big(\bar{\xi}_2, \varphi_l(\bar{\xi}_2)\big)\big| \\
&= \big|\big(\bar{\xi}_1 - \bar{\xi}_2, \varphi_l(\bar{\xi}_1) - \varphi_l(\bar{\xi}_2) + \xi_d + \varepsilon t_l\big)\big| \\
&\ge \max\{|\bar{\xi}_1 - \bar{\xi}_2|, \varepsilon t_l - C_l|\bar{\xi}_1 - \bar{\xi}_2|\} \\
&\ge 2\varepsilon
\end{aligned}
$$

since there is only the possibility that $|\bar{\xi}_1 - \bar{\xi}_2| \geq 2\varepsilon$ or $|\bar{\xi}_1 - \bar{\xi}_2| < 2\varepsilon$. Consequently,

$$|\bar{x} - \xi| \geq |x + \varepsilon t_l \eta_l - \bar{x}| - |x + \varepsilon t_l \eta_l - \xi| \geq 2\varepsilon - \varepsilon = \varepsilon \; ,$$

so $\xi \in \Omega$ and in particular $\text{supp}\, G_{l,\varepsilon,x} \subset\subset \Omega$.

In the following, choose a $\tilde{\varepsilon} > 0$ such that $\tilde{\varepsilon} < (1 + t_l)^{-1}\varepsilon_0$ for $1 \leq l \leq L$ as well as $\Omega_0 + B_{\tilde{\varepsilon}}(0) \subset\subset \Omega$ and let $0 < \varepsilon < \tilde{\varepsilon}$. Note that $\tilde{\varepsilon}$ then only depends on the domain Ω. The next step is to define $G_{l,\varepsilon}(x) = G_\varepsilon(x + \varepsilon t_l \eta_l)$ such that, for each $y \in H^r_{w, \partial_{q_1}, \ldots, \partial_{q_K}}$, $y_{l,\varepsilon}(x) = (y * G_{l,\varepsilon})(x) = \langle y, G_{l,\varepsilon,x} \rangle$ for $0 < \varepsilon < \tilde{\varepsilon}$ on each $\Omega \cap \Omega_l$. To take Ω_0 into account, let $t_0 = 0$, $\eta_0 = 0$ and set $y_{0,\varepsilon} = y * G_{0,\varepsilon} = y * G_\varepsilon$. Now, all $y_{l,\varepsilon}$ have to be assembled to an appropriate y_ε. This is of course done with a smooth partition of unity ζ_l on $\overline{\Omega}$ subordinate to the sets $\Omega_0, \ldots, \Omega_L$. We then define, as usual, the operators

$$\mathcal{M}_\varepsilon y = y_\varepsilon = \sum_{l=0}^{L} \zeta_l y_{l,\varepsilon} \; .$$

As one can easily see, this construction yields linear operators which only depend on Ω and whose images are always smooth functions, since by construction, each $\zeta_l y_{l,\varepsilon}$ is an element of $\mathcal{C}^\infty(\overline{\Omega})$ meaning that $y_\varepsilon \in \mathcal{C}^\infty(\overline{\Omega})$.

But before we prove that the first statement of the proposition holds for these y_ε, we want to obtain estimates for the L^r-norm of $w \nabla y_{l,\varepsilon}$ and $\partial_{q_k} y_{l,\varepsilon}$ restricted to the sets $\Omega \cap \Omega_l$ for $1 \leq l \leq L$. Most of the arguments from the proofs of Lemmas 4.40 and 4.42 can be reused here, we will therefore give a shortened argumentation. Consider $w \nabla y_{l,\varepsilon}$ which can be written as

$$w(x) y_{l,\varepsilon}(x) = \int_\Omega y(\xi)\big(w(x) - w(\xi)\big)\nabla G_\varepsilon(x + \varepsilon t_l \eta_l - \xi) \; \mathrm{d}\xi$$
$$+ \int_\Omega \big((w\nabla y)(\xi) + y(\xi)\nabla w(\xi)\big)G_\varepsilon(x + \varepsilon t_l \eta_l - \xi) \; \mathrm{d}\xi \; .$$

The crucial part is showing that the L^r-norm of the first term restricted to $\Omega \cap \Omega_l$ is bounded independently of ε, the second term can be estimated by standard arguments (see Lemma 4.40). Here, observe that due to Proposition 4.5 and Remark 4.6 as well as the above considerations, we have the estimates

$$|w(x + \varepsilon t_l \eta_l) - w(\xi)| \leq \|\nabla w\|_\infty \varepsilon \qquad \text{for } x \in \Omega \cap \Omega_l \; , \; |x + \varepsilon t_l \eta_l - \xi| \leq \varepsilon$$
$$|w(x) - w(x + \varepsilon t_l \eta_l)| \leq \|\nabla w\|_\infty t_l \varepsilon \qquad \text{for } x \in \Omega \cap \Omega_l$$

since $B_\varepsilon(x + \varepsilon t_l \eta_l)$ is compact in Ω as well as the line connecting x and $x + \varepsilon t_l \eta_l$. Thus,

$$\left(\int_{\Omega \cap \Omega_l} \left| \int_\Omega y(\xi)\big(w(x) - w(\xi)\big) \nabla G_\varepsilon(x + \varepsilon t_l \eta_l - \xi)\, \mathrm{d}\xi \right|^r \mathrm{d}x \right)^{1/r}$$

$$\leq (1 + t_l)\|\nabla w\|_\infty \left(\int_{\Omega \cap \Omega_l} \left| \int_\Omega \varepsilon^{-d}|y(\xi)| \left| \nabla G\Big(\frac{x - \xi}{\varepsilon} + t_l \eta_l\Big) \right| \mathrm{d}\xi \right|^r \mathrm{d}x \right)^{1/r}$$

$$\leq (1 + t_l)\|\nabla w\|_\infty \|\nabla G\|_1 \|y\|_r$$

from which follows that

$$\left(\int_{\Omega \cap \Omega_l} |w(x)\nabla y_{l,\varepsilon}(x)|^r\, \mathrm{d}x \right)^{1/r}$$

$$\leq \big((1 + t_l)\|\nabla G\|_1 + \|G\|_1\big)\|\nabla w\|_\infty \|y\|_r + \|G\|_1 \|w\nabla y\|_r \; .$$

An analog estimate for the L^r-norm of $\partial_{q_k} y$ restricted to $\Omega \cap \Omega_l$ can be obtained from the representation

$$q_k(x) \cdot \nabla y_{l,\varepsilon}(x) = \int_\Omega y(\xi)\big(q_k(x) - q_k(\xi)\big) \cdot \nabla G_\varepsilon(x + \varepsilon t_l \eta_l - \xi)\, \mathrm{d}\xi$$

$$+ \int_\Omega \big((\partial_{q_k} y)(\xi) + y(\xi)\, \mathrm{div}\, q_k(\xi)\big) G_\varepsilon(x + \varepsilon t_l \eta_l - \xi)\, \mathrm{d}\xi$$

for $x \in \Omega \cap \Omega_l$. It follows, in complete analogy to the above, that

$$\left(\int_{\Omega \cap \Omega_l} |\partial_{q_k} y_{l,\varepsilon}(x)|^r\, \mathrm{d}x \right)^{1/r}$$

$$\leq \big((1 + t_l)\|\nabla G\|_1 + d\|G\|_1\big)\|\nabla q_k\|_\infty \|y\|_r + \|G\|_1 \|\partial_{q_k} y\|_r$$

holds. Note that the above estimates also apply to $y_{0,\varepsilon}$.

We now turn to the first statement and verify that $y_\varepsilon \to y$ in $L^r(\Omega)$ as $\varepsilon \to 0$. It is clear that $y_{l,\varepsilon} \to y$ in $L^r(\Omega \cap \Omega_l)$ since $y_{l,\varepsilon}$ emerges from the convolution of y with the dilated versions of the translated mollifier $G(\cdot + t_l \eta_l)$. Hence,

$$\|y - y_\varepsilon\|_r \leq \sum_{l=0}^{L} \left(\int_{\Omega \cap \Omega_l} |y(x) - y_{l,\varepsilon}(x)|^r\, \mathrm{d}x \right)^{1/r} \to 0$$

as $\varepsilon \to 0$. Regarding the weak weighted and directional derivatives, the above estimates finally yield

$$\|w\nabla y_\varepsilon\|_r \le \sum_{l=0}^{L}\|\nabla\zeta_l\|_\infty\|w\|_\infty\left(\int_{\Omega\cap\Omega_l}|y_{l,\varepsilon}(x)|^r\,\mathrm{d}x\right)^{1/r}$$

$$+\left(\int_{\Omega\cap\Omega_l}|(w\nabla y_{l,\varepsilon})(x)|^r\,\mathrm{d}x\right)^{1/r}$$

$$\le\left(\sum_{l=0}^{L}\|\nabla\zeta_l\|_\infty\|G\|_1\|w\|_\infty\right.$$

$$+\big((1+t_l)\|\nabla G\|_1+\|G\|_1\big)\|\nabla w\|_\infty\Big)\|y\|_r$$

$$+(L+1)\|G\|_1\|w\nabla y\|_r$$

$$\le\tilde{C}_0\big(\|w\|_{Y_1^*}\big)\|y\|_r+\tilde{C}\|w\nabla y\|_r\ ,$$

with suitable $\tilde{C}_0\big(\|w\|_{Y_1^*}\big)$, $\tilde{C}$ and, analogously,

$$\|\partial_{q_k}y_\varepsilon\|_r\le\tilde{C}_1\big(\|q_k\|_{Y_d^*}\big)\|y\|_r+\tilde{C}\|\partial_{q_k}y\|_r\ .$$

Note that the $\tilde{C}_0\big(\|w\|_{Y_1^*}\big)$ and all $\tilde{C}_1\big(\|q_k\|_{Y_d^*}\big)$ remain bounded if $\|w\|_{Y_1^*}$ and $\|q_k\|_{Y_d^*}$ are bounded, respectively. The desired result is then implied by

$$\|w\nabla y_\varepsilon\|_r^r+\sum_{k=1}^{K}\|\partial_{q_k}y_\varepsilon\|_r^r\le C^r\|y\|_{H^r_{w,\partial q_1,\ldots,\partial q_K}}^r$$

with C given by the following expression (note that r' denotes the dual exponent of r)

$$C^r=(L+1)\left(\big(\tilde{C}_0(\|w\|_{Y_1^*})^{r'}+\tilde{C}^{r'}\big)^{r-1}+\sum_{k=1}^{K}\big(\tilde{C}_1(\|q_k\|_{Y_d^*})^{r'}+\tilde{C}^{r'}\big)^{r-1}\right)\ .$$

Finally, the second statement is just a consequence of the standard estimates

$$\|y_\varepsilon\|_{H^{1,r}}\le\sum_{l=0}^{L}\|y_{l,\varepsilon}\|_{H^{1,r}}\le\sum_{l=0}^{L}\|\zeta_l y_{l,\varepsilon}\|_r+\|\nabla(\zeta_l y_{l,\varepsilon})\|_r$$

$$\le\sum_{l=0}^{L}\|G\|_1\|y\|_r+\|\nabla\zeta_l\|_\infty\|G\|_1\|y\|_r+\varepsilon^{-1}\|\nabla G\|_1\|y\|_r$$

$$\le C'\varepsilon^{-1}\|y\|_r$$

where $C'>0$ is suitably chosen. ◄

After these technical considerations, the proof of the second main result of this section is immediate.

Theorem 4.45. *Let Ω be a bounded Lipschitz domain, $1 < r < \infty$, $w \in Y_1^*$ as well as $q_1, \ldots, q_K \in Y_d^*$. Then, $C^\infty(\overline{\Omega})$ is dense in $H^r_{w,\partial q_1,\ldots,\partial q_K}$ meaning in particular that $H^r_{w,\partial q_1,\ldots,\partial q_K}$ is the closure of $C^\infty(\overline{\Omega})$ under the norm (4.10).*

Proof. The result follows in analogy to the proof of Theorem 4.43. The essential step here is the combination of the first statement in Lemma 4.44 with the approximation result of Proposition 4.38. ◄

Remark 4.46. It is known that if Ω is a bounded Lipschitz domain, the set $C^\infty(\overline{\Omega})$ is also dense in $H^{1,r}(\Omega)$, see [AF03]. Since $H^{1,r}(\Omega) \hookrightarrow H^r_{w,\partial q_1,\ldots,\partial q_K}$, we can moreover interpret $H^r_{w,\partial q_1,\ldots,\partial q_K}$ as the closure of $H^{1,r}(\Omega)$ under the norm (4.10). This will become interesting for the characterization of the abstract stationary solution spaces V_p or the time-variant analogs $\mathcal{V}_p$, which is a topic of Chapter 5.

4.4.1 Calculus rules in $H^r_{w,\partial q_1,\ldots,\partial q_K}$

The possibility of smooth approximation for functions in $H^r_{w,\partial q_1,\ldots,\partial q_K}$ now allows to prove some calculus rules with respect to the weak weighted and directional derivative, namely the usual product and chain rule.

Proposition 4.47. *Let $1 < r < \infty$, $w \in Y_1^*$ and $q_1, \ldots, q_K \in Y_d^*$. Then, for each $y \in H^r_{w,\partial q_1,\ldots,\partial q_K}$ and $\zeta \in H^\infty_{w,\partial q_1,\ldots,\partial q_K}$ follows $\zeta y \in H^r_{w,\partial q_1,\ldots,\partial q_K}$ and*

$$w\nabla(\zeta y) = \zeta w \nabla y + y w \nabla \zeta \ ,$$
$$\partial_{q_k}(\zeta y) = \zeta \partial_{q_k} y + y \partial_{q_k} \zeta$$

for $k = 1, \ldots, K$. Moreover, $y \mapsto \zeta y$ is a continuous operation mapping $H^r_{w,\partial q_1,\ldots,\partial q_K}$ into itself.

Proof. The idea is to choose a $y \in C^\infty(\Omega)$ and a $\Omega' \subset\subset \Omega$ such that $y \in H^{1,\infty}(\Omega')$ in order to apply Lemmas 4.39 and 4.41 with y and ζ interchanged. This yields the existence of $w\nabla(\zeta y)$, $\partial_{q_k}(\zeta y)$ and the asserted identities in $L^r_{\mathrm{loc}}(\Omega, \mathbb{R}^d)$ and $L^r_{\mathrm{loc}}(\Omega)$, respectively.

Next, we find the continuity estimates

$$\|w\nabla(\zeta y)\|_{r,\Omega'} \leq \|w\nabla\zeta\|_\infty \|y\|_r + \|\zeta\|_\infty \|w\nabla y\|_r$$
$$\|\partial_{q_k}(\zeta y)\|_{r,\Omega'} \leq \|\partial_{q_k}\zeta\|_\infty \|y\|_r + \|\zeta\|_\infty \|\partial_{q_k} y\|_r$$

which are uniform for all $\Omega' \subset\subset \Omega$, allowing to extend the above to $L^r(\Omega, \mathbb{R}^d)$ and $L^r(\Omega)$, respectively. Finally,

$$\|\zeta y\|_{H^r_{w,\partial q_1,\ldots,\partial q_K}} \leq C\|y\|_{H^r_{w,\partial q_1,\ldots,\partial q_K}}$$

with a C with only depends on $\|\zeta\|_{H^\infty_{w,\partial q_1,\ldots,\partial q_K}}$. Hence, $y \mapsto \zeta y$ can be extended continuously to the whole space by density (Theorem 4.43).   ◄

Proposition 4.48. *Let* $1 < r < \infty$, $w \in Y_1^*$, $q_1, \ldots, q_K \in Y_d^*$ *and* $\varphi : \mathbb{R} \to \mathbb{R}$ *be a continuously differentiable function with bounded derivative. Then from* $y \in H^r_{w,\partial q_1,\ldots,\partial q_K}$ *follows* $\varphi \circ y \in H^r_{w,\partial q_1,\ldots,\partial q_K}$ *and*

$$w\nabla(\varphi \circ y) = (\varphi' \circ y)(w\nabla y) \quad , \quad \partial_{q_k}(\varphi \circ y) = (\varphi' \circ y)\partial_{q_k} y$$

for $k = 1, \ldots, K$. *The statement is still valid for* $\varphi(s) = \max\{0, s\}$, $\varphi(s) = \min\{0, s\}$ *as well as* $\varphi(s) = |s|$ *(with* $\varphi'(0) = 0$). *In particular,* $y(x) = 0$ *implies* $(w\nabla y)(x) = 0$ *and* $(\partial_{q_k} y)(x) = 0$ *almost everywhere.*

Moreover, the superposition $y \mapsto \varphi \circ y$ *is continuous in* $H^r_{w,\partial q_1,\ldots,\partial q_K} \to H^r_{w,\partial q_1,\ldots,\partial q_K}$.

Proof. The proof is an adaptation of the chain rule for classical Sobolev functions to the weak weighted and directional derivatives (confer [GT98], for example). Let $y \in \mathcal{C}^\infty(\Omega)$ so that $\varphi \circ y \in \mathcal{C}^1(\Omega)$ by the usual chain rule. The asserted identities in this case follow easily from $\nabla(\varphi \circ y) = (\varphi' \circ y)\nabla y$.

Now approximate $y \in H^r_{w,\partial q_1,\ldots,\partial q_K}$ with a sequence of $\{y_\varepsilon\}$ in $\mathcal{C}^\infty(\Omega)$ with $\varepsilon > 0$. In particular, one can find a subsequence such that we have pointwise a.e. convergence in Ω in addition to $y_\varepsilon \to y$ in $L^r(\Omega)$, $w\nabla y_\varepsilon \to w\nabla y$ in $L^r(\Omega, \mathbb{R}^d)$ and $\partial_{q_k} y_\varepsilon \to \partial_{q_k} y$ in $L^r(\Omega)$ for $k = 1, \ldots, K$. It is well-known that $\varphi \circ y_\varepsilon \to \varphi \circ y$ in $L^r(\Omega)$. Observe that $\varphi' \circ y_\varepsilon$ converges pointwise a.e. and is bounded almost everywhere. Consequently, Lebesgue's dominated-convergence theorem gives

$$(\varphi' \circ y_\varepsilon)w\nabla y_\varepsilon \to (\varphi' \circ y)w\nabla y \quad , \quad (\varphi' \circ y_\varepsilon)\partial_{q_k} y_\varepsilon \to (\varphi' \circ y)\partial_{q_k} y$$

in the respective spaces leading to the asserted identities for $w\nabla(\varphi \circ y)$ and $\partial_{q_k}(\varphi \circ y)$ by the closedness of the differential operators (see Remarks 4.31 and 4.33). This proves the chain rule in $H^r_{w,\partial q_1,\ldots,\partial q_K}$.

For the remainder, it is sufficient to consider $\varphi(s) = \max\{0, s\}$ since $\min\{0, s\} = -\max\{0, -s\}$ and $|s| = \max\{0, s\} - \min\{0, s\}$. The usual way is to approximate φ by

$$\varphi_\varepsilon(s) = \chi_{]0,\infty[}(s)\sqrt{s^2 + \varepsilon^2} - \varepsilon \quad , \quad \varphi_\varepsilon'(s) = \chi_{]0,\infty[}(s)\frac{s}{\sqrt{s^2 + \varepsilon^2}}$$

and to see that $\varphi_\varepsilon \circ y \to \varphi \circ y$ as well as $(\varphi_\varepsilon' \circ y) w \nabla y \to (\varphi' \circ y) w \nabla y$, $(\varphi_\varepsilon' \circ y)\partial_{q_k} y \to (\varphi' \circ y)\partial_{q_k} y$ in the respective spaces, again by Lebesgue's dominated-convergence theorem. Note that from $y = \max\{0, y\} + \min\{0, y\}$ then follows that $w\nabla y = 0$, $\partial_{q_k} y = 0$ almost everywhere where $y = 0$.

The continuity of $y \mapsto \varphi \circ y$ can be proven as follows. Suppose $y_l \to y$ in $H^r_{w,\partial_{q_1},\dots,\partial_{q_K}}$. First, $\varphi \circ y_l \to \varphi \circ y$ in $L^r(\Omega)$ by classical results about superposition operators (see [AZ90], for example). One can moreover assume that $w\nabla y_l \to w\nabla y$ and $\partial_{q_k} y_l \to \partial_{q_k} y$ pointwise a.e. and in $L^r(\Omega, \mathbb{R}^d)$ and $L^r(\Omega)$, respectively. Regarding the weak weighted and directional derivatives, we have

$$(\varphi' \circ y_l)(w\nabla y_l) \to (\varphi' \circ y)(w\nabla y) \quad , \quad (\varphi' \circ y_l)\partial_{q_k} y_l \to (\varphi' \circ y)\partial_{q_k} y$$

almost everywhere in Ω. This follows from the pointwise a.e. convergence of $\varphi' \circ y$ in the case where φ' is continuous. In the case where $\varphi(s) = \max\{0, s\}$, φ' is discontinuous in 0, so one also has to investigate the set where $y = 0$. The pointwise a.e. convergence is then derived from $w\nabla y_l \to |w\nabla y| = 0$ a.e. where $y = 0$, since $|(\varphi' \circ y_l)(w\nabla y_l)| \leq |w\nabla y_l| \to 0$ there. The analog conclusion is valid for $\partial_{q_k} y_l$. Furthermore, similar arguments show that this is also true for $\varphi(s) = \min\{0, s\}$ as well as $\varphi(s) = |s|$.

By Vitali's convergence theorem (cf. [Alt99], for example), we have

$$\sup_{l \in \mathbb{N}} \int_{\Omega'} |w\nabla y_l|^r \, \mathrm{d}x \to 0 \quad , \quad \text{if } |\Omega'| \to 0$$

and for each $\varepsilon > 0$ the existence of an Ω_ε such that

$$\sup_{l \in \mathbb{N}} \int_{\Omega_\varepsilon} |w\nabla y_l|^r \, \mathrm{d}x \leq \varepsilon .$$

Since φ' is bounded by, say, $C > 0$, it follows that

$$\sup_{l \in \mathbb{N}} \int_{\Omega'} |(\varphi' \circ y_l)(w\nabla y_l)|^r \, \mathrm{d}x \quad , \quad \text{if } |\Omega'| \to 0$$

as well as the fact that for each $\varepsilon > 0$ we have for $\delta = C^{-r}\varepsilon$ that

$$\sup_{l \in \mathbb{N}} \int_{\Omega_\delta} |(\varphi' \circ y_l)(w\nabla y_l)|^r \, \mathrm{d}x \leq \varepsilon$$

with the above Ω_δ. This, in turn implies $(\varphi' \circ y_l)(w\nabla y_l) \to (\varphi' \circ y)(w\nabla y)$ in $L^r(\Omega, \mathbb{R}^d)$, again by using Vitali's convergence theorem. The analog conclusions lead to $(\varphi' \circ y_l)\partial_{q_k}y_l \to (\varphi' \circ y)\partial_{q_k}y$ in $L^r(\Omega)$, thus $\varphi \circ y_l \to \varphi \circ y$ in $H^r_{w,\partial q_1,...,\partial q_K}$. ◄

Chapter 5

Time-variant spaces and application to degenerate PDEs

Until now, we have only described spaces of weak weighted and directional differentiable functions for the "stationary case" Ω. In the context of (degenerate) parabolic equations it is often useful to distinguish between the time-variable $t \in [0, T]$ and the space-variable $x \in \Omega$. This is also reflected in the choice of solution spaces which are, in the case of differential operators which are uniformly elliptic with respect to x, usually based on some Bochner integral spaces, see (3.6). Of course, a reasonable analog would allow weights w and directions $q_1, \ldots, q_K$ to vary in time which means that $H^r_{w(t), \partial q_1(t), \ldots, \partial q_K(t)}$ may be a different space for each $t \in [0, T]$, disallowing some Bochner-space formulation. Nevertheless, it will turn out that the actual definition is somewhat close to the definition of a Bochner space.

Furthermore, the theory of weak weighted and directional derivatives will be applied to analyze the solution operator of (2.4) with respect to u as well as p. It is derived that for suitable edge fields, the space $\mathcal{V}_p$ is a time-variant weighted and directional Sobolev space. We also prove a characterization of the solution space $W_p(0, T)$ which permits to obtain the uniqueness of the solution of (2.4) in $\mathcal{V}_p$. The behavior of the solution operator with respect to p is also studied. In particular, weak closedness properties which are crucial for existence of solutions of the optimal control problem (2.8) are established.

Finally, the chapter closes with some results regarding additional regularity of solution of (2.4) under additional assumptions. These results heavily rely on the given characterization of the solution space as time-variant weak weighted and directional Sobolev space.

5.1   Time-variant weights and directions

This section will be organized as follows. First, we introduce and study the
spaces of allowed time-variant weights and directions in analogy to Section 4.1.
Additionally, we are interested in weights and directions of bounded semivariation with respect to time. It is notable that, since an additional variable is
introduced, the weak* sequential continuity of superposition operators fails in
the general space but can be established for sequences where the total semivariation is bounded. Additionally, differentiability properties of non-linear
superposition operators are transferred to the time-variant setting.

   First, consider the following spaces of vector-valued functions:

Definition 5.1. For a bounded domain Ω and $k \geq 1$ we define

$$\mathcal{Y}_k = L^1(0, T; Y_k) \quad , \quad \mathcal{Y}_{\mathrm{div}} = L^1(0, T; Y_{\mathrm{div}}) .$$

   In analogy to Section 4.1, we are interested in the dual spaces.

Proposition 5.2. *The dual spaces $\mathcal{Y}_k^*$ and $\mathcal{Y}_{\mathrm{div}}^*$ are given by*

$$\mathcal{Y}_k^* = \{w \in L^\infty(]0, T[\times \Omega, \mathbb{R}^k) \mid \|w\|_{\mathcal{Y}_k^*} < \infty\}, \|w\|_{\mathcal{Y}_k^*} = \operatorname*{ess\,sup}_{t \in]0,T[} \|w(t)\|_{Y_k^*}$$

$$\mathcal{Y}_{\mathrm{div}}^* = \{q \in L^\infty(]0, T[\times \Omega, \mathbb{R}^d) \mid \|q\|_{\mathcal{Y}_{\mathrm{div}}^*} < \infty\}, \|q\|_{\mathcal{Y}_{\mathrm{div}}^*} = \operatorname*{ess\,sup}_{t \in]0,T[} \|q(t)\|_{Y_{\mathrm{div}}^*}$$

Proof. We will only examine $\mathcal{Y}_k^*$ since the proof can be carried out in complete
analogy for $\mathcal{Y}_{\mathrm{div}}^*$. Observe that $L^1(]0, T[\times \Omega, \mathbb{R}^k) \hookrightarrow \mathcal{Y}_k$ with dense continuous
embedding. Hence, each $w \in \mathcal{Y}_k^*$ can be identified with a $w \in L^\infty(]0, T[\times \Omega, \mathbb{R}^k)$
through the dual pairing

$$v \mapsto \int_0^T \int_\Omega w(t, x) \cdot v(t, x) \, \mathrm{d}x \, \mathrm{d}t .$$

Moreover, setting $v(t) = v_0 v_1(t)$ with $v_0 \in Y_k$ and $v_1 \in L^1(]0, T[)$ gives, according to the assumption

$$\left| \int_0^T \int_\Omega w(t, x) \cdot v(t, x) \, \mathrm{d}x \, \mathrm{d}t \right| \leq \|w\|_{\mathcal{Y}_k^*} \|v_0\|_{Y_k} \|v_1\|_1$$

which means that

$$\left| \int_\Omega w(t, x) \cdot v_0 \, \mathrm{d}x \right| \leq \|w\|_{\mathcal{Y}_k^*} \|v_0\|_{Y_k} \quad \text{a.e. in }]0, T[.$$

Taking the supremum over all $\{\|v_0\|_{Y_k} = 1\}$ and the essential supremum over $t \in \,]0, T[$ yields

$$\operatorname*{ess\,sup}_{t \in]0,T[} \|w(t)\|_{Y_k^*} \le \|w\|_{\mathcal{Y}_k^*} .$$

The converse direction can be seen as follows. Consider an element $w \in L^\infty(]0, T[\times \Omega, \mathbb{R}^k)$ with $\operatorname{ess\,sup}_{t \in]0,T[} \|w(t)\|_{Y_k^*} < \infty$. Then for all $v \in \mathcal{Y}_k$ which are simple with respect to t,

$$\left| \int_0^T \int_\Omega w(t,x) \cdot v(t,x) \, dx \, dt \right| \le \int_0^T \|w(t)\|_{Y_k^*} \|v(t)\|_{Y_k} \, dt$$

$$\le \operatorname*{ess\,sup}_{t \in]0,T[} \|w(t)\|_{Y_k^*} \|v\|_{\mathcal{Y}_k} ,$$

implying that $w \in \mathcal{Y}_k^*$ with $\|w\|_{\mathcal{Y}_k^*} \le \operatorname{ess\,sup}_{t \in]0,T[} \|w(t)\|_{Y_k^*}$. ◀

Remark 5.3.

(a) The space $\mathcal{Y}_k^*$ does not coincide with $L^\infty(0, T; Y_k^*)$, although they both share the same norm. This is closely connected with the failure of functions $w \in L^\infty(]0, T[\times \Omega, \mathbb{R}^k)$ regarded as $w :]0, T[\to L^\infty(\Omega, \mathbb{R}^k)$ a.e. to be measurable in the sense of Bochner, i.e. as pointwise a.e. limits of simple functions with values in $L^\infty(\Omega, \mathbb{R}^k)$.

Consequently, we are not able to integrate a function $w \in \mathcal{Y}_k^*$ in the sense of Bochner. It can be however noted that one can build the Gel'fand (or weak*-)integral of w which belongs to Y_k^*, i.e.

$$\left(\text{w-*} \int_0^T w \, dt \right)(v) = \int_0^T w(t) \cdot v \, dt$$

for each $v \in Y_k$ (see [DU77] for details).

(b) Elements $w \in \mathcal{Y}_k^*$ can be regarded as functions in $L^\infty(]0, T[\times \Omega, \mathbb{R}^k)$ for which the space gradient ∇w exists in $L^\infty(]0, T[\times \Omega, \mathbb{R}^{k \times d})$. This can be seen analogously to Proposition 4.3, by testing with $v = -\operatorname{div} z$ where z is smooth. Likewise, $w \in \mathcal{Y}_{\mathrm{div}}^*$ means, in a certain sense, that $w \in L^\infty(]0, T[\times \Omega, \mathbb{R}^d)$ as well as $\operatorname{div} w \in L^\infty(]0, T[\times \Omega)$.

(c) Since $\mathcal{Y}_k, \mathcal{Y}_{\mathrm{div}}$ are separable, we have, in analogy to Proposition 4.8, weak* sequential compactness of the dual unit balls. In other words, bounded sequences $\{w_l\}$ in $\mathcal{Y}_k^*$ and $\{q_l\}$ in $\mathcal{Y}_{\mathrm{div}}^*$ admit weakly*-convergent subsequences. As can be easily seen, a sequence satisfies $w_l \overset{*}{\rightharpoonup} w$ in $\mathcal{Y}_k^*$ if and only if

$$w_l \overset{*}{\rightharpoonup} w \text{ in } L^\infty(]0, T[\times \Omega, \mathbb{R}^k) , \quad \nabla w_l \overset{*}{\rightharpoonup} \nabla w \text{ in } L^\infty(]0, T[\times \Omega, \mathbb{R}^{k \times d}) .$$

(d) Unfortunately, a characterization of weak*-convergence in analogy to Proposition 4.11 fails, since a $w \in \mathcal{Y}_k^*$ is only "smooth with respect to x" but may have "jumps with respect to t". In particular, the norm with respect to t is not "strong enough" to enforce some compactness. This leads to the fact that the important property of weak* sequential continuity for superposition operators may also fail.

5.1.1   Superposition operators in $\mathcal{Y}_k^*$

To face the lack of weak* sequential continuity of superposition operators in $\mathcal{Y}_k^*$, which can be regarded as the lack of certain compactness, we introduce a certain property, namely the boundedness of the total semivariation, which enforces some compactness and makes it possible to establish the weak* sequential continuity. We will moreover see that this property is also useful for transferring closedness properties of the weak weighted and directional derivatives to the time-variant setting.

Recall that a function $y :]0, T[\to \mathbb{R}$ is of bounded variation if

$$\mathrm{TV}(y) = \sup_{\substack{L \geq 2 \\ 0=t_1<...t_L=T}} \sum_{l=1}^{L-1} |y(t_{l+1}) - y(t_l)| < \infty .$$

The following will be a generalization of this concept adapted to $\mathcal{Y}_k^*$ which is closely related to the definition presented in [DU77].

Definition 5.4. Let $k \geq 1$. We say that some function $w \in \mathcal{Y}_k^*$ is of *bounded semivariation* if

$$\mathrm{tv}^*(w) = \sup \left\{ \mathrm{TV}(w_v) \mid v \in Y_k \text{ with } \|v\|_{Y_k} \leq 1 \right\} < \infty ,$$

$$w_v(t) = \int_\Omega w(t, x) \cdot v(x) \, \mathrm{d}x .$$

It will turn out that sequences of uniformly bounded semivariation are "sufficiently compact" but nevertheless very general.

Lemma 5.5. *Let $\{w_l\}$ in $\mathcal{Y}_k^*$, $w \in \mathcal{Y}_k^*$ such that $w_l \overset{*}{\rightharpoonup} w$ and $\mathrm{tv}^*(w_l) \leq C$ for some $C > 0$. Then there is a subsequence $\{w_{\kappa(l)}\}$ such that $w_{\kappa(l)}(t) \overset{*}{\rightharpoonup} w(t)$ in Y_k^* for almost every $t \in]0, T[$.*

Proof. First observe that Y_k is separable, hence we can choose a dense countable subset $\{v_m\}$. Setting

$$w_{l,m}(t) = \int_\Omega w_l(t, x) \cdot v_m(x) \, \mathrm{d}x ,$$

one can see that $\mathrm{TV}(w_{l,m}) \leq C\|v_m\|_{Y_k}$, which implies that $\{w_{l,m}\}$, $l = 1, 2, \ldots$ is a sequence of functions with uniformly bounded variation on $]0, T[$ with values in $\mathbb{R}$. Thus, $\{w_{l,m}\}$, $l = 1, 2, \ldots$ is precompact in $L^1(]0, T[)$ admitting a subsequence which moreover converges pointwise almost everywhere to a z_m, i.e. on $]0, T[\backslash Z_m$ where $|Z_m| = 0$. Iterating this process over all m, diagonalizing the subsequence indices and labeling the resulting sequence with $\kappa(l)$, it is possible to obtain that

$$\lim_{l \to \infty} w_{\kappa(l),m}(t) = z_m(t) \quad \text{for } t \in]0, T[\backslash \bigcup_{m=1}^{\infty} Z_m .$$

In particular, $|w_{\kappa(l),m}(t)| \leq C_1 \|v_m\|_{Y_k}$ almost everywhere in $]0, T[$ where C_1 is a bound for $\|w_l\|_{\mathcal{Y}_k^*}$, meaning that $\int_0^T w_{\kappa(l),m}(t) v_0(t) \, dt \to \int_0^T z_m(t) v_0(t) \, dt$ for each function $v_0 \in L^1(]0, T[)$. On the other hand, according to the assumptions,

$$\int_0^T w_{\kappa(l),m}(t) v_0(t) \, dt \to \int_0^T \int_\Omega w(t, x) \cdot v_m(x) \, dx \, v_0(t) \, dt ,$$

implying $z_m(t) = \int_\Omega w(t, x) \cdot v_m(x) \, dx$ almost everywhere (without loss of generality in $]0, T[\backslash Z_m)$.

Now fix a $t \in]0, T[\backslash \bigcup_{m=1}^{\infty} Z_m$, choose an arbitrary $v \in Y_k$ and $\varepsilon > 0$. Since $\{v_m\}$ is dense in Y_k, one can choose an m such that $\|v_m - v\|_{Y_k} < \varepsilon/(2C_2)$ where C_2 is a bound for $\|w_l - w\|_{\mathcal{Y}_k^*}$. For l large enough follows

$$\left| \int_\Omega (w_{\kappa(l)} - w)(t, x) \cdot v(x) \, dx \right| \leq \left| \int_\Omega (w_{\kappa(l)} - w)(t, x) \cdot (v - v_m)(x) \, dx \right|$$

$$+ \left| \int_\Omega (w_{\kappa(l)} - w)(t, x) \cdot v_m(x) \, dx \right| \leq \frac{\varepsilon}{2} + |w_{\kappa(l),m}(t) - z_m(t)| < \varepsilon .$$

This means that $w_{\kappa(l)}(t) \overset{*}{\rightharpoonup} w(t)$ almost everywhere in $]0, T[$. ◀

Proposition 5.6. *Let* $\Theta : Y_k^* \to Y_l^*$ *a weakly* sequentially continuous mapping which is bounded on bounded sets and denote by* $\mathcal{T} : \mathcal{Y}_k^* \to \mathcal{Y}_l^*$ *the operator* $\mathcal{T}(w)(t) = \Theta(w(t))$. *Moreover, suppose* $\{w_m\}$ *in* $\mathcal{Y}_k^*$, $w \in \mathcal{Y}_k^*$ *are such that* $w_m \overset{*}{\rightharpoonup} w$ *in* $\mathcal{Y}_k^*$ *as well as* $\mathrm{tv}^*(w_m) \leq C$ *for some* $C > 0$. *Then* $\mathcal{T}(w_m) \overset{*}{\rightharpoonup} \mathcal{T}(w)$ *in* $\mathcal{Y}_l^*$ *as well as* $\mathcal{T}(w_m)(t) \overset{*}{\rightharpoonup} \mathcal{T}(w)(t)$ *in* Y_l^* *for almost every* $t \in]0, T[$.

Proof. By Lemma 5.5 we know that there is a subsequence (not relabeled) such that $w_m(t) \overset{*}{\rightharpoonup} w(t)$ for almost every $t \in]0, T[$. By the boundedness of Θ on bounded sets, one can moreover assume that $\Theta(w_m(t)) \overset{*}{\rightharpoonup} \Theta(w(t))$ a.e. in $]0, T[$

as well as $\|\Theta(w_m(t))\|_{Y_l^*} \leq C_1$ almost everywhere. Consequently, for each $v \in \mathcal{Y}_l$,

$$\int_0^T \int_\Omega \Theta(w_m(t,x)) \cdot v(t,x) \, dx \, dt \rightarrow \int_0^T \int_\Omega \Theta(w(t,x)) \cdot v(t,x) \, dx \, dt$$

by Lebesgue's dominated-convergence theorem applied to the sequence of functions $\{t \mapsto \int_\Omega \Theta(w_m(t,x)) \cdot v(t,x) \, dx\}$ in $L^1(]0,T[)$. The usual subsequence argument finally establishes the claim for the whole sequence.   ◄

Corollary 5.7. *In the situation of Proposition 4.14, the superposition operator $\mathcal{T}_\varphi : \mathcal{Y}_k^* \rightarrow \mathcal{Y}_l^*$ defined by $\mathcal{T}_\varphi(w)(t) = T_\varphi(w(t))$ is weakly* sequentially continuous in*

$$\{w \in \mathcal{Y}_k^* \mid \mathrm{tv}^*(w) \leq C\} \subset \mathcal{Y}_k^*$$

with arbitrary $C > 0$. Moreover, each sequence $\{w_m\}$ in the above set admits a subsequence for which $\mathcal{T}_\varphi(w_m)(t) \overset{*}{\rightharpoonup} \mathcal{T}_\varphi(w)(t)$ almost everywhere in $]0,T[$.

Additionally, it can easily be seen that $\mathcal{T}_\varphi$ is continuous in the strong sense on bounded sets in $\mathcal{Y}_k^*$ and maps bounded sets to bounded sets.

The differentiability of superposition operators in $\mathcal{Y}_k^*$ is also interesting for the time-variant setting.

Proposition 5.8. *Let $\Theta : Y_k^* \rightarrow Y_l^*$ be differentiable with derivative $\mathrm{D}\Theta$ which is uniformly continuous on bounded sets in Y_k^*. Then $\mathcal{T} : \mathcal{Y}_k^* \rightarrow \mathcal{Y}_l^*$ given by $\mathcal{T}(w)(t) = \Theta(w(t))$ is differentiable with derivative $\mathrm{D}\mathcal{T}(w)h = \mathrm{D}\Theta(w)h$ which is uniformly continuous on bounded sets.*

Proof. The proof is somehow analogous to the proof of the first case in Proposition 4.15, with $\mathbb{R}^k, \mathbb{R}^l$ replaced by Y_k^*, Y_l^*, respectively and $\Omega =]0,T[$. Therefore, we will give a shortened argumentation.

First remark that the uniform continuity of $\mathrm{D}\Theta$ on bounded sets implies the boundedness of $\mathrm{D}\Theta$ on bounded sets. Let $C > 0$ and choose a $0 < \delta < R$ such that whenever $w, \tilde{w} \in B_C(0) \subset Y_k^*$ with $\|w - \tilde{w}\|_{Y_k^*} < \delta$ there follows $\|\mathrm{D}\Theta(w) - \mathrm{D}\Theta(\tilde{w})\|_{\mathcal{L}(Y_k^*,Y_l^*)} \leq 1$. With an integer $M \geq C/\delta$ there holds for each $w \in B_C(0)$ that

$$\|\mathrm{D}\Theta(w) - \mathrm{D}\Theta(0)\|_{\mathcal{L}(Y_k^*,Y_l^*)}$$
$$\leq \sum_{m=1}^M \left\|\mathrm{D}\Theta\left(\frac{m}{M}w\right) - \mathrm{D}\Theta\left(\frac{m-1}{M}w\right)\right\|_{\mathcal{L}(Y_k^*,Y_l^*)} \leq M \, ,$$

implying the boundedness of $\mathrm{D}\Theta$ on $B_C(0)$.

Now it is clear that $h \mapsto D\Theta(w)h$ is bounded and continuous in $\mathcal{Y}_k^* \to \mathcal{Y}_l^*$ since

$$\operatorname*{ess\,sup}_{t \in]0,T[} \left\| D\Theta(w(t))h(t) \right\|_{Y_l^*} \leq \operatorname*{ess\,sup}_{t \in]0,T[} \left\| D\Theta(w(t)) \right\|_{\mathcal{L}(Y_k^*,Y_l^*)} \operatorname*{ess\,sup}_{t \in]0,T[} \left\| h(t) \right\|_{Y_k^*}.$$

Fix a $w \in \mathcal{Y}_k^*$. Choose a $\delta > 0$ such that for almost every $t \in]0,T[$ we have $\| D\Theta(w(t) + h(t)) - D\Theta(w(t)) \|_{\mathcal{L}(Y_k^*,Y_l^*)} < \varepsilon$ whenever $\| h \|_{\mathcal{Y}_k^*} < \delta$. Note that $s \mapsto D\Theta(w(t) + sh(t))h(t)$ is continuous for a.e. t, hence integration in Y_l^* makes sense. The remainder can thus be estimated

$$\left\| \Theta(w(t) + h(t)) - \Theta(w(t)) - D\Theta(w(t))h(t) \right\|_{Y_l^*}$$
$$= \left\| \int_0^1 \left(D\Theta(w(t) + sh(t)) - D\Theta(w(t)) \right) h(t) \, ds \right\|_{Y_l^*}$$
$$< \varepsilon \| h(t) \|_{Y_k^*},$$

yielding the stated differentiability of $\mathcal{T}$ in the point w by taking the essential supremum.

Finally, the conclusion that the operation $w \mapsto D\mathcal{T}(w)$ is uniformly continuous on bounded sets is completely analogous to Proposition 4.15. ◀

Corollary 5.9. *The superposition operator $\mathcal{T}(w)(t) = T_\varphi(w(t))$ with T_φ according to the second case of Proposition 4.15 is continuously differentiable regarded in $\mathcal{Y}_k^* \to \mathcal{Y}_l^*$.*

Remark 5.10. As can be seen in complete analogy, if $\Theta : Y_k^* \to L^\infty(\Omega, \mathbb{R}^l)$ possesses analog properties, then the corresponding $\mathcal{T} : \mathcal{Y}_k^* \to L^\infty(]0,T[\times \Omega, \mathbb{R}^l)$ is differentiable with uniformly continuous derivative (on bounded sets).

Especially, the superposition $\mathcal{T}(w)(t) = T_\varphi(w(t))$ with T_φ according to the first case of Proposition 4.15 is continuously differentiable regarded in $\mathcal{Y}_k^* \to L^\infty(]0,T[\times \Omega, \mathbb{R}^l)$.

5.2 Time-variant weighted and directional spaces

This section is concerned with notions of time-variant weak weighted and directional derivatives associated with weights and directions in $\mathcal{Y}_1^*$ and $\mathcal{Y}_d^*$, respectively. The corresponding time-variant Sobolev spaces are introduced and, in analogy to Section 4.3, strong as well as weak closedness properties are derived. Finally, it will be shown that functions on the cross product $]0,T[\times \Omega$ which are smooth with respect to the space-variable are also dense in these spaces.

In the following, the notion of the weak weighted gradient as well as the weak directional derivative will be discussed for time-variant weights and vector fields respectively. This is necessary since we cannot assume $w \in L^\infty(0, T; Y_1^*)$ and $q \in L^\infty(0, T; Y_{\text{div}}^*)$ if $w \in \mathcal{Y}_1^*$ and $q \in \mathcal{Y}_{\text{div}}^*$ respectively (see Remark 5.3), but this issue can easily be resolved by integrating on the Cartesian product $]0, T[\times \Omega$.

5.2.1   The time-variant weak weighted and directional derivative

Definition 5.11. Let $w \in \mathcal{Y}_1^*$ and $y \in L_{\text{loc}}^1(]0, T[\times \Omega)$. A function $v \in L_{\text{loc}}^1(]0, T[\times \Omega, \mathbb{R}^d)$ is defined to be the *time-variant weak weighted derivative* $v = w\nabla y$ if for every $z \in C_0^\infty(]0, T[\times \Omega, \mathbb{R}^d)$ the following equation holds:

$$\int_0^T \int_\Omega v \cdot z \, dx \, dt = -\int_0^T \int_\Omega y(w \operatorname{div} z + \nabla w \cdot z) \, dx \, dt . \qquad (5.1)$$

Furthermore, let $q \in \mathcal{Y}_{\text{div}}^*$. A $v \in L_{\text{loc}}^1(]0, T[\times \Omega)$ is defined to be the *time-variant weak directional derivative* $v = \partial_q y$, if for every $z \in C_0^\infty(]0, T[\times \Omega)$ the variational equation

$$\int_0^T \int_\Omega vz \, dx \, dt = -\int_0^T \int_\Omega y(z \operatorname{div} q + \nabla z \cdot q) \, dx \, dt \qquad (5.2)$$

is satisfied.

Note that in this definition, the differential operators are always with respect to the space coordinate x, i.e. $\nabla = \nabla_x$ and $\operatorname{div} = \operatorname{div}_x$. To avoid ambiguity, any differentiation with respect to t will be denoted by $\frac{\partial}{\partial t}$.

Remark 5.12.

(a) The time-variant weak weighted and directional derivatives according to the above definition are indeed time-variant versions of the weak weighted and directional derivatives according to Definitions 4.16 and 4.20, in the sense that $v = w\nabla y$ if and only if $v(t) = w(t)\nabla y(t)$ for almost every $t \in]0, T[$ and the analog for the weak directional derivative. Considering the weak weighted derivative, the "if part" is trivial, and the other direction can be seen by testing $w\nabla y$ with $z(t, x) = z_1(t)z_2(x)$ where $z_1 \in C_0^\infty(]0, T[)$

and $z_2 \in C_0^\infty(\Omega, \mathbb{R}^d)$. This implies

$$\int_\Omega (w\nabla y)(t) \cdot z_2 \, dx = - \int_\Omega y(t) \big(w(t) \operatorname{div} z_2 + \nabla w(t) \cdot z_2 \big) \, dx$$

a.e. in $]0, T[$ which still holds if z_2 is varied over a countable subset of $C_0^\infty(\Omega, \mathbb{R}^d)$. This subset can be chosen large enough to imply $(w\nabla y)(t) = w(t)\nabla y(t)$ for almost every $t \in]0, T[$, justifying the notation $(w\nabla y)(t) = w(t)\nabla y(t)$ which will be used in the following. The analog is also valid for the weak directional derivative.

(b) It is also easy to see that both constructions share the same strong closedness properties as the time-invariant notions, when considered on the rectangle $]0, T[\times \Omega$ (cf. Remarks 4.31 and 4.33). One has, however, to be careful when transferring the weak(*)-closedness, for example $w_k \overset{*}{\rightharpoonup} w$ in $\mathcal{Y}_1^*$, $y_k \rightharpoonup y$ in $L^r(]0, T[\times \Omega)$ and $w_k \nabla y_k \rightharpoonup \theta$ in $L^r(]0, T[\times \Omega, \mathbb{R}^d)$ does not imply $w\nabla y = \chi_{\{w \neq 0\}}\theta$, since the characterization of weak*-convergence analog to Lemma 4.9 fails, making the conclusions of Proposition 4.30 impossible. Again, this is closely connected with the non-linearity in the variational definition of the weak weighted gradient and the structure of $\mathcal{Y}_1^*$, see also Remark 5.3.

The above definition motivates to introduce time-variant weighted and directional Sobolev spaces.

Definition 5.13. Let $w \in \mathcal{Y}_1^*$ and $q_1, \ldots, q_K$ in $\mathcal{Y}_{\text{div}}^*$ for some $K \geq 0$. For $1 \leq r < \infty$ we define the space

$$\mathcal{H}_{w,\partial q_1,\ldots,\partial q_K}^r = \{ y \in L^r(0, T; L^r(\Omega)) \mid w\nabla y \in L^r(0, T; L^r(\Omega, \mathbb{R}^d)),$$
$$\partial_{q_k} y \in L^r(0, T; L^r(\Omega)), k = 1, \ldots, K \}$$

$$\|y\|_{\mathcal{H}_{w,\partial q_1,\ldots,\partial q_K}^r} = \left(\int_0^T \|y(t)\|_{H^r_{w(t),\partial q_1(t),\ldots,\partial q_K(t)}}^r \, dt \right)^{1/r}. \tag{5.3a}$$

For $r = \infty$ let

$$\mathcal{H}_{w,\partial q_1,\ldots,\partial q_K}^\infty = \{ y \in L^\infty(]0, T[\times \Omega) \mid w\nabla y \in L^\infty(]0, T[\times \Omega, \mathbb{R}^d),$$
$$\partial_{q_k} y \in L^\infty(]0, T[\times \Omega), k = 1, \ldots, K \}$$

$$\|y\|_{\mathcal{H}_{w,\partial q_1,\ldots,\partial q_K}^\infty} = \max \{ \|y\|_\infty, \|w\nabla y\|_\infty, \|\partial_{q_k} y\|_\infty, k = 1, \ldots, K \}. \tag{5.3b}$$

Remark 5.14.

(a) By Remark 5.12 and the definition, we have for each $y \in \mathcal{H}^r_{w,\partial q_1,\dots,\partial q_K}$ that

$$\|y(t)\|_r, \|w(t)\nabla y(t)\|_r, \|\partial_{q_k(t)} y(t)\|_r < \infty$$

for almost every $t \in \,]0, T[$, or, equivalently, $y(t) \in H^r_{w(t),\partial q_1(t),\dots,\partial q_K(t)}$ for almost every $t \in \,]0, T[$. Thus, the norm in Definition 5.13 indeed makes sense.

(b) Furthermore, the strong closedness of the corresponding differential operators (again, cf. Remark 5.12) imply, in analogy to Proposition 4.34, that $\mathcal{H}^r_{w,\partial q_1,\dots,\partial q_K}$ is a Banach space for $1 \leq r < \infty$ (reflexive for $1 < r < \infty$) and a Hilbert space for $r = 2$.

(c) Finally, let us remark that in analogy to Lemma 4.18, one can deduce that the identities (5.1) and (5.2) still hold for test functions $z \in \mathcal{Y}^*_d$ with compact support, in case $y \in \mathcal{H}^r_{w,\partial q_1,\dots,\partial q_K}$.

Before showing the density of smooth functions in $\mathcal{H}^r_{w,\partial q_1,\dots,\partial q_K}$ with the stronger condition $q_1, \dots, q_K \in \mathcal{Y}^*_d$, we study weak closedness of the time-variant differential operators with respect to both y as well as w and q, respectively. Again, one needs some additional assumptions if $y_k \rightharpoonup y$ in $L^r(0, T; L^r(\Omega))$, $w_k \overset{*}{\rightharpoonup} w$ in $\mathcal{Y}^*_1$ or $q_k \overset{*}{\rightharpoonup} q$ in $\mathcal{Y}^*_{\mathrm{div}}$ and $w_k \nabla y_k \rightharpoonup \theta$ or $\partial_{q_k} y_k \rightharpoonup v$, respectively. These assumptions will be pointwise a.e. convergence in $]0, T[\times \Omega$.

Proposition 5.15. *Let* $1 < r < \infty$, $\{y_k\}$ *in* $L^r(0, T; L^r(\Omega))$ *and* $y \in L^r(0, T; L^r(\Omega))$ *such that* $y_k \rightharpoonup y$.

1. *If* $\{w_k\}$ *in* $\mathcal{Y}^*_1$, $w \in \mathcal{Y}^*_1$ *with* $w_k \overset{*}{\rightharpoonup} w$ *as well as* $w_k \to w$, $\nabla w_k \to \nabla w$ *pointwise almost everywhere and* $\{w_k \nabla y_k\}$ *in* $L^r(0, T; L^r(\Omega, \mathbb{R}^d))$, $\theta \in L^r(0, T; L^r(\Omega, \mathbb{R}^d))$ *such that* $w_k \nabla y_k \rightharpoonup \theta$, *then* $w \nabla y = \theta$.

2. *If* $\{q_k\}$ *in* $\mathcal{Y}^*_{\mathrm{div}}$, $q \in \mathcal{Y}^*_{\mathrm{div}}$ *with* $q_k \overset{*}{\rightharpoonup} q$ *as well as* $q_k \to q$, $\mathrm{div}\, q_k \to \mathrm{div}\, q$ *pointwise almost everywhere and* $\{\partial_{q_k} y_k\}$ *in* $L^r(0, T; L^r(\Omega))$, $v \in L^r(0, T; L^r(\Omega))$ *such that* $\partial_{q_k} y_k \rightharpoonup v$, *then* $\partial_q y = v$.

Proof. The proof is essentially based on the observation that for each $z \in C_0^\infty(]0, T[\times \Omega)$ we have, for the respective cases, the convergence

$$-(w_k \,\mathrm{div}\, z + \nabla w_k \cdot z) \to -(w \,\mathrm{div}\, z + \nabla w \cdot z) \quad \text{in} \quad L^{r'}(0, T; L^{r'}(\Omega)) \text{ or}$$

$$-(z \,\mathrm{div}\, q_k + \nabla z \cdot q_k) \to -(z \,\mathrm{div}\, q + \nabla z \cdot q) \quad \text{in} \quad L^{r'}(0, T; L^{r'}(\Omega))$$

by the same arguments as in the proof of Proposition 4.32. Then, one easily sees that both sides of the defining integrals (5.1) or (5.2) converge, yielding the desired result. ◀

5.2.2 Density of smooth functions

The approximation properties obtained in the previous chapter can also be adapted to the time-variant setting. Here, we first restrict ourselves to the functions which are smooth with respect to the space variable. That is, we aim at proving the density of

$$\{y \in \mathcal{H}^r_{w, \partial q_1, \dots, \partial q_K} \mid y(t) \in \mathcal{C}^\infty(\Omega) \text{ for almost every } t \in \,]0, T[\} \, .$$

in the weighted and directional Sobolev space $\mathcal{H}^r_{w, \partial q_1, \dots, \partial q_K}$ for $w \in \mathcal{Y}_1^*$ and $q_1, \dots, q_K \in \mathcal{Y}_d^*$. This is sufficient for analogs to the product and chain rule of Subsection 4.4.1 to hold. To obtain that $\mathcal{H}^r_{w, \partial q_1, \dots, \partial q_K}$ is the closure of smooth functions with respect to both the time- and the space-variable, which is important for the characterization of the $\mathcal{V}_p$, the condition that Ω is a Lipschitz domain again has to be imposed. The proof of this density result will be presented subsequently.

To obtain the density of smooth functions with respect to the space-variable, we will again use the abstract approximation result of Proposition 4.36. As one can see without greater effort, the statements (restricting smoothness statements to the space-variable) are still valid if the considered domain is the cross product $]0, T[\times \Omega$ and if the mollification is as well carried out only in the direction of the space-variable.

Lemma 5.16. *Let* $1 < r < \infty$, $w \in \mathcal{Y}_1^*$, $q_1, \dots q_K \in \mathcal{Y}_d^*$ *and suppose that* $y \in L^r(0, T; L^r(\Omega))$ *satisfies* $\mathrm{supp}\, y \subset\subset \,]0, T[\times \Omega$. *Moreover, denote by* G *the standard mollifier in* $\mathbb{R}^d$, *by* G_ε *its dilated versions and by* y_ε *the function given by* $y_\varepsilon(t) = y(t) * G_\varepsilon$ *for almost every* $t \in \,]0, T[$.

Then, there is a $C > 0$, *such that from* $y \in \mathcal{H}^r_{w, \partial q_1, \dots, \partial q_K}$ *follows that* $\|y_\varepsilon\|_{\mathcal{H}^r_{w, \partial q_1, \dots, \partial q_K}} \le C \|y\|_{\mathcal{H}^r_{w, \partial q_1, \dots, \partial q_K}}$ *for sufficiently small* $\varepsilon > 0$.

Proof. First, choose $\varepsilon_0 > 0$ such that $\mathrm{supp}\, y_\varepsilon \subset\subset \,]0, T[\times \Omega$ for each $0 < \varepsilon < \varepsilon_0$. One can further assume that $\mathrm{supp}\, y(t) \subset \Omega' \subset\subset \Omega$ for almost every $t \in \,]0, T[$. Observe that $y_\varepsilon(t)$ belongs to $H^r_{w(t), \partial q_1(t), \dots, \partial q_K(t)}$ a.e. in $]0, T[$ (see Remark 5.14). Then, by virtue of Lemmas 4.40 and 4.42,

$$\|y_\varepsilon(t)\|_{H^r_{w(t)}} \le C_0 \big(\|w(t)\|_{Y_1^*} \big) \|y(t)\|_{H^r_{w(t)}} \, ,$$

$$\|y_\varepsilon(t)\|_{H^r_{\partial q_k(t)}} \le C_1 \big(\|q_k(t)\|_{Y_d^*} \big) \|y(t)\|_{H^r_{\partial q_k(t)}}$$

for each $k = 1, \dots, K$ and whenever the norms on the right-hand side is finite. By remembering how these constants are chosen in dependence of $\|w(t)\|_{Y_1^*}$ as well as $\|q_k(t)\|_{Y_d^*}$, one can convince oneself that there is a $C' > 0$ such that

$$\operatorname*{ess\,sup}_{t \in]0,T[} \max_{k=1,\dots,K} \big\{ C_0 \big(\|w(t)\|_{Y_1^*} \big), C_1 \big(\|q_k(t)\|_{Y_d^*} \big) \big\} \le C' \, .$$

Finally, recall that the norm in $\mathcal{H}^r_{w,\partial_{q_1},\ldots,\partial_{q_K}}$ is computed by taking the pointwise L^r-norm of $H^r_{w(t),\partial_{q_1}(t),\ldots,\partial_{q_K}(t)}$, thus

$$\|y_\varepsilon\|_{\mathcal{H}^r_{w,\partial_{q_1},\ldots,\partial_{q_K}}} = \left(\int_0^T \|y_\varepsilon(t)\|^r_{H^r_{w(t),\partial_{q_1}(t),\ldots,\partial_{q_K}(t)}} \, dt \right)^{1/r}$$

$$\leq (K+1)^{1/r} C' \left(\int_0^T \|y(t)\|^r_{H^r_{w(t),\partial_{q_1}(t),\ldots,\partial_{q_K}(t)}} \, dt \right)^{1/r}$$

$$\leq C \|y\|_{\mathcal{H}^r_{w,\partial_{q_1},\ldots,\partial_{q_K}}}$$

with $C \geq (K+1)^{1/r} C'$.   ◀

Theorem 5.17. *Let* $1 < r < \infty$, $w \in \mathcal{Y}^*_1$ *and* $q_1, \ldots, q_K \in \mathcal{Y}^*_d$ *for some* $K \geq 1$. *Then*

$$\{ y \in \mathcal{H}^r_{w,\partial_{q_1},\ldots,\partial_{q_K}} \mid y(t) \in \mathcal{C}^\infty(\Omega) \text{ for almost every } t \in \,]0,T[\}$$

is dense in $\mathcal{H}^r_{w,\partial_{q_1},\ldots,\partial_{q_K}}$. *In particular,* $\mathcal{H}^r_{w,\partial_{q_1},\ldots,\partial_{q_K}}$ *is the closure of the above set with respect to the norm* (5.3).

Proof. First observe that if one uses the mollification $y_\varepsilon(t) = y(t) * G_\varepsilon$ according to Lemma 5.16, the proof of Proposition 4.36 can still be carried out under the same prerequisites, since $\operatorname{supp} y_\varepsilon \subset \operatorname{supp} y + B_\varepsilon(0)$. However, one obtains that the approximation $\bar{y} \in \mathcal{D}(\Lambda)$ only satisfies $y(t) \in \mathcal{C}^\infty(\Omega)$ for almost every $t \in \,]0,T[$ instead of $y \in \mathcal{C}^\infty(]0,T[\times \Omega)$.

Now apply this to the closed differential operator

$$\Lambda : y \mapsto (w\nabla y, \partial_{q_1} y, \ldots, \partial_{q_K} y)$$

regarded in $L^r_{\mathrm{loc}}(]0,T[\times \Omega) \to L^r_{\mathrm{loc}}(]0,T[\times \Omega, \mathbb{R}^{d+K})$. The first and the second condition on Λ can be obtained from straightforward adaptations of the Lemmas 4.39 and 4.41 to the time-variant setting, while the third condition is fulfilled by virtue of Lemma 5.16. The above considerations then lead, for each $y \in L^r(]0,T[\times \Omega)$ with $w\nabla y \in L^r(]0,T[\times \Omega, \mathbb{R}^d)$ as well as $\partial_{q_k} y \in L^r(]0,T[\times \Omega)$ and $\varepsilon > 0$, to the existence of a $\bar{y}$ with $\bar{y}(t) \in \mathcal{C}^\infty(\Omega)$ a.e. and

$$\|y - \bar{y}\|^r_r + \|w\nabla y - w\nabla \bar{y}\|^r_r + \sum_{k=1}^K \|\partial_{q_k} y - \partial_{q_k} \bar{y}\|^r_r < \varepsilon$$

where the L^r-norm is taken with respect to the set $]0,T[\times \Omega$.

The claim finally follows from the equivalence of the spaces $L^r(]0,T[\times\Omega)$ and $L^r(]0,T[\times\Omega,\mathbb{R}^{d+K})$ with $L^r(0,T;L^r(\Omega))$ and $L^r(0,T;L^r(\Omega,\mathbb{R}^d))\times L^r(0,T;L^r(\Omega))^K$, respectively. ◀

Remark 5.18. In analogy to Remark 4.37, we have that for each $\Omega' \subset\subset \Omega$, $y \in \mathcal{H}^r_{w,\partial q_1,\dots,\partial q_K}$ and $\varepsilon > 0$, we can find a $\bar{y} \in \mathcal{H}^r_{w,\partial q_1,\dots,\partial q_K}$ with

$$\|y - \bar{y}\|_{\mathcal{H}^r_{w,\partial q_1,\dots,\partial q_K}} < \varepsilon \quad \text{and} \quad \bar{y}|_{]0,T[\times\Omega'} \in L^r(0,T;H^{1,r}(\Omega')) \ .$$

This can be seen as follows: First take the approximation $\tilde{y}$ from Theorem 5.17 and cut it off with respect to t, i.e. $\bar{y} = \chi_{[\delta,T-\delta]}\tilde{y}$ for some small $\delta > 0$. Then, $\bar{y}|_{]0,T[\times\Omega'} \in L^r(0,T;H^{1,r}(\Omega'))$ since $\bar{y}$ coincides with $\tilde{y}$ on a compact subset of $]0,T[\times\Omega$, thus its restriction can, again by construction, be written as a finite linear combination of elements in $L^r(0,T;H^{1,r}(\Omega'))$ (confer Proposition 4.36 and its proof as well as Remark 4.37).

It is still open whether $\bar{y}$ approximates y in $\mathcal{H}^r_{w,\partial q_1,\dots,\partial q_K}$. For see this, we remark that choosing $\tilde{y}$ such that $\|y - \tilde{y}\|_{\mathcal{H}^r_{w,\partial q_1,\dots,\partial q_K}} < \frac{\varepsilon}{2}$ and $\delta > 0$ such that

$$\left(\int_{]0,\delta]\cup[T-\delta,T[} \|\tilde{y}(t)\|^r_{H^r_{w(t),\partial q_1(t),\dots,\partial q_K(t)}} \, \mathrm{d}t\right)^{1/r} < \frac{\varepsilon}{2}$$

(which is possible since $t \mapsto \|\tilde{y}(t)\|^r_{H^r_{w(t),\partial q_1(t),\dots,\partial q_K(t)}}$ is an integrable function) yields $\|y - \bar{y}\|_{\mathcal{H}^r_{w,\partial q_1,\dots,\partial q_K}} < \varepsilon$ by a simple use of the triangle inequality.

Remark 5.19. The approximation result for the time-variant weak and directional Sobolev spaces allows us to state analogs to the calculus rules of Subsection 4.4.1. In the situation of Theorem 5.17, we have that $y \in \mathcal{H}^r_{w,\partial q_1,\dots,\partial q_K}$ and $\zeta \in \mathcal{H}^\infty_{w,\partial q_1,\dots,\partial q_K}$ implies that the product satisfies $\zeta y \in \mathcal{H}^r_{w,\partial q_1,\dots,\partial q_K}$ with the identities analog to Proposition 4.47. Likewise, for $\varphi : \mathbb{R} \to \mathbb{R}$ continuously differentiable with bounded derivative, we have $\varphi \circ y \in \mathcal{H}^r_{w,\partial q_1,\dots,\partial q_K}$ with identities analog to Proposition 4.48 and continuity of the respective operations. Both statements can be shown in analogy to the proofs of the respective propositions.

Remembering Theorem 4.44, one is again able to prove stronger results if Ω is a bounded Lipschitz domain.

Theorem 5.20. *Let Ω be a bounded Lipschitz domain, $1 < r < \infty$, $w \in \mathcal{Y}_1^*$ and $q_1,\dots,q_K \in \mathcal{Y}_d^*$. Then, $C^\infty([0,T] \times \overline{\Omega})$ is dense in $\mathcal{H}^r_{w,\partial q_1,\dots,\partial q_K}$.*

Proof. Let $\varepsilon > 0$ and a $y \in \mathcal{H}^r_{w, \partial_{q_1}, \ldots, \partial_{q_K}}$ be given. The first step is to construct a $\bar{y}_1 \in L^r(0, T; H^{1,r}(\Omega))$ which is close to y such that $\bar{y}_1(t) \in \mathcal{C}^\infty(\overline{\Omega})$ for almost every $t \in]0, T[$. The construction carried out in Lemma 4.44 yields a $\delta_0 > 0$ (called $\tilde{\varepsilon}$ there) which does only depend on Ω and, for almost every $t \in]0, T[$, a family $\{y_\delta(t)\}$ where $0 < \delta < \delta_0$ such that $y_\delta(t) \in \mathcal{C}^\infty(\overline{\Omega})$ (called y_ε in Lemma 4.44) and

$$\|y_\delta(t)\|_{H^r_{w(t), \partial_{q_1}(t), \ldots, \partial_{q_K}(t)}}$$
$$\leq C\big(\|w(t)\|_{Y_1^*}, \|q_1(t)\|_{Y_d^*}, \ldots, \|q_K(t)\|_{Y_d^*}\big) \|y(t)\|_{H^r_{w(t), \partial_{q_1}(t), \ldots, \partial_{q_K}(t)}}$$

with C remaining bounded whenever the norms $\|w(t)\|_{Y_1^*}$ and $\|q_k(t)\|_{Y_d^*}$ are bounded, which is a.e. the case since $\|w\|_{\mathcal{Y}_1^*} < \infty$ as well as $\|q_k\|_{\mathcal{Y}_d^*} < \infty$ for $1 \leq k \leq K$. Moreover, $y_\delta \in L^r(0, T; H^{1,r}(\Omega))$ by the second statement in Lemma 4.44, so one can find a linear combination $\bar{y}_1$ of some of the y_δ such that $\|y - \bar{y}_1\|_{\mathcal{H}^r_{w, \partial_{q_1}, \ldots, \partial_{q_K}}} < \varepsilon/2$. This can be seen by applying a straightforward analog of Proposition 4.38 to the time-variant differential operator $\Lambda y = (w \nabla y, \partial_{q_1} y, \ldots, \partial_{q_K} y)$ considered on $L^r(]0, T[\times \Omega)$ with the difference that one only assumes $y_\varepsilon \in \mathcal{D}(\Lambda)$ and concludes $\bar{y}_1 \in \mathcal{D}(\Lambda)$ instead of $y_\delta, \bar{y}_1 \in \mathcal{C}^\infty([0, T] \times \overline{\Omega})$.

Now observe that $]0, T[\times \Omega$ is a Lipschitz domain if Ω is a Lipschitz domain. This can be deduced from the result stating that bounded Lipschitz domains are exactly the bounded domains satisfying the uniform cone condition (see [Gri85], Theorem 1.2.2.2) and the fact that the uniform cone condition is preserved when taking the Cartesian product (see [Hoc89], Satz 3.1). Hence, $\mathcal{C}^\infty([0, T] \times \overline{\Omega})$ is dense in $H^r_{\partial_{e_2}, \ldots, \partial_{e_{d+1}}}$ considered with respect to the domain $]0, T[\times \Omega$, since one is able to apply Theorem 4.45. This yields a $\bar{y}_2 \in \mathcal{C}^\infty([0, T] \times \overline{\Omega})$ such that

$$\left(\int_0^T \int_\Omega |\bar{y}_2 - \bar{y}_1|^r + \sum_{i=1}^d \left| \frac{\partial \bar{y}_2}{\partial x_i} - \frac{\partial \bar{y}_1}{\partial x_i} \right|^r \, dx \, dt \right)^{1/r} < \frac{\varepsilon}{2C'}$$

where $C' > 0$ is chosen such that $\|\cdot\|_{\mathcal{H}^r_{w, \partial_{q_1}, \ldots, \partial_{q_K}}} \leq C' \|\cdot\|_{H^r_{\partial_{e_2}, \ldots, \partial_{e_{d+1}}}}$ (having the appropriate identifications in mind), for instance

$$C' = \max \left\{ 1, \left(\|w\|_\infty^r + \sum_{k=1}^K \|q_k\|_\infty^r \right)^{1/r} \right\}.$$

Together with the above, one gets the desired estimate

$$\|\bar{y}_2 - y\|_{\mathcal{H}^r_{w, \partial_{q_1}, \ldots, \partial_{q_K}}} \leq C' \|\bar{y}_2 - \bar{y}_1\|_{H^r_{\partial_{e_2}, \ldots, \partial_{e_{d+1}}}} + \|\bar{y}_1 - y\|_{\mathcal{H}^r_{w, \partial_{q_1}, \ldots, \partial_{q_K}}} < \varepsilon. \quad \blacktriangleleft$$

5.3 Application to degenerate parabolic PDEs

In the following, we demonstrate how the theory of (time-variant) weighted and directional Sobolev spaces can be applied to the weak theory for degenerate parabolic partial differential equations of the type (2.4). It will be shown that for bounded Lipschitz domains Ω, a suitable edge-intensity function σ as well as vector fields $p \in \mathcal{Y}_d^*$, $\|p\|_{\mathcal{Y}_d^*} \leq 1$, a weight $w \in \mathcal{Y}_1^*$ as well as some directions $q_1, \dots, q_K \in \mathcal{Y}_d^*$ can be found such that $\mathcal{V}_p = \mathcal{H}^2_{w, \partial q_1, \dots, \partial q_K}$. Unfortunately, the employed techniques do not seem to work for the characterization of the solution space $W_p(0, T)$ in terms of (time-variant) weighted and directional Sobolev spaces. By replacing the requirement that Ω is a Lipschitz domain with the condition that p does not degenerate at the boundary (which will be explained later), one is, however, able to obtain such a characterization. This will also imply $\bar{W}_p(0, T) = W_p(0, T)$ and solve the uniqueness problem for weak solutions of (2.4) in $\mathcal{V}_p$.

Moreover, many conclusions which are commonly used in the theory of parabolic PDEs with uniformly elliptic operators with respect to the space-variable can then be drawn. The characterization also is, together with the weak closedness properties, the key to gain insight about the behavior of the solution operator for (2.4) with respect to p by studying the associated weighted and directional derivatives. This is because these solutions and their associated weak weighted and directional derivatives can now be interpreted as elements of the fixed spaces $L^2\big(0, T; L^2(\Omega, \mathbb{R}^d)\big)$ and $L^2\big(0, T; L^2(\Omega)\big)^K$ for all feasible p, i.e. even if the space $\mathcal{V}_p$ itself varies. Hence, one can say that the technique of weak weighted and directional derivatives pulls a whole class of Hilbert spaces back to one fixed Hilbert space.

However, there is a price we have to pay for this powerful tool. As can be seen below, the characterization requires that the vector field p is in $\mathcal{Y}_d^*$ and not only measurable (and bounded). Furthermore, we restrict ourselves to special edge-intensity functions σ which are smooth by definition.

We will first demonstrate how the operation of the diffusion tensor associated with a vector field p, i.e. $y \mapsto D_p^2 \nabla y$ can be written in terms of weak weighted and directional derivatives. Therefore, for each p, a weight w and a set of directions q_k are associated.

The idea behind this is the following. Fix a $p \in \mathbb{R}^d$ with $|p| \leq 1$. By definition, D_p^2 can be decomposed into

$$D_p^2 = \begin{cases} (1 - \sigma(|p|))I + \sigma(|p|) \sum_{i=1}^{d-1} p_i \otimes p_i & \text{if } p \neq 0 \\ I & \text{if } p = 0 \end{cases} \tag{5.4a}$$

where p_i are orthonormal vectors spanning the orthogonal complement of $\frac{p}{|p|}$.

This also reads as

$$D_p^2 = w^2 I + \sum_{i=1}^{d-1} q_i \otimes q_i \tag{5.4b}$$

where $w = \sqrt{1 - \sigma(|p|)}$, $q_i = \sqrt{\sigma(|p|)}p_i$ and p_i are an arbitrary set of orthonormal vectors in the case $p = 0$.

Now let $p \in L^\infty(]0,T[\times \Omega, \mathbb{R}^d)$ be given and perform this construction for almost every $p(t,x)$. For functions $y, z \in C^\infty([0,T] \times \overline{\Omega})$ we have

$$\int_0^T \langle D_{p(t)}^2 \nabla y(t)^\mathrm{T}, \nabla z(t)^\mathrm{T} \rangle_{L^2} \, dt$$

$$= \int_0^T \langle w(t)\nabla y(t), w(t)\nabla z(t) \rangle_{L^2} + \sum_{i=1}^{d-1} \langle \partial_{q_i(t)} y(t), \partial_{q_i(t)} z(t) \rangle_{L^2} \, dt . \tag{5.4c}$$

Hence, we express the differential operator with respect to the space-variable in (2.4) in terms of (weak) weighted and directional derivatives and, by adding $\langle y, z \rangle_{L^2}$, in terms of the scalar product in $\mathcal{V}_p$, as can be seen in Definition 3.12. Moreover, the norm in $\mathcal{V}_p$ coincides with (5.3) for $r = 2$ which suggests equality of the spaces $\mathcal{V}_p$ and $\mathcal{H}^2_{w,\partial q_1,\dots,\partial q_{d-1}}$.

In order to apply Theorem 5.20 and to obtain equality of $\mathcal{V}_p$ with a weighted and directional Sobolev space, it is necessary, nevertheless, to verify that the construction yields feasible w and q_i. Therefore, the mappings $p \mapsto w$ as well as $p \mapsto q_i$ have to be studied. It turns out that for general dimensions $d \geq 1$, it is necessary to construct the directions q_k in a different way than the above. This will be examined in the following lemmas. We begin with specifying the edge-intensity functions and vector fields we are considering now.

Condition 5.21. *Let* $\sigma : [0, \infty[\to [0,1]$ *be a strictly monotone increasing and twice continuously differentiable function satisfying*

$$\sigma(0) = 0 , \qquad\qquad \sigma(1) = 1 ,$$
$$\sigma'(0) = 0 , \qquad\qquad \sigma'(1) = 0 ,$$
$$\sigma''(0) = 0 , \qquad\qquad \sigma''(1) = 0 ,$$

as well as $\sigma(s) = 1$ *for* $s > 1$.

Condition 5.22. *Let* $p \in \mathcal{Y}_d^*$ *be such that* $\|p\|_{\mathcal{Y}_d^*} \leq 1$.

Example 5.23. Possible choices for σ in $[0,1]$ are for instance

$$\sigma(|p|) = |p|^3 \left(6|p|^2 - 15|p| + 10\right) ,$$

$$\sigma(|p|) = c^{-1} \int_0^{|p|} e^{\frac{1}{s(s-1)}} \, ds \quad \text{with} \quad c = \int_0^1 e^{\frac{1}{s(s-1)}} \, ds ,$$

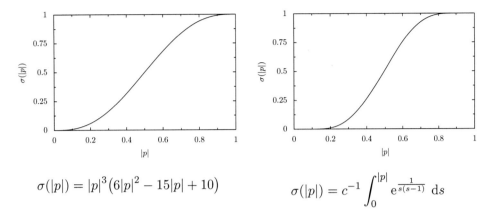

$$\sigma(|p|) = |p|^3\left(6|p|^2 - 15|p| + 10\right) \qquad \sigma(|p|) = c^{-1}\int_0^{|p|} e^{\frac{1}{s(s-1)}}\,ds$$

Figure 5.1: Sample edge-intensity functions fulfilling Condition 5.21.

see Figure 5.1 for an illustration.

Lemma 5.24. *Let σ fulfill Condition 5.21.*

1. *Then $s \mapsto \sqrt{\sigma(s)}$ and $s \mapsto \sqrt{1 - \sigma(s)}$ are continuously differentiable functions with vanishing derivatives at 0 and 1.*

2. *Moreover, $\sqrt{\sigma(s)}/s$ and $\sqrt{1 - \sigma(s)}/(1 - s)$ can be extended to continuous functions with $\lim_{s\to 0} \sqrt{\sigma(s)}/s = \lim_{s\to 1} \sqrt{1 - \sigma(s)}/(1 - s) = 0$.*

Proof. For $s \in \,]0, 1[$, the first claim can be verified by differentiation:

$$\frac{\partial}{\partial s}\sqrt{\sigma(s)} = \frac{\sigma'(s)}{2\sqrt{\sigma(s)}} \quad , \quad \frac{\partial}{\partial s}\sqrt{1 - \sigma(s)} = \frac{-\sigma'(s)}{2\sqrt{1 - \sigma(s)}} \,,$$

while $\frac{\partial}{\partial s}\sqrt{\sigma(s)}\big|_{s=1} = 0$ as well as $\frac{\partial}{\partial s}\sqrt{1 - \sigma(s)}\big|_{s=0} = 0$ according to the assumptions. From Condition 5.21 also follows $|\sigma(s)| < \varepsilon s^2$ and $|1 - \sigma(s)| < \varepsilon(1 - s)^2$ for each $\varepsilon > 0$ and in a neighborhood of 0 and 1, respectively, hence $\frac{\partial}{\partial s}\sqrt{\sigma(s)}\big|_{s=0} = \frac{\partial}{\partial s}\sqrt{1 - \sigma(s)}\big|_{s=1} = 0$.

It remains to show that $\frac{\partial}{\partial s}\sqrt{\sigma(s)}$ is also continuous in 0. Note that

$$\frac{\sigma'(s)^2}{2} = \int_0^s \sigma''(\tau)\sigma'(\tau)\,d\tau$$

$$\leq \left(\max_{\tau\in[0,s]}|\sigma''(\tau)|\right)\int_0^s \sigma'(\tau)\,d\tau = \left(\max_{\tau\in[0,s]}|\sigma''(\tau)|\right)\sigma(s),$$

since, by assumption, σ is strictly monotone increasing, meaning $\sigma' \geq 0$. By continuity, it is possible to choose an s small enough such that we obtain $\max_{\tau \in [0,s]} |\sigma''(\tau)| < 2\varepsilon^2$ for each $\varepsilon > 0$. This implies

$$\frac{\sigma'(s)^2}{4\sigma(s)} < \varepsilon^2 \quad \Rightarrow \quad \frac{\sigma'(s)}{2\sqrt{\sigma(s)}} < \varepsilon$$

again by $\sigma, \sigma' \geq 0$. Analogously, one proves that $\frac{\partial}{\partial s}\sqrt{1 - \sigma(s)}$ is continuous in 1. Finally, the second claim is just a consequence of the first. ◀

With the help of the above, it is possible to show that $p \mapsto w(p) = \sqrt{1 - \sigma(|p|)}$ is mapping $\mathcal{Y}_d^* \to \mathcal{Y}_d^*$.

Lemma 5.25. *In the situation of Condition 5.21, the weight function $p \mapsto \sqrt{1 - \sigma(|p|)}$ mapping $\mathbb{R}^d \to \mathbb{R}$ is continuously differentiable with bounded derivative.*

The superposition operator $\mathcal{T}_w(p)(t,x) = \sqrt{1 - \sigma(|p(t,x)|)}$ is mapping continuously between $\mathcal{Y}_d^* \to \mathcal{Y}_1^*$ and takes bounded sets to bounded sets.

Proof. Note that the function $\varphi : \mathbb{R}^d \to \mathbb{R}$ defined by

$$\varphi(v) = \begin{cases} \sqrt{1 - \sigma(|v|)} & \text{if } |v| \leq 1 \\ 0 & \text{if } |v| > 1 \end{cases}$$

and with which each $p \in \mathcal{Y}_d^*$ is superposed, is Lipschitz continuous. To see this, observe (with the help of Lemma 5.24) that φ is continuously differentiable with derivative

$$\nabla\varphi(v) = \begin{cases} \dfrac{-\sigma'(|v|)}{2|v|\sqrt{1 - \sigma(|v|)}} v & \text{if } 0 < |v| < 1 \\ 0 & \text{else .} \end{cases}$$

Proposition 4.14 then establishes that the superposition with φ maps continuously between $Y_d^* \to Y_1^*$ and is bounded on bounded sets. The statement on $\mathcal{T}_w$ in $\mathcal{Y}_d^*$ then follows by applying Corollary 5.7. ◀

Remark 5.26. From Corollary 5.7 also follows that $\mathcal{T}_w$ is weakly* sequentially continuous on sets of bounded semivariation. Moreover, if $p_l \overset{*}{\rightharpoonup} p$ in $\mathcal{Y}_d^*$ and the semivariation of $\{p_l\}$ is bounded, then there is a subsequence such that $\mathcal{T}_w(p_l)(t) \overset{*}{\rightharpoonup} \mathcal{T}_w(p_l)(t)$ in Y_1^* for almost every $t \in\,]0, T[$.

We now want to establish a connection between the vector field p and the directions spanning the respective orthogonal complement. Here, for general

$d \geq 1$, we need in fact more than $d - 1$ mappings, due to a technical difficulty which is explained as follows.

Recall that we seek a smooth mapping assigning each p a set of orthogonal directions all having the length $\sqrt{\sigma(|p|)}$ and spanning the orthogonal complement of p if $p \neq 0$. If we restrict ourselves to $|p| = 1$, then this amounts, because of $\sigma(1) = 1$, to finding mappings q_k which map the unit sphere in $\mathbb{R}^d$ into itself and such that $q_k(p) \cdot p = 0$ for all $|p| = 1$, meaning that the mappings have to be tangential. But in the case $d = 3$ (or, more general, in odd dimensions), such a mapping q_k cannot be continuous: a continuous tangential mapping on S^2 has to vanish somewhere (the hairy ball theorem, see e.g. [EG79]), i.e. there is a $p \in S^2$ such that $|q_k(p)| = 0 \neq 1 = \sigma(1)$.

By choosing discontinuous mappings q_k, however, one is possibly confronted with the failure of associated superposition operators being well-defined. Moreover, the smoothness of vector fields p cannot be transformed to the superposed $q_k(p)$ if q_k is not smooth itself. Nevertheless, by defining some $\tilde{q}_k$ locally on the unit sphere, cutting off smoothly and extending suitably to the whole $\mathbb{R}^d$, a workaround can be found. Such a construction is the topic of the next lemmas.

Lemma 5.27. *For $d \geq 2$, there exist smooth mappings $\tilde{q}_k : S^{d-1} \to S^{d-1}$, $k = 1, \ldots, K$ such that for each $p \in S^{d-1}$ there holds*

$$I - p \otimes p = \sum_{k=1}^{K} \tilde{q}_k(p) \otimes \tilde{q}_k(p) . \tag{5.5}$$

Proof. Let (φ_l, U_l) for $l = 1, \ldots, L$ an atlas of the sphere $S^{d-1} \subset \mathbb{R}^d$ and $\psi_l : S^{d-1} \to \mathbb{R}$ such that ψ_l^2 forms a C^∞-partition of unity subordinate to U_l. Then there exists, for each l, a C^∞-orthonormal frame $p_1^l, \ldots, p_{d-1}^l : U_l \to S^{d-1}$ of the tangential space, which is the orthogonal complement of span$\{p\}$ for each $|p| = 1$. Then define $\tilde{q}_i^l(p) = \psi_l(p)p_i^l(p)$ with zero extension outside of U_l for $i = 1, \ldots, d - 1$ and $l = 1, \ldots, L$ and renumber the double indices to $\tilde{q}_k$ where $k = 1, \ldots, L(d-1)$. Note that since each $p_1^l(p), \ldots, p_{d-1}^l(p)$ is an orthonormal basis of span$\{p\}^\perp$, we have for $p \in U_l$ that

$$I - p \otimes p = \sum_{i=1}^{d-1} p_i^l(p) \otimes p_i^l(p) .$$

Since $\sum_{l=1}^{L} \psi_l^2(p) = 1$ for all $|p| = 1$ we deduce

$$I - p \otimes p = \sum_{l=1}^{L} \psi_l^2(p)(I - p \otimes p) = \sum_{l=1}^{L} \sum_{i=1}^{d-1} \psi_l(p)p_i^l(p) \otimes \psi_l(p)p_i^l(p)$$

$$= \sum_{k=1}^{K} \tilde{q}_k(p) \otimes \tilde{q}_k(p) . \qquad \blacktriangleleft$$

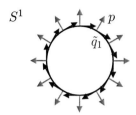

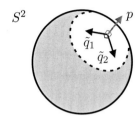

(a) Illustration of $\tilde{q}_1$ for $d = 2$

(b) Illustration of local perpendicular directions $\tilde{q}_1$, $\tilde{q}_2$ for $d = 3$.

Figure 5.2: Illustration of the vector fields $\tilde{q}_k$ spanning the directions perpendicular to each $|p| = 1$. (a) In the case $d = 2$, one mapping $\tilde{q}_1$ is sufficient. The vectors p with $|p| = 1$ are depicted as arrows which are perpendicular to the circle while the $\tilde{q}_1(p)$ are shown as tangential arrows. (b) In the case $d = 3$, one has to construct the perpendicular directions locally together with a partition of unity. The region within the dashed lines depicts such a local neighborhood. Again, the arrow labeled with p indicates a unit direction where $\tilde{q}_1(p)$, $\tilde{q}_2(p)$ are indicated by two corresponding tangential arrows. Moreover, $\tilde{q}_1$, $\tilde{q}_2$ are smooth with respect to p, i.e. $\tilde{q}_1(p)$ and $\tilde{q}_2(p)$ do not deviate too much from the values the illustration hints at.

Remark 5.28. In the case $d = 2$, a single $q_1 : S^1 \to S^1$ is sufficient to obtain (5.5): just set $\tilde{q}_1(p) = \left(\begin{smallmatrix} 0 & -1 \\ 1 & 0 \end{smallmatrix}\right)p$ which is a rotation by $\frac{\pi}{2}$. For higher even dimensions, one can decide when a construction of $d-1$ tangential vector fields with the desired properties is possible [Ada62]. It turns out that this is only the case for $d = 2, 4, 8$. Thus, for all other d, a construction like the one carried out in Lemma 5.27 is necessary. See Figure 5.2 for an illustration of $\tilde{q}_1$ for $d = 2$ as well as the construction of Lemma 5.27 for $d = 3$.

With the help of σ, it is possible to extend the $\tilde{q}_k$ to the whole $\mathbb{R}^d$ in a continuously differentiable way.

Lemma 5.29. *Let σ fulfill Condition 5.21. The $\tilde{q}_k : S^{d-1} \to S^{d-1}$ of Lemma 5.27 can be extended to some continuously differentiable $q_k : \mathbb{R}^d \to \mathbb{R}^d$ such that for each $|p| \leq 1$ the $q_k(p)$ suffice the identity*

$$\sigma(|p|)\left(I - \tfrac{p}{|p|} \otimes \tfrac{p}{|p|}\right) = \sum_{k=1}^{K} q_k(p) \otimes q_k(p) \ . \tag{5.6}$$

The corresponding superposition operators $\mathcal{Y}_d^* \to \mathcal{Y}_d^*$ acting pointwise a.e. via $\mathcal{T}_{q_k}(p)(t, x) = q_k\big(p(t, x)\big)$ are continuous and bounded on bounded sets.

Proof. We extend the $\tilde{q}_k$ as follows. First, project the argument p to S^{d-1}, then apply q_k and multiply the result by $\sqrt{\sigma(|p|)}$. This gives

$$q_k(p) = \sqrt{\sigma(|p|)}\tilde{q}_k\left(\frac{p}{|p|}\right) ,$$

which can be extended by 0 in 0. The q_k are obviously differentiable with derivative

$$\nabla q_k(p) = \frac{\sqrt{\sigma(|p|)}}{|p|}\nabla\tilde{q}_k\left(\frac{p}{|p|}\right)\left(I - \frac{p}{|p|}\otimes\frac{p}{|p|}\right) + \frac{\sigma'(|p|)}{2\sqrt{\sigma(|p|)}}\tilde{q}_k\left(\frac{p}{|p|}\right)\otimes\frac{p}{|p|}$$

for $p \neq 0$. For $p = 0$ the derivative satisfies $\nabla q_k(0) = 0$ since $\sqrt{\sigma(|p|)}/|p| \to 0$ for $|p| \to 0$, see Lemma 5.24. Moreover, ∇q_k is continuous, even in 0, again by Lemma 5.24. The identity (5.6) then follows from the definition of the q_k and (5.5).

Finally, again by Corollary 5.7, each superposition operator $\mathcal{T}_{q_k} : \mathcal{Y}_d^* \to \mathcal{Y}_d^*$ is continuous and bounded on bounded sets. ◄

Remark 5.30. Again, $\mathcal{T}_{q_k}$ is also weakly* sequentially continuous in the respective spaces, if the semivariation is bounded, as can be seen by Corollary 5.7. Moreover, for each sequence $p_l \overset{*}{\rightharpoonup} p$ of bounded semivariation there exists a subsequence for which $\mathcal{T}_{q_k}(p_l)(t) \overset{*}{\rightharpoonup} \mathcal{T}_{q_k}(p)(t)$ in Y_d^* almost everywhere in $]0,T[$.

Collecting these prerequisites, we can turn towards the proof of the first main result of this section, the characterization of the weak solution spaces $\mathcal{V}_p$ in terms of weak and directional Sobolev spaces.

Theorem 5.31. *Let Ω be a bounded Lipschitz domain as well as σ and p satisfy Conditions 5.21 and 5.22, respectively. Then, in the sense of Hilbert space isometry,*

$$\mathcal{V}_p = \mathcal{H}^2_{w,\partial q_1,\ldots,\partial q_K} ,$$

$$\langle y, z\rangle_{\mathcal{V}_p} = \langle y, z\rangle_{L^2} + \langle w\nabla y, w\nabla z\rangle_{L^2} + \sum_{k=1}^K \langle \partial_{q_k} y, \partial_{q_k} z\rangle_{L^2}$$

where, for $k = 1,\ldots,K$,

$$w = \sqrt{1 - \sigma(|p|)} \quad , \quad q_k = \sqrt{\sigma(|p|)}\tilde{q}_k\left(\frac{p}{|p|}\right) \tag{5.7}$$

with $\tilde{q}_k$ constructed in Lemma 5.27.

In the case that Ω is not a Lipschitz domain, one still has that $\mathcal{V}_p \subset \mathcal{H}^2_{w,\partial q_1,\ldots,\partial q_K}$ in the sense of Hilbert-space isometry.

Proof. From Lemmas 5.25 and 5.29 we know that $w \in Y_1^*$ and $q_k \in Y_d^*$, thus $\mathcal{H}^2_{w,\partial q_1,\dots,\partial q_K}$, according to Definition 5.13, makes sense. Verify, for $y \in L^2(0,T;H^1(\Omega))$, that the scalar products in $\mathcal{V}_p$ and $\mathcal{H}^2_{w,\partial q_1,\dots,\partial q_K}$ coincide, implying $\mathcal{V}_p \subset \mathcal{H}^2_{w,\partial q_1,\dots,\partial q_K}$ in the sense of isometries. This follows in analogy to (5.4): With $w = \sqrt{1 - \sigma(|p|)}$ and (5.6) we have

$$D_p^2 = w^2 I + \sum_{k=1}^{K} q_k \otimes q_k$$

and consequently the desired equality of scalar products. The equality of spaces in the case where Ω is a Lipschitz domain then is a consequence of the density result of Theorem 5.20.   ◄

Having a characterization of $\mathcal{V}_p$ immediately gives a characterization of the associated $\bar{W}_p(0,T)$, where weak solutions of (2.4) in the sense of Definition 3.17 are contained, see Remark 3.18. Recalling the existence results in Chapter 3, one could also ask for a characterization of the possibly "smaller" space $W_p(0,T)$ such that the weak solution spaces of (2.4) are completely described in terms of weak and directional Sobolev spaces. Even more attractive would be a result $W_p(0,T) = \bar{W}_p(0,T)$ since uniqueness in $\mathcal{V}_p$ of a weak solution according to Definition 3.17 would follow (see Remark 3.18 and Theorem 3.25).

Recall, that the spaces $W_p(0,T)$ are constructed by closure of sufficiently smooth functions with respect to a norm, so an approach which suggests itself in order to prove $W_p(0,T) = \bar{W}_p(0,T)$ is again to prove density for smooth functions. Unfortunately and maybe unexpectedly, it turns out that there is a technical difficulty: We have to approximate with $L^2(0,T;H^1(\Omega))$-functions simultaneously in $\mathcal{H}^2_{w,\partial q_1,\dots,\partial q_K}$ and $\mathcal{H}^{2*}_{w,\partial q_1,\dots,\partial q_K}$ which does not seem to work without further assumptions on p. The reason for this difficulties is mainly the behavior of the approximations close to the boundary, so one way to circumvent these difficulties is to introduce the following notion.

Definition 5.32. Let $M_\partial < 1$ and $p \in \mathcal{Y}_d^*$ with $\|p\|_{\mathcal{Y}_d^*} \leq 1$. The vector field p is called *non-degenerate at the boundary* if

$$|p(t,x)| \leq \min\left\{1, M_\partial + M_x^{-1}\operatorname{dist}(x,\partial\Omega)\right\}$$

for almost every $(t,x) \in \,]0,T[\,\times\,\Omega$.

Roughly speaking, p being non-degenerate at the boundary means that $|p| < 1$ on $]0,T[\,\times\,\partial\Omega$ uniformly in t and x. Hence, functions in $\mathcal{V}_p$ are H^1-functions close to $\partial\Omega$, so one can reduce our considerations to the interior parts of Ω.

In the following, fix $M_\partial < 1$ but close to 1 to which we refer implicitly whenever we mention that p is non-degenerate at the boundary. Under this additional assumption we have, for appropriate p, a density result for $\bar{W}_p(0,T)$ which immediately implies $\bar{W}_p(0,T) = W_p(0,T)$.

Proposition 5.33. *Let σ and p according to Conditions 5.21 and 5.22, respectively, be given and denote by w and $q_1, \ldots, q_K$ the associated weight and directions according to (5.7).*

If p is non-degenerate at the boundary, then, in the sense of isometries (with the usual norms),

$$V_p = \mathcal{H}^2_{w,\partial q_1,\ldots,\partial q_K}$$

as well as

$$W_p(0,T) = \bar{W}_p(0,T) = \left\{ y \in \mathcal{H}^2_{w,\partial q_1,\ldots,\partial q_K} \;\middle|\; \frac{\partial y}{\partial t} \in \mathcal{H}^{2*}_{w,\partial q_1,\ldots,\partial q_K} \right\}.$$

Moreover, the set

$$H^1(\,]0,T[\,\times\Omega) = \left\{ y \in L^2\big(0,T;H^1(\Omega)\big) \;\middle|\; \frac{\partial y}{\partial t} \in L^2\big(0,T;L^2(\Omega)\big) \right\} \qquad (5.8)$$

is dense in each $W_p(0,T)$.

Proof. We first prove that the condition that p is non-degenerate at the boundary implies $V_p = \mathcal{H}^2_{w,\partial q_1,\ldots,\partial q_K}$ without any regularity assumptions on the domain. Therefore, introduce the sets

$$\Omega_1 = \{x \in \Omega \mid \text{dist}(x,\partial\Omega) \geq \tfrac{1}{3}M_x(1-M_\partial)\}$$
$$\Omega_2 = \{x \in \Omega \mid \text{dist}(x,\partial\Omega) \geq \tfrac{2}{3}M_x(1-M_\partial)\}$$

which satisfy $\Omega_2 \subset\subset \Omega_1 \subset\subset \Omega$, so we can find, on the one hand, a smooth cutoff-function η_1 such that $\eta_1 \in \mathcal{C}_0^\infty(\Omega)$ as well as $\eta_1 = 1$ on Ω_1 and, on the other hand, an $\eta_2 \in \mathcal{C}_0^\infty(\Omega_1)$ such that $\eta_2 = 1$ on Ω_2. Denote by $\eta'_m = 1 - \eta_m$ for $m = 1, 2$ which vanish on Ω_m, respectively. Now split a function $y \in \mathcal{H}^2_{w,\partial q_1,\ldots,\partial q_K}$ into an interior part and a boundary part, i.e. $y = y_1 + y_2$ with $y_1 = \eta_1 y$, $y_2 = \eta'_1 y$, yielding functions in $\mathcal{H}^2_{w,\partial q_1,\ldots,\partial q_K}$ by the usual calculus rules (see Remark 5.19 as well as Lemma 4.47).

The interior part has compact support in Ω with respect to the space variable, hence one can choose a $\delta_0 > 0$ such that for each $0 < \delta < \delta_0$ the function $y_{1,\delta}(t) = y_1(t) * G_\delta$ is smooth with compact support in Ω with respect to x. Here, G_δ denote once again the dilated versions of the standard mollifier in $\mathbb{R}^d$. Hence, by Lemma 5.16 and the continuity of multiplication with a smooth function, the sequence $\{y_{1,\delta}\}$ is bounded in $\mathcal{H}^2_{w,\partial q_1,\ldots,\partial q_K}$ with bound $C_1\|y\|_{\mathcal{H}^2_{w,\partial q_1,\ldots,\partial q_K}}$.

Examining y_2, note that for almost every $(t,x) \in\ ]0,T[\ \times \Omega\backslash\Omega_1$, we have $|p(t,x)| \leq \frac{1}{3}(2M_\partial + 1)$ from which follows

$$w(t,x) \geq \sqrt{1 - \sigma\big(\tfrac{1}{3}(2M_\partial + 1)\big)} = w_1 > 0 \quad \text{a.e. in }]0,T[\ \times \Omega\backslash\Omega_1 \ .$$

Consequently, ∇y_2 exists on $]0,T[\ \times \Omega\backslash\Omega_1$ (see Remark 5.12 and Proposition 4.19) meaning that $y_2 \in L^2\big(0,T;H^1(\Omega)\big)$ with norm estimate

$$\|y_2\|_2^2 + \|\nabla y_2\|_2^2 \leq \|y_2\|_2^2 + w_1^{-2}\|w\nabla y_2\|_2^2$$
$$\leq (1 + w_1^{-2})\|y_2\|_{\mathcal{H}^2_{w,\partial q_1,\ldots,\partial q_K}}^2 \leq C_2^2\|y\|_{\mathcal{H}^2_{w,\partial q_1,\ldots,\partial q_K}}^2$$

for a suitable $C_2 > 0$.

Together, $y_\delta = y_{1,\delta} + y_2$ satisfy

$$\|y_\delta\|_{L^2(0,T;H^1)} \leq \big(\|G\|_1^2\|\eta_1 y\|_2^2 + \delta^{-2}\|\nabla G\|_1^2\|\nabla(\eta_1 y)\|_2^2\big)^{1/2}$$
$$+ C_2\|y\|_{\mathcal{H}^2_{w,\partial q_1,\ldots,\partial q_K}}$$
$$\leq C_3\delta^{-1}\|y\|_2 + C_2\|y\|_{\mathcal{H}^2_{w,\partial q_1,\ldots,\partial q_K}} \ ,$$
$$\|y_\delta\|_{\mathcal{H}^2_{w,\partial q_1,\ldots,\partial q_K}} \leq C_1\|y\|_{\mathcal{H}^2_{w,\partial q_1,\ldots,\partial q_K}} + \|y_2\|_{\mathcal{H}^2_{w,\partial q_1,\ldots,\partial q_K}}$$
$$\leq C_4\|y\|_{\mathcal{H}^2_{w,\partial q_1,\ldots,\partial q_K}} \ ,$$

hence it is constituting a sequence in $L^2\big(0,T;H^1(\Omega)\big)$ which is bounded in $\mathcal{H}^2_{w,\partial q_1,\ldots,\partial q_K}$. One can easily see that $y_\delta \to y$ in $L^2\big(0,T;L^2(\Omega)\big)$ as $\delta \to 0$, so the argumentation used in the previous density theorems also gives us an approximating element here: The boundedness of $\{y_\delta\}$ implies the existence of a weakly converging subsequence, labeled by δ_l. By continuous embedding $\mathcal{H}^2_{w,\partial q_1,\ldots,\partial q_K} \hookrightarrow L^2\big(0,T;L^2(\Omega)\big)$, its limit has to coincide with y. Mazur's Lemma then states that there are convex combinations $\bar{y}_l = \sum_{l'=1}^{l} \zeta_{l,l'} y_{\delta_{l'}}$ where $\zeta_{l,l'} \geq 0$ and $\sum_{l'=1}^{l} \zeta_{l,l'} = 1$ such that $\{\bar{y}_l\}$ converges to strongly to y. Choosing l large enough, we can achieve that for each $\varepsilon > 0$ there is an l such that

$$\|\bar{y}_l - y\|_{\mathcal{H}^2_{w,\partial q_1,\ldots,\partial q_K}} < \varepsilon \ ,$$

hence $L^2\big(0,T;H^1(\Omega)\big)$ is dense in $\mathcal{H}^2_{w,\partial q_1,\ldots,\partial q_K}$ implying, by the definition of V_p, that $V_p = \mathcal{H}^2_{w,\partial q_1,\ldots,\partial q_K}$ and thus the first statement.

An immediate consequence is that

$$\bar{W}_p(0,T) = \left\{y \in \mathcal{H}^2_{w,\partial q_1,\ldots,\partial q_K} \ \Big|\ \frac{\partial y}{\partial t} \in \mathcal{H}^{2*}_{w,\partial q_1,\ldots,\partial q_K}\right\} \ ,$$

so let $y \in \bar{W}_p(0,T)$. We are going to show that y can be approximated with functions in $H^1(]0,T[\times \Omega)$. Denote by $\mathcal{M}_\delta$ the approximation operator used above, i.e.

$$(\mathcal{M}_\delta y)(t) = \big(\eta_1 y(t)\big) * G_\delta + \eta_1' y(t) \ .$$

According to the above argumentation, each $\mathcal{M}_\delta$ maps $\mathcal{H}^2_{w,\partial q_1,\dots,\partial q_K}$ linearly and continuously into itself. We want to show that this also applies to the L^2-adjoint mapping which reads as

$$(\mathcal{M}_\delta^* z)(t) = \eta_1\big(z(t) * G_\delta\big) + \eta_1' z(t) \ ,$$

since G is symmetric. For potentially smaller $\delta > 0$, which are chosen such that $\operatorname{supp}\eta_1 + B_\delta(0) \subset\subset \Omega$, the norm of the convolution cut off with η_1 is again bounded:

$$\|\eta_1(z * G_\delta)\|_{\mathcal{H}^2_{w,\partial q_1,\dots,\partial q_K}} \le C_5 \|z\|_{\mathcal{H}^2_{w,\partial q_1,\dots,\partial q_K}}$$

with some $C_5 > 0$, see Lemma 5.16 (and Proposition 4.47). Furthermore, with the product rule one can also estimate the respective norm of $\eta_1' z$, so $\mathcal{M}_\delta^*$ maps $\mathcal{H}^2_{w,\partial q_1,\dots,\partial q_K}$ linearly and continuously into itself with norm estimate bounded independent of δ by, say, C_6.

Thus, one can extend $\mathcal{M}_\delta$ to $\mathcal{H}^{2*}_{w,\partial q_1,\dots,\partial q_K}$ by defining $\langle \mathcal{M}_\delta y, z \rangle_{\mathcal{H}^* \times \mathcal{H}} = \langle y, \mathcal{M}_\delta^* z \rangle_{\mathcal{H}^* \times \mathcal{H}}$. It is easy to see that $\mathcal{M}_\delta$ maps $\mathcal{H}^{2*}_{w,\partial q_1,\dots,\partial q_K}$ linearly and continuously into itself, again with norm estimate C_6.

Observe that for test functions $z(t,x) = z_1(t)z_2(x)$ with $\mathcal{C}_0^\infty(]0,T[)$ and $z_2 \in H^1(\Omega)$, each $\mathcal{M}_\delta^*$ commutes with differentiation with respect to t, i.e. $\frac{\partial \mathcal{M}_\delta^* z}{\partial t} = \mathcal{M}_\delta^* \frac{\partial z}{\partial t}$. Hence, if $y \in \bar{W}_p(0,T)$, the distributional derivative $\frac{\partial \mathcal{M}_\delta y}{\partial t}$ obeys,

$$\Big\langle \frac{\partial \mathcal{M}_\delta y}{\partial t}, z \Big\rangle_{\mathcal{H}^* \times \mathcal{H}} = -\Big\langle \mathcal{M}_\delta y, \frac{\partial z}{\partial t} \Big\rangle_{L^2} = -\Big\langle y, \mathcal{M}_\delta^* \frac{\partial z}{\partial t} \Big\rangle_{L^2} = -\Big\langle y, \frac{\partial \mathcal{M}_\delta^* z}{\partial t} \Big\rangle_{L^2}$$

$$= \Big\langle \frac{\partial y}{\partial t}, \mathcal{M}_\delta^* z \Big\rangle_{\mathcal{H}^* \times \mathcal{H}} = \Big\langle \mathcal{M}_\delta \frac{\partial y}{\partial t}, z \Big\rangle_{\mathcal{H}^* \times \mathcal{H}} \ ,$$

so $\mathcal{M}_\delta$ also commutes with differentiation with respect to the time-variable if the latter is in $\mathcal{H}^{2*}_{w,\partial q_1,\dots,\partial q_K}$. Hence, the sequence $\{\mathcal{M}_\delta y\}$ is also bounded in $\bar{W}_p(0,T)$:

$$\|\mathcal{M}_\delta y\|^2_{\mathcal{H}^2_{w,\partial q_1,\dots,\partial q_K}} + \Big\| \frac{\partial \mathcal{M}_\delta y}{\partial t} \Big\|^2_{\mathcal{H}^{2*}_{w,\partial q_1,\dots,\partial q_K}}$$

$$\le C_4^2 \|y\|^2_{\mathcal{H}^2_{w,\partial q_1,\dots,\partial q_K}} + C_6^2 \Big\| \frac{\partial y}{\partial t} \Big\|^2_{\mathcal{H}^{2*}_{w,\partial q_1,\dots,\partial q_K}} \ ,$$

so by remarking that $\mathcal{M}_\delta y \to y$ in $L^2(0, T; L^2(\Omega))$ as $\delta \to 0$ and by concluding analogously to the above, one can find a convex combination $\bar{y}_1$ of finitely many $\mathcal{M}_{\delta_i} y$ such that $\|\bar{y}_1 - y\|_{\bar{W}_p(0,T)} < \frac{\varepsilon}{2}$.

Now, consider the restriction of $\eta_2' \bar{y}_1$ to $]0, T[\times \Omega \backslash \Omega_2$. We introduce the space

$$H_{0_2}^1(\Omega \backslash \Omega_2) = \overline{\{y \in \mathcal{C}^\infty(\Omega \backslash \Omega_2) \cap H^1(\Omega \backslash \Omega_2) \mid y = 0 \text{ on } \partial \Omega_2\}}^{\|\cdot\|_{H^1}}$$

which is the subspace of functions in $H^1(\Omega \backslash \Omega_2)$ with zero boundary values on $\partial \Omega_2$. Note that the zero extension operator is linear and continuous between $H_{0_2}^1(\Omega \backslash \Omega_2) \to H^1(\Omega)$. So, on the one hand, $\eta_2' \bar{y}_1 \in L^2(0, T; H_{0_2}^1(\Omega \backslash \Omega_2))$ with

$$\|\eta_2' \bar{y}_1\|_{L^2(0,T;H_{0_2}^1(\Omega \backslash \Omega_2))} \leq (1 + w_2^{-2})^{1/2} \|\eta_2' \bar{y}_1\|_{\mathcal{H}^2_{w,\partial q_1,\ldots,\partial q_K}} \leq C_7 \|y\|_{\mathcal{H}^2_{w,\partial q_1,\ldots,\partial q_K}}$$

where w_2 is chosen such that $w \geq w_2 > 0$ a.e. on $]0, T[\times \Omega \backslash \Omega_2$, for instance

$$w_2 = \sqrt{1 - \sigma\left(\tfrac{1}{3}(2 + M_\partial)\right)} > 0 \ ,$$

and a suitable C_7. On the other hand, $\frac{\partial \eta_2' \bar{y}_1}{\partial t} \in L^2(0, T; H_{0_2}^1(\Omega \backslash \Omega_2)^*)$ since for each $z_1 \in \mathcal{C}_0^\infty(]0, T[)$ and $z_2 \in H_{0_2}^1(\Omega \backslash \Omega_2)$ we have, by zero extension,

$$-\int_0^T \langle \eta_2' \bar{y}_1(t), z_2 \rangle_{L^2(\Omega \backslash \Omega_2)} \frac{\partial z_1}{\partial t}(t) \, dt = -\int_0^T \langle \eta_2' \bar{y}_1(t), z_2 \rangle_{L^2(\Omega)} \frac{\partial z_1}{\partial t}(t) \, dt$$

$$= \left\langle \frac{\partial \eta_2' \bar{y}_1}{\partial t}, z \right\rangle_{\mathcal{H}^* \times \mathcal{H}}$$

$$\leq C_8 \left\| \frac{\partial y}{\partial t} \right\|_{\mathcal{H}^{2*}_{w,\partial q_1,\ldots,\partial q_K}} \|z\|_{\mathcal{H}^2_{w,\partial q_1,\ldots,\partial q_K}}$$

$$\leq C_8 \left\| \frac{\partial y}{\partial t} \right\|_{\mathcal{H}^{2*}_{w,\partial q_1,\ldots,\partial q_K}} \|z\|_{L^2(0,T;H_{0_2}^1(\Omega \backslash \Omega_2))}$$

where $z(t, x) = z_1(t) z_2(x)$ and C_8 is a continuity constant arising from the multiplication with η_2'. This is a classical setting from which we know that one can achieve approximation with elements in $\mathcal{C}^\infty(0, T; H_{0_2}^1(\Omega \backslash \Omega_2))$ by extension and convolution with respect to the time-variable (see [Sho96], for example). In particular, it is possible to find a $\bar{y}_2 \in L^2(0, T; H_{0_2}^1(\Omega \backslash \Omega_2))$ with $\frac{\partial y}{\partial t} \in L^2(0, T; L^2(\Omega \backslash \Omega_2))$ such that

$$\|\bar{y}_2 - \eta_2' \bar{y}_1\|_{L^2(0,T;H_{0_2}^1(\Omega \backslash \Omega_2))}^2$$

$$+ \left\| \frac{\partial}{\partial t}(\bar{y}_2 - \eta_2' \bar{y}_1) \right\|_{L^2(0,T;H_{0_2}^1(\Omega \backslash \Omega_2)^*)}^2 < \frac{\varepsilon^2}{4(1 + w_2^2)} \ .$$

Thus, $\bar{y} = \eta_2 \bar{y}_1 + \bar{y}_2 = \bar{y}_1 + \bar{y}_2 - \eta_2' \bar{y}_1$ satisfies

$$
\begin{aligned}
\|\bar{y} - y\|_{\bar{W}_p(0,T)} &\leq \|\bar{y}_1 - y\|_{\bar{W}_p(0,T)} + \|\bar{y}_2 - \eta_2' \bar{y}_1\|_{\bar{W}_p(0,T)} \\
&\leq \frac{\varepsilon}{2} + (1 + w_2^{-2})^{1/2} \Big(\|\bar{y}_2 - \eta_2' \bar{y}_1\|^2_{L^2(0,T;H^1(\Omega\setminus\Omega_2))} \\
&\quad + \Big\| \frac{\partial}{\partial t}(\bar{y}_2 - \eta_2' \bar{y}_1) \Big\|^2_{L^2(0,T;H^1(\Omega\setminus\Omega_2)^*)} \Big)^{1/2} \\
&< \varepsilon ,
\end{aligned}
$$

which means that $\bar{y}$ indeed approximates y. To finish the proof, verify the smoothness properties of $\bar{y}$: The function $\eta_2 \bar{y}_1$ only has its support in Ω_1 and coincides with a finite convex combination of $(\eta_1 y) * G_\delta$ there. For each of the above $(\eta_1 y) * G_\delta$, we can easily verify that

$$
\|(\eta_1 y) * G_\delta\|_{L^2(0,T;H^1)} \leq C_9 \delta^{-1} \|\eta_1 y\|_2 \leq C_{10} \delta^{-1} \|y\|_{\mathcal{H}^2_{w,\partial q_1,\dots,\partial q_K}} .
$$

Considering the derivative $\frac{\partial}{\partial t}\big((\eta_1 y) * G_\delta\big)$, we remark the that the adjoint operation $z(t) \mapsto \eta_1\big(z(t) * G_\delta\big)$ possesses the property that

$$
\|\eta_1(z * G_\delta)\|_{\mathcal{H}^2_{w,\partial q_1,\dots,\partial q_K}} \leq C_{11} \|z * G_\delta\|_{L^2(0,T;H^1)} \leq C_{12} \delta^{-1} \|z\|_2
$$

which leads to

$$
\Big| \Big\langle \frac{\partial}{\partial t}\big((\eta_1 y) * G_\delta\big), z \Big\rangle_{\mathcal{H}^* \times \mathcal{H}} \Big| \leq C_{12} \delta^{-1} \Big\| \frac{\partial y}{\partial t} \Big\|_{\mathcal{H}^{2*}_{w,\partial q_1,\dots,\partial q_K}} \|z\|_2 ,
$$

noting that $(\eta_1 y) * G_\delta$ commutes with differentiation with respect to t by the same argumentation already used above to prove an analog for $\mathcal{M}_\delta$. It follows that $\frac{\partial}{\partial t}\big((\eta_1 y) * G_\delta\big) \in L^2\big(0,T;L^2(\Omega)\big)$ and consequently $\eta_2 \bar{y}_1$.

Finally, by construction, $\bar{y}_2 \in \mathcal{C}^\infty\big(0,T;H^1_{02}(\Omega\setminus\Omega_2)\big)$ and can be extended by zero to a $\bar{y}_2 \in L^2\big(0,T;H^1(\Omega)\big)$ which satisfies $\frac{\partial \bar{y}_2}{\partial t} \in L^2\big(0,T;L^2(\Omega)\big)$. Thus, $\bar{y} = \eta_2 \bar{y}_1 + \bar{y}_2$ indeed possesses the asserted smoothness property.

The statement $W_p(0,T) = \bar{W}_p(0,T)$ then follows from the density of $H^1(]0,T[\times \Omega)$: For each given $y \in \bar{W}_p(0,T)$ and $\varepsilon > 0$ one chooses a $\bar{y}_1 \in H^1(]0,T[\times\Omega)$ such that $\|\bar{y} - y\|_{\bar{W}_p(0,T)} < \varepsilon/2$. Interpreting $H^1(]0,T[\times\Omega)$ according to (5.8), constructing an extension as well as convolving with respect to the time-variable, one can find a $\bar{y} \in \mathcal{C}^\infty\big(0,T;H^1(\Omega)\big)$ such that

$$
\|\bar{y} - \bar{y}_1\|_{\bar{W}_p(0,T)} \leq \|\bar{y} - \bar{y}_1\|_{H^1(]0,T[\times\Omega)} < \varepsilon/2 ,
$$

so $\bar{y} \in \mathcal{C}^\infty\big(0,T;H^1(\Omega)\big) \subset \mathcal{AC}\big(0,T;H^1(\Omega)\big) \subset W_p(0,T)$ approximates y up to ε which was chosen arbitrarily, yielding that $W_p(0,T) = \bar{W}_p(0,T)$ by definition of $W_p(0,T)$ according to (3.9). ◀

Remark 5.34. It immediately follows that the functions in $\mathcal{C}^\infty(]0,T[\times\Omega)$ which also belong to $H^1(]0,T[\times\Omega)$ are a dense subset of $\bar{W}_p(0,T)$.

Moreover, if, in addition to the assumptions in Proposition 5.33, Ω is a Lipschitz domain, then one can deduce, in analogy to Theorem 5.20, that $\mathcal{C}^\infty([0,T]\times\overline{\Omega})$ is also a dense subset of $\bar{W}_p(0,T)$.

With these considerations, one can formulate the second main result of this section, a uniqueness theorem in the space $\mathcal{V}_p = \mathcal{H}^2_{w,\partial q_1,\ldots,\partial q_K}$ and weak formulation only in terms of weighted and directional weak derivatives:

Theorem 5.35. *Let σ fulfill Condition 5.21, p fulfill Condition 5.22 as well as be non-degenerate at the boundary and denote by w and $q_1,\ldots,q_K$ the associated weight and directions according to (5.7).*

Then, a $y \in \mathcal{H}^2_{w,\partial q_1,\ldots,\partial q_K}$ is the unique weak solution of (2.4) in the sense of Definition 3.17 if and only if for each $z \in L^2(0,T;H^1(\Omega))$ with $\frac{\partial z}{\partial t} \in L^2(0,T;L^2(\Omega))$ and $z(T) = 0$, the variational equation

$$-\left\langle \frac{\partial z}{\partial t}, y \right\rangle_{L^2} + \langle w\nabla y, w\nabla z\rangle_{L^2} + \sum_{k=1}^{K}\langle \partial_{q_k} y, \partial_{q_k} z\rangle_{L^2}$$

$$= \langle u, z\rangle_{\mathcal{H}^*\times\mathcal{H}} + \langle y_0, z(0)\rangle_{L^2} \quad (5.9)$$

is satisfied.

Proof. First observe that the variational formulation (5.9) is indeed equivalent to (3.10). By Proposition 5.33, $\mathcal{V}_p = \mathcal{H}^2_{w,\partial q_1,\ldots,\partial q_K}$ in the sense of isometrical isomorphisms, hence one can replace, in (3.10), the scalar product and dual pairing in $\mathcal{V}_p$ with the scalar product and dual pairing in $\mathcal{H}^2_{w,\partial q_1,\ldots,\partial q_K}$, respectively, which yields (5.9) (the dual pairing $\mathcal{V}_p^* \times \mathcal{V}_p$ with $\frac{\partial z}{\partial t}$ can be replaced by the scalar product in L^2, since $\frac{\partial z}{\partial t} \in L^2(0,T;L^2(\Omega))$). Furthermore, it is easy to see that the set

$$\left\{ z \in L^2(0,T;H^1(\Omega)) \;\middle|\; \frac{\partial z}{\partial t} \in L^2(0,T;L^2(\Omega)), \; z(T) = 0 \right\}$$

is a subspace of $\bar{W}_p(0,T)$. Since, again by Proposition 5.33, $W_p(0,T) = \bar{W}_p(0,T)$, it also has to be contained in $\{z \in W_p(0,T) \mid z(T) = 0\}$, which is a closed subspace of $W_p(0,T)$. The inclusion is moreover dense, since both sets contain the (relatively) dense subset $\{z \in \mathcal{AC}(0,T;H^1(\Omega)) \mid z(T) = 0\}$.

Hence, it is sufficient to test with functions $z \in L^2(0,T; H^1(\Omega))$ with $\frac{\partial z}{\partial t} \in L^2(0,T; L^2(\Omega))$ and $z(T) = 0$.

The uniqueness statement finally follows from the observation that each $y \in \mathcal{V}_p$ which satisfies (5.9) is in $\bar{W}_p(0,T) = W_p(0,T)$, see Remark 3.18, and the fact that the solution is unique in $W_p(0,T)$, confer Theorem 3.25. This leaves no possibility for the existence of a different solution. ◄

Remark 5.36. It seems like the condition that p is non-degenerate at the boundary is not the best possible for obtaining the equivalence $W_p(0,T) = \bar{W}_p(0,T)$. In view of Theorem 5.31, for example, one can conjecture that Ω being a Lipschitz domain and $\|p\|_{\mathcal{Y}_d^*} \leq 1$ is sufficient. Such an approach however requires choosing an appropriate family of smoothing operators $\{\mathcal{M}_\delta\}$: In the proof of Proposition 5.33 it becomes apparent that $\{\mathcal{M}_\delta\}$ as well as the family of L^2-adjoints $\{\mathcal{M}_\delta^*\}$ have to be bounded considered between $\mathcal{V}_p \to \mathcal{V}_p$ as $\delta \to 0$ in order to approximate both y and $\frac{\partial y}{\partial t}$. Choosing $\{\mathcal{M}_\delta\}$ as time-variant versions of the operations constructed in Lemma 4.44 which turned out to be successful in Theorem 5.31, however, will not expose these properties: As one can see with after some computations, $\mathcal{M}_\delta^* z$ will always be zero close to the boundary of Ω, so it will not approximate each $z \in \mathcal{V}_p$ for general p (remember that in the classical setting, $H_0^1(\Omega)$ is in the most cases a proper subspace of $H^1(\Omega)$ and so is $L^2(0,T; H_0^1(\Omega)) \subset L^2(0,T; H^1(\Omega))$).

So, in order to prove $W_p(0,T) = \bar{W}_p(0,T)$, one has to come up with a better approximation strategy. Alternatively, one can follow an approach which leads to restrictions on p different from the requirement that p is non-degenerate at the boundary and which is also inspired by the proof of Proposition 5.33: Impose more regularity on p with respect to the time-variable. The prerequisite that p is non-degenerate at the boundary realizes this in a certain sense. It implies that close to the boundary, the norm in $W_p(0,T)$ is essentially H^1 with respect to space and the dual for the derivative with respect to time, which can be interpreted as the function space at $t \in [0,T]$ being constant with respect to time. So, we can smooth in the direction of t and get approximating elements. This is probably still possible if the function space associated with $t \in [0,T]$ varies in a smooth way.

Besides its importance to prove existence of solutions for (2.8) (as we will see in Chapter 6) and to obtain uniqueness in $\mathcal{V}_p$, the result that the weak solution of (2.4) is in a weighted and directional Sobolev space can be used to manipulate the solution e.g. by means of superposition operators. Such operations are the key to obtain additional properties for the solutions of the degenerate parabolic equation.

5.4   Boundedness and additional regularity

This section is devoted to the derivation of boundedness and additional regularity for the weak solutions of (2.4). In particular, our aim is to show conditions under which it can be ensured that the evaluations of $y(t_n)$, as they appear in the discrepancy term of the optimal control problem (2.8), are smooth images in the sense that $y(t_n) \in V_{p(t_n)}$. It will turn out that it is sufficient, besides smoothness, to restrict the vector fields p in order to control the behavior of $\frac{\partial y}{\partial t}$. We moreover show that $y_0 \in L^\infty(\Omega)$ as well as $u \in L^\infty(]0, T[\times \Omega)$ leads to a $y \in L^\infty(]0, T[\times \Omega)$, a statement which might also become interesting in the context of constrained optimization. Also, a kind of maximum principle for the homogeneous equation can be established.

Throughout this section, it will be assumed that σ satisfies Condition 5.21 as well as p satisfies Condition 5.22 and is non-degenerate at the boundary, so that the density of $\mathcal{C}^\infty(]0, T[\times \Omega)$ is given (see Remark 5.34) and we are able to apply Theorem 5.35.

5.4.1 L^∞-estimates

We will first present the L^∞-result and maximum principle.

Proposition 5.37. *Let Ω be a Lipschitz domain, σ and p according to Conditions 5.21 and 5.22, respectively, and let p be non-degenerate at the boundary.*

Suppose that $y_0 \in L^\infty(\Omega)$ and $u \in L^\infty(]0, T[\times \Omega)$. Then the weak solution y of (2.4) belongs to $L^\infty(]0, T[\times \Omega)$ with $\|y\|_\infty \leq \|y_0\|_\infty + T\|u\|_\infty$.

Moreover, if $u = 0$ and $y_0(x) \in [c_1, c_2]$ a.e. in Ω for some constants $c_1 \leq c_2$, then $y(t, x) \in [c_1, c_2]$ for almost every $(t, x) \in]0, T[\times \Omega$.

Proof. Choose some $C_1, C_2 \in \mathbb{R}$. It is clear that $t \mapsto y(t) - (C_1 + tC_2)$ belongs to $\mathcal{V}_p$ and by Remark 5.19 (see also Proposition 4.48) we know that

$$z(t, x) = \big(y(t, x) - (C_1 + tC_2)\big)_+ = \max\{0, y(t, x) - (C_1 + tC_2)\}$$

is also an element of $\mathcal{V}_p$. Hence, we can use (3.12) in Remark 3.27 to deduce that

$$\left\langle \frac{\partial y}{\partial t}(t), z(t) \right\rangle_{V(t)^* \times V(t)} + \langle D_{p(t)} \nabla y(t)^{\mathrm{T}}, D_{p(t)} \nabla z(t)^{\mathrm{T}} \rangle_{L^2} = \langle u, z \rangle_{V(t)^* \times V(t)} \quad (*)$$

for almost every $t \in]0, T[$. We now investigate each term in $(*)$ separately.

Regarding the first term, our claim is that

$$\left\langle \frac{\partial y}{\partial t}(t), z(t) \right\rangle_{V(t)^* \times V(t)} = \frac{1}{2} \frac{\partial}{\partial t} \|z(t)\|_2^2 + C_2 \langle \mathbf{1}, z(t) \rangle \quad (**)$$

for almost every $t \in {]0, T[}$, where $\mathbf{1}$ denotes the function which is constant 1 on Ω. This is remarkable since $\frac{\partial z}{\partial t} \notin V_p^*$ in general, but the derivative of the squared norm does still exist. First suppose that $y \in C^\infty([0, T] \times \overline{\Omega})$, hence by the classical chain rule for weakly differentiable functions it follows that

$$\frac{\partial}{\partial t}\left(y - (C_1 + tC_2)\right)_+ = \begin{cases} 0 & \text{if } y \le C_1 + tC_2 \\ \frac{\partial y}{\partial t} - C_2 & \text{if } y > C_1 + tC_2 \end{cases}$$

almost everywhere in ${]0, T[} \times \Omega$. Consequently,

$$\frac{1}{2}\frac{\partial}{\partial t}\left\|\left(y(t) - (C_1 + tC_2)\right)_+\right\|_2^2 = \left\langle \frac{\partial y}{\partial t}(t) - C_2,\ \left(y(t) - (C_1 + tC_2)\right)_+ \right\rangle_{L^2} \quad (***)$$

since $\frac{\partial}{\partial t}\left(y - (C_1 + tC_2)\right)_+ = 0$ a.e. where $y \le C_1 + tC_2$. Now suppose that $y \in W_p(0, T)$ and choose a sequence $y_l \to y$ in $W_p(0, T)$ with $y_l \in C^\infty({]0, T[} \times \Omega)$, see Remark 5.34. Then, due to the continuity of the operation, $\left(y_l - (C_1 + tC_2)\right)_+ \to \left(y - (C_1 + tC_2)\right)_+$ in V_p (again Remark 5.19). Thus, the right-hand side of $(***)$ as a function of $t \in {]0, T[}$ converges, as $l \to \infty$, to

$$\left\langle \frac{\partial y_l}{\partial t}(t) - C_2,\ \left(y_l(t) - (C_1 + tC_2)\right)_+ \right\rangle_{L^2}$$
$$\to \left\langle \frac{\partial y}{\partial t}(t) - C_2,\ \left(y(t) - (C_1 + tC_2)\right)_+ \right\rangle_{V(t)^* \times V(t)}$$

in $L^1({]0, T[})$. Moreover,

$$\left\|\left(y_l(t) - (C_1 + tC_2)\right)_+\right\|_2^2 \to \left\|\left(y(t) - (C_1 + tC_2)\right)_+\right\|_2^2$$

in $L^1({]0, T[})$, so by the closedness of the weak derivative we get $(**)$ after some rearrangements.

Going back to $(*)$, the second term can be estimated

$$\langle D_{p(t)}\nabla y(t)^{\mathrm{T}},\ D_{p(t)}\nabla z(t)^{\mathrm{T}}\rangle_{L^2} = \langle D_{p(t)}\nabla z(t)^{\mathrm{T}},\ D_{p(t)}\nabla z(t)^{\mathrm{T}}\rangle_{L^2} \ge 0$$

using that $D_p\nabla z^{\mathrm{T}} = 0$ almost everywhere where $z = 0$ and $D_p\nabla y^{\mathrm{T}} = D_p\nabla z^{\mathrm{T}}$ else. This is again a consequence of the characterization of solution spaces (Theorem 5.31) as well as the established calculus rules (Proposition 4.48 and Remark 5.19). Combining this with $(*)$ gives

$$\frac{1}{2}\frac{\partial}{\partial t}\|z(t)\|_2^2 \le \langle u(t) - C_2,\ z(t)\rangle_{L^2}$$

for almost every $t \in {]0, T[}$ and consequently,

$$\max_{t \in [0, T]} \|z(t)\|_2^2 \le \|z(0)\|_2^2 + \int_0^T 2\langle u(t) - C_2,\ z(t)\rangle_{L^2}\ dt\ .$$

Now, with $C_1 = \|y_0\|_\infty$ and $C_2 = \|u\|_\infty$, the above implies $z = 0$ a.e. in $]0, T[\times \Omega$, meaning $y \leq C_1 + TC_2$ a.e. The same argument applied to $-y$ with data $-y_0$ and $-u$ leads to $y \geq -C_1 - TC_2$. This establishes the L^∞-estimate. The case where $u = 0$ and $c_1 \leq y_0 \leq c_2$ a.e. in Ω, we can choose $C_1 = c_2$ and $C_2 = 0$ to conclude $y \leq c_2$ almost everywhere and $C_1 = -c_1$ and $C_2 = 0$ to establish that $-y \leq -c_1$.   ◀

5.4.2   Additional regularity with respect to time

The following subsection deals with sufficient conditions for weak solutions y of (2.4) to satisfy $y(t_n) \in V_{p(t_n)}$ where $t_1, \ldots, t_N$ are the control times introduced in Chapter 2. The first step is to define a space associated with this kind of "regularity".

Definition 5.38. Let $0 < t_1 < t_2 < \ldots t_N = T$, σ be given according to Condition 5.21 and $p \in \mathcal{Y}_d^*$ with $\|p\|_{\mathcal{Y}_d^*} \leq 1$ and $p(t)$ continuous from the left at each t_n with values in Y_d^*.

Then the space $\mathcal{V}_p(t_1, \ldots, t_n)$ is defined as the closure of the set of functions in $W_p(0, T)$ for which $y(t_n) \in V_{p_-(t_n)}$, where $p_-(t_n)$ denotes the left limit of $p(t)$ at t_n. It is equipped with the norm

$$\|y\|^2_{\mathcal{V}_p(t_1, \ldots, t_N)} = \|y\|^2_{\mathcal{V}_p} + \sum_{n=1}^{N} \|y(t_n)\|^2_{V_{p_-(t_n)}} .$$

It is clear that $\mathcal{V}_p(t_1, \ldots, t_N)$ is a Hilbert space.

Now since the actual $t_1, \ldots, t_N$ do not matter and all t_n can be treated the same, we turn to a slightly different framework. Our aim is to obtain $y(t^*) \in V_{p(t^*)}$ where $y(t^*)$ is the weak solution of (2.4) at some time $t^* \in]0, T]$ with some data $u \in L^2(0, T; L^2(\Omega))$. To outline the idea of proof roughly, one first shows that $\frac{\partial y}{\partial t} \in L^2(t^* - \delta, t^*; L^2(\Omega))$ for some $\delta > 0$. This implies higher regularity with respect to space for almost every $y(t)$ with $t \in [t^* - \delta, t^*]$. With an argument similar to the one used in Proposition 3.16, we see that each function possessing the above regularity is contained in the space $V_p(t^*)$.

Recall that in the theory of evolution equations with uniformly elliptic operator in space, we have to require that the derivative of the coefficients with respect to time is bounded in the L^2-sense for some interval $[t^* - \delta, t^*]$ in order to get $\frac{\partial y}{\partial t} \in L^2(t^* - \delta, t^*; L^2(\Omega))$. Here, in the general degenerate case, this property may fail even under the smoothness condition $\frac{\partial p}{\partial t} \in L^\infty(]0, T[\times \Omega)$, thus further requirements have to be made.

Roughly speaking, two issues may cause that $\frac{\partial y}{\partial t}$ is not contained in the space $L^2(t^* - \delta, t^*; L^2(\Omega))$: First, it may happen that for some $t < t^*$ and some $x \in \Omega$ that $|p(t,x)| = 1$ holds and that the function $|p(\tilde{t}, x)|$ is strictly decreasing for $\tilde{t}$ in a neighborhood of t. Since for $|p(t,x)| = 1$ the equation degenerates, the solution $y(t)$ is not necessarily smooth (see Proposition 3.9, for example) and the situation is comparable to a uniformly elliptic parabolic equation with initial value in $L^2(\Omega)$: we only have $\frac{\partial y}{\partial t} \in L^2(0, \delta; H^1(\Omega)^*)$ but not $\frac{\partial y}{\partial t} \in L^2(0, \delta; L^2(\Omega))$.

Second, if we have $|p(t,x)| = 1$ for some $(t,x) \in \,]t^* - \delta, t^*[\times \Omega$ and the derivative $\frac{\partial p}{\partial t}(t,x)$ has components perpendicular to $p(t,x)$, it might happen that the solution space $V_{p(t)}$ varies with time in such a way that no inclusion relation holds. Consequently, smooth components in one instant might become jumps in the next, which also corresponds to non-smooth initial values in a certain way and the property $\frac{\partial y}{\partial t} \in L^2(t^* - \delta, t^*; L^2(\Omega))$ may fail as well.

To resolve these issues, maybe the easiest way is to impose that $V_{p(t)}$ does not change in, say, $]t^* - \delta, t^*[$. This can for example be ensured by requiring that $\frac{\partial p}{\partial t} = 0$ for these t. Other conditions are possible, but for the sake of simplicity we restrict ourselves to p being constant on small intervals.

The following proposition shows sufficient conditions for

$$\frac{\partial y}{\partial t} \in L^2\big(t_n - \delta_n/2, t_n; L^2(\Omega)\big) \ .$$

It uses the usual argument which involves "testing" with $\frac{\partial y}{\partial t}$. Such an approach turns out to be successful even in the case of a degenerate equation where the weak solution is only an element of $W_p(0,T)$. Here, it turns out that the already-established results about the convergence of Galerkin approximations for weak solutions of (2.4), see Lemma 3.23, are again useful.

Proposition 5.39. *In the situation of Definition 5.38, let p be such that $p(t)$ is constant on $[t_n - \delta_n, t_n]$ for some $\delta_n > 0$ and $u \in L^2(0, T; L^2(\Omega))$. Then, the weak solution y of (2.4) (in $W_p(0,T)$) satisfies $\frac{\partial y}{\partial t} \in L^2\big(t_n - \delta_n/2, t_n; L^2(\Omega)\big)$ for $n = 1, \dots, N$ with norm estimate*

$$\int_{t_n - \delta_n/2}^{t_n} \left\| \frac{\partial y}{\partial t}(\tau) \right\|_2^2 \, \mathrm{d}\tau \leq C_n \big(\|u\|_2^2 + \|y_0\|_2^2 \big)$$

where $C_n > 0$ only depends on δ_n.

Proof. For the proof, we use the Galerkin solutions y_l derived in Lemma 3.23 which approximate, according to Theorem 3.25, the unique solution y in $W_p(0,T)$.

In the following, fix an n and derive a $L^2\big(t_n - \delta_n, t_n; L^2(\Omega)\big)$-estimate for $\sqrt{t - (t_n - \delta_n)} \frac{\partial y_l}{\partial t}$. Note that for almost every $t \in [t_n - \delta_n, t_n]$, one can test the

Galerkin solutions with $\frac{\partial y_l}{\partial t}(t)$, hence, by (3.11),

$$\left\|\frac{\partial y_l}{\partial t}(t)\right\|_2^2 = -\left\langle D_{p(t)}\nabla y_l(t)^{\mathrm{T}}, D_{p(t)}\nabla\frac{\partial y_l}{\partial t}(t)^{\mathrm{T}}\right\rangle_{L^2} + \left\langle u(t), \frac{\partial y_l}{\partial t}(t)\right\rangle_{L^2}$$
$$= -\frac{1}{2}\frac{\partial}{\partial t}\left\|D_{p(t)}\nabla y_l(t)^{\mathrm{T}}\right\|_2^2 + \left\langle u(t), \frac{\partial y_l}{\partial t}(t)\right\rangle_{L^2}$$

since $\frac{\partial}{\partial t}$ can be interchanged with $D_{p(t)}\nabla$. This is due to the fact that $y \mapsto D_{p(t)}\nabla y$ considered on $\mathrm{span}\{\varphi_1,\dots,\varphi_l\}$ is a linear mapping between finite-dimensional spaces which is moreover independent of t, since $p(t)$ is constant on $[t_n - \delta_n, t_n]$. Now, integrating the identity over $[t, t_n]$, with t still in $[t_n - \delta_n, t_n]$, rearranging the terms and estimating $-\frac{1}{2}\|D_{p(t_n)}\nabla y_l(t_n)^{\mathrm{T}}\|_2^2 \le 0$ gives

$$\int_t^{t_n}\left\|\frac{\partial y_l}{\partial t}(s)\right\|_2^2 \mathrm{d}s \le \frac{1}{2}\left(\int_t^{t_n}\|u(s)\|_2^2 + \left\|\frac{\partial y_l}{\partial t}(s)\right\|_2^2 \mathrm{d}s + \left\|D_{p(t)}\nabla y_l(t)^{\mathrm{T}}\right\|_2^2\right)$$

and, after another integration over $[t_n - \delta_n, t_n]$, further rearrangements as well as the application of Fubini's theorem,

$$\int_{t_n-\delta_n}^{t_n}(t - (t_n - \delta_n))\left\|\frac{\partial y_l}{\partial t}(t)\right\|_2^2 \mathrm{d}t \le \delta_n \int_{t_n-\delta_n}^{t_n}\|u(t)\|_2^2 + \left\|D_{p(t)}\nabla y_l(t)^{\mathrm{T}}\right\|_2^2 \mathrm{d}t$$
$$\le \delta_n C\big(\|u\|_2^2 + \|y_0\|_2^2\big)$$

with a suitable $C > 0$. The latter estimate is a consequence of the continuity estimate for the Galerkin solutions y_l in the $\mathcal{V}_p$-norm (cf. Lemma 3.23). Finally,

$$\int_{t_n-\delta_n/2}^{t_n}\left\|\frac{\partial y_l}{\partial t}(t)\right\|_2^2 \mathrm{d}t \le \frac{2}{\delta_n}\int_{t_n-\delta_n}^{t_n}(t - (t_n - \delta_n))\left\|\frac{\partial y_l}{\partial t}(t)\right\|_2^2 \mathrm{d}t$$
$$\le 2C\big(\|u\|_2^2 + \|y_0\|_2^2\big),$$

which gives the desired statement through weak convergence as $l \to \infty$, again by Lemma 3.23.   ◀

Proposition 5.40. *Let Ω be a bounded Lipschitz domain. Under the prerequisites of Proposition 5.39, there also holds $y(t_n) \in V_{p(t_n)}$ for $n = 1,\dots,N$ with norm estimate*

$$\|y\|_{\mathcal{V}_p(t_1,\dots,t_N)}^2 \le C\big(\|y_0\|_2^2 + \|u\|_2^2\big).$$

Proof. It is clear that $p(t)$ is continuous from the left in t_n for $n = 1,\dots,N$ since $p(t)$ is constant on $[t_n - \delta_n, t_n]$. Hence, we will construct a suitable space which is continuously embedded in $\mathcal{V}_p(t_1,\dots,t_N)$ according to Definition 5.38 and show that y indeed belongs to that space.

Therefore, first consider the adjoint operation for the differential operator $y \mapsto (w\nabla y, \partial_{q_1} y, \ldots, \partial_{q_K} y)$ which is closely connected with the differential operator $y \mapsto D_p \nabla y^{\mathrm{T}}$, see Theorem 5.31. Let $\varphi_0 \in L^1_{\mathrm{loc}}(\Omega, \mathbb{R}^d)$, $\varphi_1, \ldots, \varphi_K \in L^1_{\mathrm{loc}}(\Omega)$ and define the weak adjoint operation as follows:

$$v = -\operatorname{div}(w\varphi_0) + \sum_{k=1}^{K} \partial^*_{q_k} \varphi_k$$

$$\Leftrightarrow \int_\Omega vz \, \mathrm{d}x = \int_\Omega \varphi_0 \cdot w\nabla z \, \mathrm{d}x + \sum_{k=1}^{K} \int_\Omega \varphi_k(q_k \cdot \nabla z) \, \mathrm{d}x \quad \forall z \in C_0^\infty(\Omega) \ . \tag{$*$}$$

Observe that for smooth functions $\varphi_0, \varphi_1, \ldots, \varphi_K$, this operation amounts to

$$(\varphi_0, \varphi_1, \ldots, \varphi_K) \mapsto -\operatorname{div}\left(w\varphi_0 + \sum_{k=1}^{K} q_k\varphi_k\right) \ .$$

Since $p(t)$ is constant on $[t_n - \delta_n, t_n]$, the associated $w(t)$ and $q_1(t), \ldots, q_K(t)$ are as well constant. Hence, we can define, for each t_n, a space of "higher regularity" for $y \in V_{p(t_n)}$:

$$V^2_{p(t_n)} = \left\{ y \in V_{p(t_n)} \ \middle| \ -\operatorname{div}\left(w(w\nabla y)\right) + \sum_{k=1}^{K} \partial^*_{q_k}(\partial_{q_k} y) \in L^2(\Omega) \right\}$$

$$\|y\|^2_{V^2_{p(t_n)}} = \|y\|^2_{V_{p(t_n)}} + \left\| -\operatorname{div}\left(w(w\nabla y)\right) + \sum_{k=1}^{K} \partial^*_{q_k} \partial_{q_k} y \right\|^2_2 \ .$$

Note that for smooth functions y, the norm of $V^2_{p(t_n)}$ also reads as

$$\|y\|^2_{V^2_{p(t_n)}} = \|y\|^2_2 + \left\| D_{p(t_n)} \nabla y^{\mathrm{T}} \right\|^2_2 + \left\| \operatorname{div}\left(D^2_{p(t_n)} \nabla y^{\mathrm{T}}\right) \right\|^2_2 \ ,$$

so the $V^2_{p(t_n)}$ can roughly be seen as an analog to $y, \Delta y \in L^2(\Omega)$ for the degenerate case, i.e. as a space of function which are smoother than elements in $V_{p(t_n)}$ except for possible jumps indicated by $p(t_n)$. The time-invariant spaces $V^2_{p(t_n)}$ can easily be "assembled" to an appropriate time-variant space

$$V^2_p = \left\{ y \in V_p \ \middle| \ y|_{[t_n - \delta_n/2, t_n]} \in L^2\left(t_n - \delta_n/2, t_n; V^2_{p(t_n)}\right) \text{ for } n = 1, \ldots, N \right\}$$

$$\|y\|^2_{V^2_p} = \|y\|^2_{V_p} + \sum_{n=1}^{N} \int_{t_n - \delta_n/2}^{t_n} \|y(\tau)\|^2_{V^2_{p(t_n)}} \, \mathrm{d}\tau \ ,$$

which serves as a foundation for the solution space

$$W_p^2(0,T) = \left\{ y \in \mathcal{V}_p^2 \ \Big| \ \frac{\partial y}{\partial t}\Big|_{[t_n-\delta_n/2, t_n]} \in L^2\big(t_n - \delta_n/2, t_n; L^2(\Omega)\big) \right.$$

$$\left. \text{for } n = 1, \ldots, N \right\},$$

$$\|y\|_{W_p^2}^2 = \|y\|_{\mathcal{V}_p^2}^2 + \sum_{n=1}^{N} \int_{t_n-\delta_n/2}^{t_n} \left\| \frac{\partial y}{\partial t}(\tau) \right\|_2^2 \, d\tau$$

for which we want to prove that the weak solution y of (2.4) associated with u and y_0 according to Proposition 5.39 is contained in with continuous dependence on the data. The claim then follows from the embedding $W_p^2(0,T) \cap W_p(0,T) \hookrightarrow \mathcal{V}_p(t_1, \ldots, t_N)$ which we will prove first.

For this purpose, pick a y as follows:

$$y \in \mathcal{AC}\big(0,T; H^1(\Omega)\big) \cap \bigcap_{1 \le n \le N} L^2\big(t_n - \delta_n/2, t_n; V_{p(t_n)}^2\big)$$

$$y(t) = \frac{2\big(t - (t_n - \delta_n)\big)}{\delta_n} y(2t_n - \delta_n - t) \quad \text{if } t \in [t_n - \delta_n, t_n - \delta_n/2] . \qquad (**)$$

The condition on the intervals $[t_n - \delta_n, t_n - \delta_n/2]$ here means that y is essentially symmetric with respect to $t_n - \delta_n/2$; some weight makes sure that $y(t_n - \delta_n) = 0$ and that the transition in $t_n - \delta_n/2$ does not break the property $y \in \mathcal{AC}\big(0,T; H^1(\Omega)\big)$ (compare with the extension operator in Proposition 3.16). Moreover, one can see that $y \in \mathcal{V}_p(t_1, \ldots, t_N)$ as well as $y \in L^2\big(t_n - \delta_n, t_n - \delta_n/2; V_{p(t_n)}^2\big)$. Thus, with some differential calculus, we obtain (for almost every $t \in [t_n - \delta_n, t_n]$ and each $n = 1, \ldots, N$, which are omitted in the following)

$$\frac{1}{2}\frac{\partial}{\partial t}\left(\|w\nabla y\|_2^2 + \sum_{k=1}^{K} \|\partial_{q_k} y\|_2^2 \right) = \left\langle \frac{\partial}{\partial t}(w\nabla y), w\nabla y \right\rangle_{L^2} + \sum_{k=1}^{K} \left\langle \frac{\partial}{\partial t}(\partial_{q_k} y), \partial_{q_k} y \right\rangle_{L^2}$$

$$= \left\langle w\nabla \frac{\partial y}{\partial t}, w\nabla y \right\rangle_{L^2} + \sum_{k=1}^{K} \left\langle \partial_{q_k} \frac{\partial y}{\partial t}, \partial_{q_k} y \right\rangle_{L^2}$$

$$= \left\langle \frac{\partial y}{\partial t}, -\operatorname{div}\big(w(w\nabla y)\big) + \sum_{k=1}^{K} \partial_{q_k}^* \partial_{q_k} y \right\rangle_{L^2}$$

$$\le \left\| \frac{\partial y}{\partial t} \right\|_2 \left\| -\operatorname{div}\big(w(w\nabla y)\big) + \sum_{k=1}^{K} \partial_{q_k}^* \partial_{q_k} y \right\|_2$$

$$\le \frac{1}{2}\left(\|y\|_{V_{p(t_n)}^2}^2 + \left\| \frac{\partial y}{\partial t} \right\|_2^2 \right) .$$

Integration and the fact that $y(t_n - \delta_n) = 0$ leads to the estimate

$$
\|y(t_n)\|^2_{V_{p(t_n)}} = \|y(t_n)\|^2_2 + \|w\nabla y(t_n)\|^2_2 + \sum_{k=1}^{K} \|\partial_{q_k} y(t_n)\|^2_2
$$

$$
\leq C_1 \|y\|^2_{W_p} + \int_{t_n-\delta_n}^{t_n} \|y(\tau)\|^2_{V^2_{p(t_n)}} + \left\| \frac{\partial y}{\partial t}(\tau) \right\|^2_2 \, d\tau
$$

$$
\leq C_1 \|y\|^2_{W_p} + C_2 \int_{t_n-\delta_n/2}^{t_n} \|y(\tau)\|^2_{V^2_{p(t_n)}} + \left\| \frac{\partial y}{\partial t}(\tau) \right\|^2_2 \, d\tau
$$

for some $C_2 > 0$, where the embedding result of Proposition 3.16 and the symmetry (**) have been applied. Summing up, one finds that

$$
\sum_{n=1}^{N} \|y(t_n)\|^2_{V_{p(t_n)}} \leq N C_1 \|y\|^2_{W_p} + C_2 \|y\|^2_{W_p^2} \tag{***}
$$

which leads to the desired estimate. Now, each $y \in W_p^2(0,T) \cap W_p(0,T)$ with $y \in \mathcal{AC}(0,T;H^1(\Omega))$ can be mapped to a $\tilde{y}$ such that (**) holds for $\tilde{y}$ and $\tilde{y} = y$ on each $[t_n - \delta_n/2, t_n]$: Choose a smooth function $\eta \in C^\infty([0,T])$ with $\eta(t_n - \delta_n) = 0$ for each $1 \leq n \leq N$ and $\eta(t) = 1$ on each $[t_n - \delta_n/2, t_n]$. Then, one can easily verify that $y \mapsto \tilde{y}$ defined by

$$
\tilde{y}(t) = \begin{cases} \frac{2(t-(t_n-\delta_n))}{\delta_n} y(2t_n - \delta_n - t) & \text{if } t \in [t_n - \delta_n, t_n - \delta_n/2] \\ \eta(t)y(t) & \text{else} \end{cases}
$$

is a continuous mapping between $W_p^2(0,T) \cap W_p(0,T)$ and itself. Moreover, $\tilde{y}$ satisfies (**) for each of the above y. Hence, (***) holds for $\tilde{y}$, meaning that, since $\tilde{y}(t_n) = y(t_n)$,

$$
\sum_{n=1}^{N} \|y(t_n)\|^2_{V_{p(t_n)}} \leq C_3 \left(\|y\|^2_{W_p} + \|y\|^2_{W_p^2} \right) .
$$

The desired continuous embedding now follows from the density of the y chosen according to the above with respect to the space $W_p^2(0,T) \cap W_p(0,T)$.

It remains to show that under the conditions of Proposition 5.39, the solution indeed belongs to $W_p^2(0,T)$. Let y be the weak solution of (2.4) with data $u \in L^2(0,T;L^2(\Omega))$ and $y_0 \in L^2(\Omega)$. Then, by Proposition 5.39, the additional regularity with respect to time $\frac{\partial y}{\partial t} \in L^2(t_n - \delta_n/2, t_n; L^2(\Omega))$ for each $n = 1, \dots, N$ follows. Plugging this into the variant of the weak formulation (3.12) derived in Remark 3.27, taking (5.9) into account and recalling the definition of the weak adjoint differential operator $(*)$, one obtains

$$\langle w(t)\nabla y(t),\, w(t)\nabla z\rangle_{L^2} + \sum_{k=1}^{K}\langle \partial_{q_k(t)}y(t),\, \partial_{q_k(t)}z\rangle_{L^2} = \Big\langle u(t) - \frac{\partial y}{\partial t}(t),\, z\Big\rangle_{L^2}$$

for all $z \in C_0^\infty(\Omega)$ and almost every $t \in \bigcup_{1\leq n\leq N}[t_n - \delta_n/2, t_n]$, meaning that, considering the respective restrictions to $[t_n - \delta_n/2, t_n]$,

$$- \operatorname{div}\big(w(w\nabla y)\big) + \sum_{k=1}^{K}\partial_{q_k}^* \partial_{q_k} y = \frac{\partial y}{\partial t} - u \in L^2\big(t_n - \delta_n/2, t_n; L^2(\Omega)\big)$$

for each $n = 1,\ldots,N$. But this implies that $y \in \mathcal{V}_p^2$ and consequently $y \in W_p^2(0,T)$ in addition to $y \in W_p(0,T)$. One finally gets the continuity of the solution mapping with respect to the $W_p^2(0,T)$-norm from the estimate proven in Proposition 5.39:

$$\|y\|_{W_p^2}^2 = \|y\|_{\mathcal{V}_p}^2 + \sum_{n=1}^{N}\int_{t_n-\delta_n/2}^{t_n}\Big\|\frac{\partial y}{\partial t}(\tau)\Big\|_2^2$$

$$+ \Big\| - \operatorname{div}\big(w(w\nabla y)(\tau)\big) + \sum_{k=1}^{K}\partial_{q_k}^* \partial_{q_k} y(\tau)\Big\|_2^2 \,\mathrm{d}\tau$$

$$\leq \|y\|_{\mathcal{V}_p}^2 + \sum_{n=1}^{N}\int_{t_n-\delta_n/2}^{t_n} 3\Big\|\frac{\partial y}{\partial t}(\tau)\Big\|_2^2 + 2\|u(\tau)\|_2^2 \,\mathrm{d}\tau$$

$$\leq C_4\big(\|y_0\|_2^2 + \|u\|_2^2\big)$$

where also the estimate from Theorem 3.25 is used. This theorem also ensures $\|y\|_{W_p}^2 \leq C_5\big(\|y_0\|_2^2 + \|u\|_2^2\big)$, hence, with the help of the continuous embedding proven above, we get the asserted estimate

$$\|y\|_{\mathcal{V}_p(t_1,\ldots,t_N)}^2 \leq C\big(\|y_0\|_2^2 + \|u\|_2^2\big) . \qquad \blacktriangleleft$$

Remark 5.41. Note that for the conclusion

$$- \operatorname{div}\big(w(w\nabla y)\big) + \sum_{k=1}^{K}\partial_{q_k}^* \partial_{q_k} y \in L^2\big(t_n - \delta_n/2, t_n; L^2(\Omega)\big) ,$$

the variational formulation of the "adjoint differential operator" was necessary since this fact was derived directly from the weak formulation of (2.4) which is also a variational formulation. Therefore, the characterization provided by Theorem 5.31 (and the resulting weak formulation (5.9)) was necessary in the sense that otherwise, the conclusion could not be drawn. This points out the range of application for such a result.

Remark 5.42. To summarize the regularity results roughly, we get that the weak solution of (2.4) satisfies

$$\frac{\partial y}{\partial t} \in L^2\big(t_n - \delta_n/2, t_n; L^2(\Omega)\big) \quad , \quad y(t_n) \in V_{p(t_n)}$$

with continuous L^2-dependence on y_0 and u, as soon as $p(t)$ is constant in the interval $[t_n - \delta_n, t_n]$. This also means that, if we discretize with respect to t, i.e. the t_n represent the nodes of some discretization, and choose p as constant on the intervals between two nodes, we get that each solution is in $V_{p(t_n)}$ at each node. Such results then can for example be useful to obtain error estimates for numerical realizations but also shows $y(T) = y(t_N)$ has to be an image which is smooth with the exception of jumps along the edges indicated by $p(T)$.

Chapter 6

Analysis of the optimization problem

After having provided a thorough analysis of the degenerate parabolic equations of the type (2.4), i.e. showed existence and uniqueness of weak solutions, identified the solution spaces and analyzed their behavior with respect to the parameter p, we can now collect these results to prove existence of optimal control problems associated with the equation. Such problems are usually stated as minimization problems over a *control* u and a *state* y which are related by the so-called *state equation*. In the situation of (2.8), the optimization problem also includes the parameter identification of p, which can be interpreted as an additional control variable. The two control variables (u, p) are connected with y by the partial differential equation (2.4). Thus, we consider the following general optimal control problem:

$$\min_{\substack{u \in L^2(0,T;L^2(\Omega)) \\ \|p\|_{\mathcal{Y}_d^*} \leq 1}} \Psi(y) + \Phi_u(u) + \Phi_p(p) \tag{6.1}$$

$$\frac{\partial y}{\partial t} = \operatorname{div}\left(D_p^2 \nabla y^{\mathrm{T}}\right) + u \quad , \quad \frac{\partial y}{\partial_p \nu} = 0 \quad , \quad y = y_0 \ .$$

First, the existence of optimal controls and associated states in the respective function spaces is shown. It turns out that such a result can be established by requiring some weak compactness for u but strong compactness for p in a certain sense. Moreover, it will be necessary to assume that each admissible p is non-degenerate at the boundary in the sense of Definition 5.32. As we will see, the necessity of this strong compactness originates from the difficulties proving the closedness of the solution operator under weak (and weak*) convergence (cf. Subsection 5.1.1 and Proposition 5.15). The requirement that p is non-degenerate at the boundary has its origin in the alternate weak formulation of Theorem 5.35 which will be employed in the existence theorem. Thus, the conditions for existence will seem quite restrictive at the first glance.

In the subsequent section, as far as it is possible, necessary conditions for optimality of some pair (u, p) are derived for a particular choice of the penalty functionals Ψ, Φ_u and Φ_p. Especially, the adjoint of the equation (2.4) is computed, which turns out to be of the same type. The situation here is unfortunately not satisfactory: The operator assigning each (u, p) the weak solution of (2.4) lacks differentiability for the interesting case where $\|p\|_\infty = 1$, i.e. where a degeneracy does occur. The first-order necessary conditions can nevertheless be useful for approximating some stationary points or computing approximate numerical solutions.

6.1   Existence of optimal solutions

We now want to show the existence of a solution of the control problem (2.8) in a general framework. The usual way to obtain existence of minimizers for a given functional is the direct method in the calculus of variations. This method proceeds as follows: First, it is observed that the functional is bounded by below, hence admitting an infimum. A sequence whose functional values converge to the infimum is then chosen. For suitable functionals (namely coercive functionals), such sequences admit a subsequence which converges in a certain sense. If, additionally, the objective functional is lower semi-continuous with respect to this convergence, then the limit is indeed a minimizer.

One main topic about the direct method is choosing the appropriate spaces and convergence notions such that minimizing sequences have a convergent subsequence for preferably general functionals. The functional for which one wants to obtain a minimizing element still has to be lower semi-continuous with respect to the convergence notion associated with these spaces. In the theory of optimal control, the state y is treated separately from the control (u, p) and the underlying solution operator $y = \mathcal{S}(u, p)$ (or constraint, in general) is studied. This has the advantage that the objective functionals Ψ, Φ_u and Φ_p in (6.1) for y, u and p, respectively, can be investigated separately from $\mathcal{S}$. In particular, one can impose coercivity as well as lower semi-continuity and obtains minimizers y^*, u^* and p^*. For the recombination, it remains to show that y^* is the state associated with (u^*, p^*) which amounts to the closedness of the solution operator $\mathcal{S}$. Considering the problem (6.1), this is where the most effort has to be spent and the theory developed in Chapters 4–5 has another powerful application.

We will obtain that a solution of (6.1) exists if some general and somehow abstract conditions on the functionals are satisfied. We will later give concrete examples of functionals for which the conditions can be verified successfully. The abstract conditions read as follows:

Condition 6.1.

1. Let $\Psi : \mathcal{C}^*(0,T;L^2(\Omega)) \to \mathbb{R}$ be a weakly lower semi-continuous (or convex and lower semi-continuous) functional which penalizes the state y and is bounded from below.

2. Let $\Phi_u : L^2(0,T;L^2(\Omega)) \to \mathbb{R} \cup \{\infty\}$ be a proper, coercive, weakly lower semi-continuous functional penalizing the right-hand side u such that it is bounded from below on bounded sets.

3. Let the penalty of the edge function $\Phi_p : \mathcal{Y}_d^* \to \mathbb{R} \cup \{\infty\}$ be proper, weakly* lower semi-continuous and bounded from below on the unit ball in $\mathcal{Y}_d^*$ and such that $\Phi_p(p) < \infty$ implies that p is non-degenerate at the boundary (with some M_∂ which may depend on $\Phi_p(p)$).

   Furthermore, let Φ_p such that for each $C \in \mathbb{R}$ we have, for some $C' > 0$,

   $$\{p \in \mathcal{Y}_d^* \mid \Phi_p(p) \le C\} \subset \{p \in \mathcal{Y}_d^* \mid \mathrm{tv}^*(p) \le C'\}$$

   as well as for each sequence $\{p_l\}$ with $\Phi_p(p_l) \le C$ and $\|p_l\|_{\mathcal{Y}_d^*} \le 1$ there follows that $\nabla p_l(t) \in L^\infty(\Omega,\mathbb{R}^{d \times d})$ is relatively compact in $L^1(\Omega,\mathbb{R}^{d \times d})$ for almost every $t \in]0,T[$.

The third condition might need a bit of an explanation. Without such a condition, we can only deduce the weak*-convergence of a sequence $\{p_l\}$ in $\mathcal{Y}_d^*$ associated with a minimizing sequence. In order to employ Theorem 5.35 and to put the setting in the context of weighted and directional weak derivatives, the limit p^* has to be non-degenerate at the boundary. Moreover, as already mentioned, weak*-convergence in $\mathcal{Y}_d^*$ causes problems regarding the lack of weak* sequential continuity of the superposition operators according to (5.7). By requiring that p is of bounded semivariation in the sense of Definition 5.4, the weak* sequential continuity can be ensured with the help of Corollary 5.7, see Remarks 5.26 and 5.30. Moreover, in order to apply Proposition 5.15, we also have to make sure that the gradient and divergence of the considered weights and directions, respectively, converge pointwise almost everywhere in the domain $]0, T[\times \Omega$. This will follow from the relative compactness of $\nabla p(t)$ in $L^1(\Omega, \mathbb{R}^{d \times d})$ for almost every $t \in]0, T[$, as we will see in the following lemma.

Of course, one can think of other conditions on Φ_p, as long as they ensure that weak*-convergence of p_l can be transferred to the associated weights and directions and the requirements of Proposition 5.15 are fulfilled.

Lemma 6.2. *Let Φ_p be a functional according to Condition 6.1 and $\{p_l\}$ in $\mathcal{Y}_d^*$ a sequence such that $\|p_l\|_{\mathcal{Y}_d^*} \leq 1$, $p_l \xrightarrow{*} p$ as well as $\Phi_p(p_l) \leq C$ for some $C > 0$.*

Then, for a subsequence, the $w_l(t)$ and $q_{k,l}(t)$ associated with $p_l(t)$ according to (5.7) converge weakly* for almost every t to the $w(t)$ and $q_k(t)$ associated with $p(t)$ (in Y_1^* and Y_d^*, respectively). In particular, $w_l \xrightarrow{*} w$, $q_{k,l} \xrightarrow{*} q_k$ in the respective spaces as well as pointwise almost everywhere.

Moreover, we have that $\nabla w_l \to \nabla w$ as well as $\operatorname{div} q_{k,l} \to \operatorname{div} q_k$ pointwise almost everywhere in $]0, T[\times \Omega$.

Proof. To establish the pointwise a.e. weak*-convergence for a subsequence (with respect to t), apply Lemma 5.5 to get that almost every $p_l(t) \xrightarrow{*} p(t)$ in Y_d^* since Φ_p according to Condition 6.1 enforces the required prerequisites for $\{p_l\}$. Remarks 5.26 and 5.30 then imply $w_l(t) \xrightarrow{*} w(t)$ in Y_1^* as well as $q_{k,l}(t) \xrightarrow{*} q_k(t)$ in Y_d^* for almost every $t \in]0, T[$. The weak*-convergence follows immediately whereas the pointwise a.e. convergence can be obtained from applying Proposition 4.11 to almost every $\{w_l(t)\}$ and $\{q_{k,l}(t)\}$, respectively, giving pointwise convergence with respect to x and consequently a.e. on the cylinder $]0, T[\times \Omega$.

Proving the remaining part requires a bit more work. Note that we already have $p_l(t) \xrightarrow{*} p(t)$ almost everywhere. This means in particular that $p_l(t) \to p(t)$ in $L^1(\Omega, \mathbb{R}^d)$ for almost every t as well as $\nabla p_l(t) \xrightarrow{*} \nabla p(t)$ in $L^\infty(\Omega, \mathbb{R}^{d \times d})$ (see Lemma 4.9). By assumption on the Φ_p, almost every $\nabla p_l(t)$ lies within a relatively compact set of $L^1(\Omega, \mathbb{R}^{d \times d})$.

Fix a $t \in]0, T[$ for which the above is true. Our claim is now that $\nabla p_l(t) \to \nabla p(t)$ in $L^1(\Omega, \mathbb{R}^{d \times d})$. If this is not the case, there is an $\varepsilon > 0$ and another subsequence, still denoted by $\nabla p_l(t)$, such that

$$\|\nabla p_l(t) - \nabla p(t)\|_1 \geq \varepsilon .$$

Since $\{\nabla p_l(t)\}$ still is relatively compact in $L^1(\Omega, \mathbb{R}^{d \times d})$, yet another subsequence converges to $\nabla p(t)$ in $L^1(\Omega, \mathbb{R}^{d \times d})$ by the closedness of the gradient operator, a contradiction. Thus, $p_l(t) \to p(t)$ in $H^{1,1}(\Omega, \mathbb{R}^d)$ for almost every t.

Using classical results on the continuity of superposition operators mapping $H^{1,1}(\Omega, \mathbb{R}^d)$ to $H^{1,1}(\Omega)$ as well as $H^{1,1}(\Omega, \mathbb{R}^d)$ we can deduce, with the help of Lemmas 5.25 and 5.29, that the weights and directions associated with p_l and

p, respectively, according to (5.7) satisfy $w_l(t) \to w(t)$ in $H^{1,1}(\Omega)$ as well as $q_{k,l}(t) \to q_k(t)$ in $H^{1,1}(\Omega, \mathbb{R}^d)$ for almost every $t \in]0, T[$. Hence, by choosing an appropriate C_1 (which is possible since the respective sequences converge weakly* and are thus bounded) we get, for almost every $t \in]0, T[$,

$$\|w_l(t)\|_1 + \|\nabla w_l(t)\|_1 \leq |\Omega|(\|w_l(t)\|_\infty + \|\nabla w_l(t)\|_\infty)$$
$$\leq (1 + M_x^{-1})|\Omega| \|w_l\|_{\mathcal{Y}_d^*} \leq C_1$$

and analogously
$$\|q_{k,l}(t)\|_1 + \|\nabla q_{k,l}(t)\|_1 \leq C_1 .$$

By Lebesgue's dominated convergence theorem (for Bochner spaces) there follows $w_l \to w$ in $L^1(0, T; H^{1,1}(\Omega))$ and $q_{k,l} \to q_k$ in $L^1(0, T; H^{1,1}(\Omega, \mathbb{R}^d))$. In particular, again going to suitable subsequences, $\nabla w_l \to \nabla w$ as well as $\operatorname{div} q_{k,l} \to \operatorname{div} q_k$ pointwise a.e. in $]0, T[\times \Omega$, what was to show. ◄

Theorem 6.3. *Let σ satisfy Condition 5.21 and $y_0 \in L^2(\Omega)$. Under the assumptions of Condition 6.1, the optimal control problem (6.1) possesses at least one solution $(u^*, p^*) \in L^2(0, T; L^2(\Omega)) \times \mathcal{Y}_d^*$.*

Proof. Denote by $\mathcal{S}(u, p) = y$ the weak solution operator of the partial differential equation (2.4), by

$$F(u, p) = \Psi(\mathcal{S}(u, p)) + \Phi_u(u) + \Phi_p(p) + I_{\{\|p\|_{\mathcal{Y}_d^*} \leq 1\}}(p)$$

the reduced functional to be minimized and by $\mathcal{Z} = L^2(0, T; L^2(\Omega)) \times \mathcal{Y}_d^*$ the space in which the minimization problem is posed. Our aim is to show existence of minimizers for F. This will be done in three steps. First, we derive the existence of a minimizing sequence $\{(y_l, u_l, p_l)\}$ with $y_l = \mathcal{S}(u_l, p_l)$ which converges weakly(*), then show that the limit (y^*, u^*, p^*) also satisfies $y^* = \mathcal{S}(u^*, p^*)$. Finally, it is verified that the limit is indeed a minimizer for F.

First note that F satisfies $F(u, p) \geq \Phi_u(u) + C_1 + C_2$ where C_1 and C_2 are the bounds from below for Ψ and Φ_p (on the unit ball of $\mathcal{Y}_d^*$). Then, is it easy to see that $\|u\|_2 \to \infty$ implies $F(u, p) \to \infty$ regardless of p. The other way around, if $\{F(u_l, p_l)\}$ is bounded from above then $\|u_l\|_2$ is also bounded and of course $\|p_l\|_{\mathcal{Y}_d^*} \leq 1$, enforced by the indicator functional.

Furthermore, for each $u \in L^2(0, T; L^2(\Omega))$ and $p \in \mathcal{Y}_d^*$ with $\|p\|_{\mathcal{Y}_d^*} \leq 1$, there exists a weak solution of (2.4) in $\mathcal{C}(0, T; L^2(\Omega))$ by Theorem 3.25 and Corollary

3.26, hence $\Psi \circ S$ is finite for all u and all admissible p. Since Φ_u and Φ_p are proper, there is at least one (u, p) for which $F(u, p) < \infty$ and one can find a minimizing sequence $\{(u_l, p_l)\}$ in $\mathcal{Z}$, i.e. $\lim_{l \to \infty} F(u_l, p_l) = \inf_{(u,p) \in \mathcal{Z}} F(u, p)$ with $-\infty$ allowed. Now, apply the conclusions of the last paragraph to obtain that $\|u_l\|_2 \leq C_3$ for some $C_3 > 0$. Since Φ_u is also bounded from below on bounded sets, say by $C_4 \in \mathbb{R}$, we obtain $F(u_l, p_l) \geq C_1 + C_2 + C_4$ for all $l \in \mathbb{N}$, hence the infimum is actually finite.

Since $\{u_l\}$ is bounded in $L^2(0, T; L^2(\Omega))$, it admits a weakly convergent subsequence $\{u_l\}$, not relabeled, with limit u^*. Furthermore, $\|p_l\|_{\mathcal{Y}_d^*} \leq 1$ implies the existence of a weak*-convergent subsequence $p_l \overset{*}{\rightharpoonup} p^*$ in $\mathcal{Y}_d^*$ (see Remark 5.3). Applying Theorem 3.40 now yields that $y_l = S(u_l, p_l)$ possesses a weakly converging subsequence in $\mathcal{C}^*(0, T; L^2(\Omega))$ with limit y^*.

We now turn to the second step and show that for yet another subsequence of $\{(u_l, p_l)\}$, the corresponding solutions y_l of the partial differential equation also converge to the solution of the limit problem, i.e. $S(u^*, p^*) = y^*$. We recall that for a fixed l, y_l satisfies the weak formulation for (2.4) (also confer Definition 3.17), i.e.

$$- \langle y_l(t), \frac{\partial z}{\partial t}(t) \rangle_{V_l^* \times V_l} + \int_0^T \langle D_{p_l(t)} \nabla y_l(t)^{\mathrm{T}}, D_{p_l(t)} \nabla z(t)^{\mathrm{T}} \rangle_{L^2} \, dt$$

$$= \int_0^T \langle u_l(t), z(t) \rangle_{L^2} \, dt + \langle y_0, z(0) \rangle_{L^2} \quad (*)$$

for each $z \in W_{p_l}(0, T)$ with $z(T) = 0$. This is where the characterization of the solution spaces come into account. With the help of Lemma 6.2, we conclude that, again restricting to a subsequence, the corresponding weights and directions, satisfy $w_l \overset{*}{\rightharpoonup} w^*$ in $\mathcal{Y}_1^*$ and $q_{k,l} \overset{*}{\rightharpoonup} q_k^*$ in $\mathcal{Y}_d^*$, respectively (here, w_l and w^* as well as $q_{k,l}$ and q_k^* are associated with p_l and p^* by (5.7)). There moreover holds that $\nabla w_l \to \nabla w^*$ and $\mathrm{div}\, q_{k,l} \to \mathrm{div}\, q_k^*$ almost everywhere in $]0, T[\times \Omega$. With the help of the fact that $V_{p_l} \subset \mathcal{H}^2_{w_l, \partial q_{l,1}, \dots, \partial q_{l,K}}$ according to Theorem 5.31 (which is applicable since σ, p_l fulfill Conditions 5.21 and 5.22, respectively) the variational equation $(*)$ also implies that

$$- \langle y_l, \frac{\partial z}{\partial t} \rangle_{L^2} + \langle w_l \nabla y_l, w_l \nabla z \rangle_{L^2} + \sum_{k=1}^K \langle \partial_{q_{k,l}} y_l, \partial_{q_{k,l}} z \rangle_{L^2}$$

$$= \langle u_l, z \rangle_{L^2} + \langle y_0, z(0) \rangle_{L^2} \quad (**)$$

for each l as well as $z \in L^2(0, T; H^1(\Omega))$, $\frac{\partial z}{\partial t} \in L^2(0, T; L^2(\Omega))$ and $z(T) = 0$. We now pass to the limit as $l \to \infty$ and want to obtain convergence of the variational equation. With the result of Theorem 5.35 applied to p^* (which has to be non-degenerate at the boundary according to Condition 6.1), this will imply that $(**)$ in the limit for all test functions according to the above is sufficient for $\mathcal{S}(u^*, p^*) = y^*$.

From the a-priori estimate in Theorem 3.25 follows that the norm $\|y_l\|_{\mathcal{V}_{p_l}} \leq C_5$ for some $C_5 > 0$ which does not depend on l. In particular, we have

$$\|y_l\|_2 \leq C_5 \quad , \quad \|w_l \nabla y_l\|_2 \leq C_5 \quad \text{and} \quad \|\partial_{q_{k,l}} y_l\|_2 \leq C_5 .$$

Hence, there is yet another subsequence such that $y_l \rightharpoonup y^*$, $w_l \nabla y_l \rightharpoonup \theta$ and $\partial_{q_{k,l}} y_l \rightharpoonup v_k$ in the respective L^2-spaces. The equality of the weak $\mathcal{C}^*(0, T; L^2(\Omega))$-limit y^* and the weak $L^2(0, T; L^2(\Omega))$-limit of $\{y_l\}$ follows from the continuous injection $\mathcal{C}^*(0, T; L^2(\Omega)) \hookrightarrow L^2(0, T; L^2(\Omega))$. Since $w_l \overset{*}{\rightharpoonup} w^*$ in $\mathcal{Y}_1^*$, $q_{k,l} \overset{*}{\rightharpoonup} q_k^*$ in $\mathcal{Y}_d^*$ and we have the pointwise a.e. convergence of $w_l \to w^*$, $\nabla w_l \to \nabla w^*$ and $q_{k,l} \to q_k^*$, $\text{div } q_{k,l} \to \text{div } q_k^*$, the respective closedness properties (see Proposition 5.15) yield $\theta = w^* \nabla y^*$ as well as $v_k = \partial_{q_k^*} y^*$.

Furthermore, the characterization of the respective weak*-convergence in Lemma 4.9 ensures that $w_l(t) \to w^*(t)$ in $L^\infty(\Omega)$ and $q_{k,l}(t) \to q_k^*(t)$ in $L^\infty(\Omega, \mathbb{R}^d)$ for almost every $t \in \ ]0, T[$, hence $w_l(t) \nabla z(t) \to w^*(t) \nabla z(t)$ as well as $\partial_{q_{k,l}(t)} z(t) \to \partial_{q_k^*(t)} z(t)$ a.e. in the respective L^2-spaces, leading to $w_l \nabla z \to w^* \nabla z$ in $L^2(0, T; L^2(\Omega, \mathbb{R}^d))$ and $\partial_{q_{k,l}} z \to \partial_{q_k^*} z$ for $L^2(0, T; L^2(\Omega))$ by the dominated convergence theorem of Lebesgue for Bochner spaces. Combining strong and weak convergence yields

$$-\langle y^*, \frac{\partial z}{\partial t} \rangle + \langle w^* \nabla y^*, w^* \nabla z \rangle + \sum_{k=1}^{K} \langle \partial_{q_k^*} y^*, \partial_{q_k^*} z \rangle = \langle u^*, z \rangle + \langle y_0, z(0) \rangle$$

when taking the limit in $(**)$. Therefore, y^* is the state associated with (u^*, p^*), or $\mathcal{S}(u^*, p^*) = y^*$ in $L^2(0, T; L^2(\Omega))$. Again, the continuous embedding argument from above also allows to conclude that the equation is true even in $\mathcal{C}(0, T; L^2(\Omega))$.

It remains to show that the triple (y^*, u^*, p^*) is a minimizer of the functional. But Condition 6.1 states that the functionals Ψ, Φ_u and $\Phi_p + I_{\{\|p\|_{\mathcal{Y}^*} \leq 1\}}$ are lower semi-continuous with respect to the convergence we have for y_l, u_l and p_l, respectively. Thus,

$$F(u^*, p^*) \leq \liminf_{l \to \infty} \ \Psi(y_l) + \Phi_u(u_l) + \Phi_p(p_l) = \min_{(u,p) \in \mathcal{Z}} F(u, p) . \quad \blacktriangleleft$$

Remembering Section 2.2, we derived in (2.6) some possible functionals which may serve as objective and regularization functionals in the optimal control problem (2.8). It was mentioned that the functional Φ_p has to contain some additional regularization terms in order to obtain existence. In view of Theorem 6.3 (and Condition 6.1), it becomes apparent how to choose these regularization terms. The following two lemmas state a possible choice for the regularization of p and establish Condition 6.1 for (2.6). Regarding the penalization for u, we furthermore show the feasibility of some power of weighted sequence norms with respect to the coefficients associated with an orthonormal basis $\{\varphi_j\}$ of $L^2(\Omega)$. In preparation to the result that verifies Condition 6.1 for this particular choice, we need a weakly* lower semi-continuity statement for the functionals associated with the total (semi)-variation.

Lemma 6.4. *The functionals*

$$p \mapsto \mathrm{tv}^*(p) \quad , \quad q \mapsto \operatorname*{ess\,sup}_{t \in]0,T[} \mathrm{TV}\big(q(t)\big)$$

where

$$\mathrm{TV}(q) = \sup \left\{ \int_\Omega q \cdot \mathrm{div}\, v \; \mathrm{d}x \;\middle|\; v \in \mathcal{C}_0^1(\Omega, \mathbb{R}^{d \times d \times d}) \,,\; \|v\|_\infty \leq 1 \right\}, \qquad (6.2\mathrm{a})$$

$$\|v\|_\infty = \sup_{x \in \Omega} \left(\sum_{i_1,i_2,i_3=1}^{d} v_{i_1,i_2,i_3}(x)^2 \right)^{1/2}, \qquad (6.2\mathrm{b})$$

in $\mathcal{Y}_d^*$ and $L^\infty(]0,T[\times \Omega, \mathbb{R}^{d \times d})$, respectively, are proper and weakly* lower semi-continuous. Moreover, for each $0 < M_\partial < 1$, the set

$$\mathcal{P}_\partial = \{ p \in \mathcal{Y}_d^* \mid |p(t,x)| \leq \min\, \{1, M_\partial + M_x^{-1} \,\mathrm{dist}(x, \partial\Omega)\}$$
$$\text{for a.e. } (t,x) \in\;]0,T[\times \Omega \} \quad (6.2\mathrm{c})$$

is non-empty and weakly* sequentially closed in $\mathcal{Y}_d^*$.

Proof. Consider the functional tv^* first. It is proper since $\mathrm{tv}^*(0) = 0$. Let $p_l \overset{*}{\rightharpoonup} p$ in $\mathcal{Y}_d^*$ which in particular means that, for each $v \in Y_d$ the functions

$$\bar{p}_{v,l}(t) = \int_\Omega p_l(t,x) \cdot v(x) \; \mathrm{d}x \quad , \quad \bar{p}_v(t) = \int_\Omega p(t,x) \cdot v(x) \; \mathrm{d}x$$

satisfy $\bar{p}_{v,l} \overset{*}{\rightharpoonup} \bar{p}_v$ in $L^\infty(]0,T[)$ and, by continuous embedding, they also converge weakly in $L^1(]0,T[)$. Now since the total variation on $]0,T[$ is weakly lower semi-continuous (see [AV94], for example), it follows

$$\mathrm{TV}(\bar{p}_v) \leq \liminf_{l \to \infty} \mathrm{TV}(\bar{p}_{v,l})$$

and consequently

$$
\mathrm{tv}^*(p) = \sup_{\|v\|_{Y_d} \leq 1} \mathrm{TV}(\bar{p}_v) \leq \sup_{\|v\|_{Y_d} \leq 1} \liminf_{l \to \infty} \mathrm{TV}(\bar{p}_{v,l})
$$

$$
\leq \liminf_{l \to \infty} \sup_{\|v\|_{Y_d} \leq 1} \mathrm{TV}(\bar{p}_{v,l}) = \liminf_{l \to \infty} \mathrm{tv}^*(p_l)
$$

by the definition of the total semivariation (compare Definition 5.4). Hence, tv^* is weakly* lower semi-continuous.

Now consider $q \mapsto \mathrm{ess\,sup}_{t \in]0,T[} \mathrm{TV}(q)$ in $L^\infty(]0,T[\times \Omega, \mathbb{R}^{d \times d})$. Note that it is proper (admitting 0 at 0) and convex. Let a sequence satisfy $q_l \overset{*}{\rightharpoonup} q$ in $L^\infty(]0,T[\times \Omega, \mathbb{R}^{d \times d})$ which means in particular that $q_l \rightharpoonup q$ in $L^1(]0,T[\times \Omega, \mathbb{R}^{d \times d})$. Thus, it suffices to show that the functional

$$
q \mapsto \mathrm{ess\,sup}_{t \in]0,T[} \mathrm{TV}(q) \quad , \quad q \in L^1(]0,T[\times \Omega, \mathbb{R}^{d \times d})
$$

is lower semi-continuous (since it is convex). Assume that $q_l \to q$ in the space $L^1(0,T; L^1(\Omega, \mathbb{R}^{d \times d}))$ and without loss of generality that moreover $q_l(t) \to q(t)$ in $L^1(\Omega, \mathbb{R}^{d \times d})$ for a.e. $t \in]0,T[$ (see [Alt99], Lemma 1.18, for example). The functional TV itself is lower semi-continuous which can be seen as follows. Denote by

$$
D = \{ \mathrm{div}\, v \mid v \in \mathcal{C}_0^1(\Omega, \mathbb{R}^{d \times d \times d}) , \ \|v\|_\infty \leq 1 \} ,
$$

hence

$$
\mathrm{TV}(\tilde{q}) = \sup_{\tilde{v} \in D} \left\{ \liminf_{l \to \infty} \int_\Omega \tilde{q}_l \cdot \tilde{v} \, \mathrm{d}x \right\}
$$

$$
\leq \liminf_{l \to \infty} \sup_{\tilde{v} \in D} \left\{ \int_\Omega \tilde{q}_l \cdot \tilde{v} \, \mathrm{d}x \right\} = \liminf_{l \to \infty} \mathrm{TV}(\tilde{q}_l)
$$

whenever $\tilde{q}_l \to \tilde{q}$ in $L^1(\Omega, \mathbb{R}^{d \times d})$. Consequently, with an analog argumentation and Fatou's lemma,

$$
\mathrm{ess\,sup}_{t \in]0,T[} \liminf_{l \to \infty} \mathrm{TV}(q_l(t)) \leq \liminf_{l \to \infty} \mathrm{ess\,sup}_{t \in]0,T[} \mathrm{TV}(q_l(t)) ,
$$

leading to the desired lower semi-continuity.

Finally, $\mathcal{P}_\partial$ is obviously non-empty. Remember that $L^1(]0,T[\times \Omega, \mathbb{R}^d) \hookrightarrow \mathcal{Y}_d$ densely, so we have the dense embedding $\mathcal{Y}_d^* \hookrightarrow L^\infty(]0,T[\times \Omega, \mathbb{R}^d)$ which is weak* sequentially continuous. Thus, if $p_l \overset{*}{\rightharpoonup} p$ in $\mathcal{Y}_d^*$ with each $p_l \in \mathcal{P}_\partial$, then also $p_l \overset{*}{\rightharpoonup} p$ in $L^\infty(]0,T[\times \Omega, \mathbb{R}^d)$. By weak* sequential closedness in the latter space, it is clear that

$$
|p(t,x)| \leq \min \{1, M_\partial + M_x^{-1} \mathrm{dist}(x, \partial\Omega)\} \quad \text{a.e. in }]0,T[\times \Omega ,
$$

hence $p \in \mathcal{P}_\partial$. ◀

Lemma 6.5. *Let Ω satisfy the cone condition (see, e.g. [AF03]), $0 < M_\partial < 1$, $0 < t_1 < \ldots, t_N = T$, $y_1, \ldots, y_N \in L^2(\Omega)$, $\{\varphi_j\}$, $j = 1, 2, \ldots$ be an orthonormal basis of $L^2(\Omega)$, $0 < \omega_0 \leq \omega_j$, $1 \leq r \leq 2$ and $\lambda_u, \lambda_p, \mu_1, \mu_2 > 0$ some given regularization parameters. The functionals*

$$\Psi(y) = \sum_{n=1}^{N} \frac{\|y(t_n) - y_n\|^2}{2} \, , \tag{6.3a}$$

$$\Phi_u(u) = \lambda_u \int_0^T \sum_{j=1}^{\infty} \omega_j |\langle u(t), \varphi_j \rangle|^r \; \mathrm{d}t \, , \tag{6.3b}$$

$$\Phi_p(p) = \lambda_p \int_0^T \int_\Omega \sigma\big(|p(t,x)|\big) \; \mathrm{d}x \; \mathrm{d}t + \mu_1 \operatorname{tv}^*(p)$$
$$+ \mu_2 \operatorname*{ess\,sup}_{t \in]0,T[} \operatorname{TV}\big(\nabla p(t)\big) + I_{\mathcal{P}_\partial}(p) \tag{6.3c}$$

fulfill Condition 6.1.

Proof. The functional Ψ is convex and continuous in $C^*\big(0, T; L^2(\Omega)\big)$ since it is a composition of point evaluation and taking the squared norm in a Hilbert space (recall that $C^*\big(0, T; L^2(\Omega)\big)$ and $C\big(0, T; L^2(\Omega)\big)$ share the same norm and that point evaluation is still continuous in $C^*\big(0, T; L^2(\Omega)\big)$). In particular, Ψ is weakly lower semi-continuous (as a consequence of being convex and lower semi-continuous, see [ET76], for example). Furthermore, $\Psi \geq 0$ which means that it is also bounded from below.

Regarding Φ_u, note that this functional can be written as the integral over the functional

$$\Phi_{u,t}(u) = \sum_{j=1}^{\infty} \omega_j |\langle u, \varphi_j \rangle|^r$$

with respect to t. Thus, as we will see, if $\Phi_{u,t}$ is proper, convex, lower semi-continuous and non-negative on $L^2(\Omega)$, Φ_u will inherit these properties on $L^2\big(0, T; L^2(\Omega)\big)$. But $\Phi_{u,t}$ can be regarded as an integral over the r-th power of the modulus of the coefficients of $u(t)$ in the sequence space ℓ^2. The mapping $c \mapsto |c|^r$ is convex and lower semi-continuous, which implies the lower semi-continuity of $\Phi_{u,t}$ with the help of the isometrical isomorphism

$$\mathcal{I}: L^2(\Omega) \to \ell^2 \, , \quad u \mapsto \{\langle u, \varphi_j \rangle\}_j \, .$$

We will now show the coercivity of $\Phi_{u,t}$. Again, we can restrict ourselves to the coefficients which we denote by $\tilde{u} \in \ell^2$ and the functional

$\tilde{\Phi}(\tilde{u}) = \sum_{j=1}^{\infty} \omega_j |\tilde{u}_j|^r$ which acts on the coefficients. Now, with the inequality $|\tilde{u}_j|^2 \leq |\tilde{u}_j|^r \|\tilde{u}\|_2^{2-r}$ and as well as $0 < \omega_0 \leq \omega_j$ we get

$$\omega_0 \sum_{j=1}^{\infty} |\tilde{u}_j|^2 \leq \|\tilde{u}\|_2^{2-r} \sum_{j=1}^{\infty} \omega_j |\tilde{u}_j|^r .$$

Hence, $\omega_0 \|\tilde{u}\|_2^r \leq \tilde{\Phi}(\tilde{u})$. Pulling the setting back to $u \in L^2(\Omega)$, this means $\omega_0 \|u\|_2^r \leq \Phi_{u,t}(u)$ showing that $\Phi_{u,t}(u) \to \infty$ whenever $\|u\|_2 \to \infty$.

It remains to show that

$$\Phi(u) = \lambda_u \int_0^T \Phi_{u,t}(u(t)) \, dt$$

is also proper, convex, lower semi-continuous and coercive. The non-trivial part is merely the coercivity and lower semi-continuity. Former can easily be deduced from the estimate on $\Phi_{u,t}$:

$$\omega_0 \|u\|_2^r \leq \int_0^T \Phi_{u,t}(u(t)) \, dt .$$

The lower semi-continuity follows from Fatou's lemma and of course from the lower semi-continuity of $\Phi_{u,t}$. If $u_l \to u$ in $L^2(0, T; L^2(\Omega))$, then there exists a subsequence for which $u_l(t) \to u(t)$ in $L^2(\Omega)$ almost everywhere, hence

$$\Phi_u(u) \leq \int_0^T \liminf_{l \to \infty} \lambda_u \Phi_{u,t}(u_l(t)) \, dt$$

$$\leq \liminf_{l \to \infty} \lambda_u \int_0^T \Phi_{u,t}(u_l(t)) \, dt = \liminf_{l \to \infty} \Phi_u(u_l)$$

since $\Phi_{u,t}(u(t)) \geq 0$. This holds independently for each subsequence, thus also for the whole sequence.

Turning towards Φ_p, it is easy to see that this functional is proper and bounded from below. We use Lemma 6.4 to establish the weak* lower semi-continuity of

$$\tilde{\Phi}_p(p) = \lambda_p \int_0^T \int_{\Omega} \sigma(|p(t,x)|) \, dx \, dt + \mu_1 \, tv^*(p)$$

as follows. Since we are only interested in the limes inferior, we can assume, without loss of generality, that the sequence $p_l \overset{*}{\rightharpoonup} p$ in $\mathcal{Y}_d^*$ also satisfies $tv^*(p_l) \leq C$ for some $C > 0$. Hence, by Lemma 5.5, $p_l(t) \overset{*}{\rightharpoonup} p(t)$ almost everywhere in

$]0, T[$ and $p_l(t) \to p(t)$ in $\mathcal{C}(\Omega)$ a.e. due to Lemma 4.9. Thus, the integral in $\tilde{\Phi}_p$ converges leading to

$$\tilde{\Phi}_p(p) \leq \lim_{l\to\infty} \lambda_p \int_0^T \int_\Omega \sigma\big(|p_l(t,x)|\big) \; dx \; dt + \liminf_{l\to\infty} \; \mu_1 \, \mathrm{tv}^*(p_l)$$
$$= \liminf_{l\to\infty} \tilde{\Phi}_p(p_l)$$

by virtue of Lemma 6.4. Also, by the particular choice of $\tilde{\Phi}_p$, it is immediate that for each $C \in \mathbb{R}$, $\Phi_p(p) \leq C$ implies $\mathrm{tv}^*(p) \leq C\mu_1^{-1} = C'$, hence

$$\{p \in \mathcal{Y}_d^* \mid \Phi_p(p) \leq C\} \subset \{p \in \mathcal{Y}_d^* \mid \mathrm{tv}^*(p) \leq C'\} \;.$$

Now consider $p \mapsto \mathrm{ess\,sup}_{t\in]0,T[} \mathrm{TV}\big(\nabla p(t)\big)$ and observe that this functional is proper and weakly* lower semi-continuous, again due to Lemma 6.4 and the fact that $p_l \overset{*}{\rightharpoonup} p$ in $\mathcal{Y}_d^*$ implies $\nabla p_l \overset{*}{\rightharpoonup} \nabla p$ in $L^\infty(]0,T[\times \Omega, \mathbb{R}^{d\times d})$ (see Remark 5.3). Next, we want to obtain the relative compactness of $\nabla p_l(t)$ for a.e. $t \in]0,T[$ whenever $\mathrm{ess\,sup}_{t\in]0,T[} \mathrm{TV}\big(\nabla p_l(t)\big) \leq C$ for some $C > 0$ as well as $\|p_l\|_{\mathcal{Y}_d^*} \leq 1$. We restrict ourselves to the $t \in]0,T[$ where $\mathrm{TV}\big(\nabla p(t)\big) \leq C$ and $\|\nabla p(t)\|_\infty \leq M_x^{-1}$ (see Propositions 5.2 and 4.3), which is still a set of full measure. Note that since $\|p_l\|_{\mathcal{Y}_d^*} \leq 1$, we have

$$\|\nabla p_l(t)\|_1 + \mathrm{TV}\big(\nabla p_l(t)\big) \leq |\Omega| M_x^{-1} + C$$

which means that $\nabla p(t)$ is relatively compact in $L^1(\Omega, \mathbb{R}^{d\times d})$ by the compact embedding $\mathrm{BV}(\Omega) \hookrightarrow L^1(\Omega, \mathbb{R}^{d\times d})$ for bounded domains Ω satisfying the cone condition, see [Giu84]. (Due to the equivalence of norms in finite dimensional spaces, the statement can be verified component-wise.) This applies to each of the above t, hence $\nabla p_l(t)$ is a.e. relatively compact in $L^1(\Omega, \mathbb{R}^{d\times d})$.

Finally, combining the functional to

$$\Phi_p(p) = \tilde{\Phi}_p(p) + \mu_2 \, \mathrm{ess\,sup}_{t\in]0,T[} \; \mathrm{TV}\big(\nabla p(t)\big) + I_{\mathcal{P}_\partial}(p)$$

it turns out, by using the above results and the properties of $\mathcal{P}_\partial$ stated in Lemma 6.4, that it indeed fulfills the requirements of Condition 6.1 (obviously, $p \in \mathcal{P}_\partial$ implies that p is non-degenerate at the boundary). ◀

Eventually, by combining the general existence theorem (Theorem 6.3) and the previous lemma, we obtain the existence of at least one solution of the optimal control problem stated in (2.8) with the functionals according to (2.6).

Theorem 6.6. *Let Ω satisfy the cone condition, $M_x > 0$, $0 < M_\partial < 1$, σ according to Condition 5.21, $y_0 \in L^2(\Omega)$ and $0 < t_1 < \ldots < t_N = T$, $y_1, \ldots, y_N$, $\{\varphi_j\}$, $\{\omega_j\}$, r as well as $\lambda_u, \lambda_p, \mu_1, \mu_2$ be given according to Lemma 6.5. Then, the optimal control problem*

$$
\min_{\substack{u \in L^2(0,T;L^2(\Omega)) \\ \|p\|_{\mathcal{Y}_d^*} \leq 1}} \sum_{n=1}^{N} \frac{\|y(t_n) - y_n\|_2^2}{2} + \lambda_u \int_0^T \sum_{j=1}^{\infty} \omega_j |\langle u(t), \varphi_j \rangle_{L^2}|^r \ dt
$$

$$
+ \lambda_p \int_0^T \int_\Omega \sigma\big(|p(t,x)|\big) \ dx \ dt + \mu_1 \operatorname{tv}^*(p)
$$

$$
+ \mu_2 \operatorname*{ess\,sup}_{t \in]0,T[} \operatorname{TV}\big(\nabla p(t)\big) + I_{\mathcal{P}_\partial}(p) \tag{6.4a}
$$

$$
\frac{\partial y}{\partial t} = \operatorname{div}\big(D_p^2 \nabla y^{\mathsf{T}}\big) + u \quad in \ ]0,T[\times \Omega
$$

$$
\frac{\partial y}{\partial_p \nu} = 0 \qquad\qquad on \ ]0,T[\times \partial\Omega \tag{6.4b}
$$

$$
y(0) = y_0 \qquad\qquad on \ \{0\} \times \Omega
$$

possesses at least one solution $(u^*, p^*) \in L^2\big(0,T;L^2(\Omega)\big) \times \mathcal{Y}_d^*$.

6.2   First-order optimality conditions

The aim of this section is to derive some first-order necessary conditions for optimal solutions of (6.1) in the special case of the functionals given by (6.3). It will turn out that the weak solution operator $\mathcal{S}$ of (2.4) is only differentiable in points where the solution is smooth enough. In particular, the differentiability with respect to the edge field p can only be established in points where it satisfies $\|p\|_\infty < 1$, i.e. where the solution y belongs to $W(0,T)$, see Remark 3.28. This is unsatisfactory since the evolution equation is "designed" in a way to produce non-smooth solutions, leaving a gap for the case where p is such that $\|p\|_\infty = 1$, i.e. the equation does degenerate at some points. Nevertheless, computing the derivative of the solution and deriving first-order necessary conditions for the special case can provide more insight to the solution operator and the optimization problem. Especially, the particular form of the derivative of $\mathcal{S}$ gives a hint why the differentiability might fail for the degenerate case. Moreover, when one tries to solve a discretized version of the optimization problem in finitely many dimensions, see Chapter 7, discrete first-order necessary conditions can be derived, which turn out to be very similar to the conditions presented in the following.

6.2.1   Differentiability of the control-to-state mapping

Let us investigate the derivative of the solution operator of (2.4) for vector fields with do not cause the differential equation to degenerate. This can be done with the help of standard calculus for the Fréchet derivative in general Banach spaces. Note that the solution operator is linear and continuous in u such that we only have to examine the behavior with respect to p.

Remark 6.7. Let $E : W(0,T) \times L^2(\Omega) \times L^2(0,T;H^1(\Omega)) \to \mathbb{R}$ and an initial-value operator E_1 as well as a differential operator E_2 be given. We say that E is the *weak form* of the initial-value/differential operator (E_1, E_2), if

$$E\big(y, z(0), z\big) = -\langle E_1 y,\, z(0)\rangle_{L^2} + \int_0^T \langle (E_2 y)(t),\, z(t)\rangle_{L^2}\ dt$$

for each $y \in C^\infty([0,T] \times \overline{\Omega})$ and $z \in W(0,T)$ with $z(T) = 0$. Here, we implicitly incorporate homogeneous Neumann boundary conditions on $]0,T[\times \partial\Omega$ by omitting a corresponding boundary integral.

Defining E by

$$E\big(y, z(0), z\big) = -\langle y(0),\, z(0)\rangle_{L^2} + \int_0^T \langle \frac{\partial y}{\partial t}(t),\, z(t)\rangle_{L^2}$$
$$+ \langle D_{p(t)} \nabla y(t)^{\mathrm{T}},\, D_{p(t)} \nabla z(t)^{\mathrm{T}}\rangle_{L^2}\ dt$$

and performing some integration by parts with respect to the space-variable analogously to the derivation of (3.4) from (3.1), one can see that E is the weak form of $y \mapsto \big(y(0), \frac{\partial y}{\partial t} - \operatorname{div}(D_p^2 \nabla y^{\mathrm{T}})\big)$ (with homogeneous Neumann boundary conditions on the respective boundary) in the above sense.

Lemma 6.8. *Let σ satisfy Condition 5.21, $\mathcal{E}$ be the operator assigning p the weak form of*

$$y \mapsto \Big(y(0), \frac{\partial y}{\partial t} - \operatorname{div}(D_p^2 \nabla y^{\mathrm{T}})\Big)$$

for $p \in L^\infty(]0,T[\times \Omega, \mathbb{R}^d)$ with $\|p\|_\infty < 1$, see also Remarks 6.7 and 3.28, and consider it as a mapping between the open sets

$$\{p \in L^\infty(]0,T[\times \Omega, \mathbb{R}^d) \mid \|p\|_\infty < 1\}$$
$$\to \mathcal{L}\big(W(0,T), L^2(\Omega) \times L^2(0,T;H^1(\Omega)^*)\big)\ .$$

Then $\mathcal{E}$ is continuously Fréchet-differentiable with derivative

$$(\mathrm{D}\mathcal{E}(p)q)(y, z) = \int_0^T \Big\langle \Big(\frac{\partial D_{p(t)}^2}{\partial p}q(t)\Big)\nabla y(t)^{\mathrm{T}}, \nabla z(t)^{\mathrm{T}}\Big\rangle_{L^2} \, dt$$

for $(y, z(0), z) \in W(0, T) \times L^2(\Omega) \times L^2(0, T; H^1(\Omega)^*)$ which corresponds to the weak form of the operator

$$y \mapsto \Big(0, -\operatorname{div}\Big(\Big(\frac{\partial D_p^2}{\partial p}q\Big)\nabla y^{\mathrm{T}}\Big)\Big)$$

(implicitly incorporating the boundary mapping $-\nu \cdot \big(\frac{\partial D_p^2}{\partial p}q\big)\nabla y^{\mathrm{T}}$) where

$$\frac{\partial D_p^2}{\partial p}q = -q \cdot \tfrac{p}{|p|}\sigma'(|p|)\tfrac{p}{|p|} \otimes \tfrac{p}{|p|} - \sigma(|p|)P_p\tfrac{q}{|p|} \otimes \tfrac{p}{|p|} - \sigma(|p|)\tfrac{p}{|p|} \otimes \tfrac{q}{|p|}P_p, \quad (6.5)$$

with $\tfrac{p}{|p|}$ extended by 0 in 0 and

$$P_p = \begin{cases} \big(I - \tfrac{p}{|p|} \otimes \tfrac{p}{|p|}\big) & \text{if } p \neq 0 \\ I & \text{if } p = 0 \,. \end{cases}$$

Proof. Note that for $\|p\|_\infty < 1$ we can decompose $\mathcal{E}$ into a superposition part $p \mapsto D_p^2$ from $L^\infty(]0, T[\times \Omega, \mathbb{R}^d) \to L^\infty(]0, T[\times \Omega, \mathbb{R}^{d \times d})$ and the weak form of the differentiation part $D \mapsto \big(\delta_0, \frac{\partial}{\partial t} - \operatorname{div}(D\nabla^{\mathrm{T}})\big)$ mapping between the spaces $L^\infty(]0, T[\times \Omega, \mathbb{R}^{d \times d}) \to \mathcal{L}\big(W(0, T), L^2(\Omega) \times L^2(0, T; H^1(\Omega)^*)\big)$.

By standard results for superposition operators, the superposition part is continuously differentiable with derivative taken pointwise verifying $q \mapsto \frac{\partial D_p^2}{\partial p}q$ according to (6.5). This can be proven with the help of standard calculus rules, the identities

$$\nabla(|p|) = \frac{1}{|p|}p^{\mathrm{T}} \quad , \quad \nabla\Big(\frac{p}{|p|}\Big) = \frac{1}{|p|}\Big(I - \frac{p}{|p|} \otimes \frac{p}{|p|}\Big)$$

and Lemma 5.24 which allows to extend (6.5) continuously in 0.

The differentiation part can be written as

$$D \mapsto \Big(z \mapsto -\langle y(0), z(0)\rangle_{L^2} + \int_0^T \Big\langle \frac{\partial y}{\partial t}(t), z(t)\Big\rangle_{(H^1)^* \times H^1} \, dt$$

$$+ \int_0^T \langle D(t)\nabla y(t)^{\mathrm{T}}, \nabla z(t)^{\mathrm{T}}\rangle_{L^2} \, dt\Big)$$

which is affine linear and continuous since

$$\int_0^T \langle D(t)\nabla y(t)^{\mathrm{T}}, \nabla z(t)^{\mathrm{T}}\rangle_{L^2} \, dt \le \|D\|_\infty \|y\|_{W(0,T)} \|z\|_{L^2(0,T;H^1)} \, .$$

Thus, it is (continuously) differentiable with the linear part as derivative, i.e. $Q \mapsto (0, -\operatorname{div}(Q\nabla^{\mathrm{T}}))$ in the weak form. Thus, the composition is also continuously differentiable while the chain rule yields the asserted identity. ◀

Knowing the derivative of the forward operator $\mathcal{E}$ with respect to p allows to establish differentiability and to compute the derivative of the solution operator $(u, p) \mapsto \mathcal{S}(u, p)$. For this purpose, we also introduce the mapping $\mathcal{S}_0$ which maps $u \in L^2(0, T; L^2(\Omega))$ and $p \in \mathcal{Y}_d^*$ to the weak solution of (2.4) with the initial value $y_0 = 0$.

Lemma 6.9. *Let σ satisfy Condition 5.21. Then, the weak solution operator $\mathcal{S}$ of (2.4) considered in both variables, i.e.*

$$\mathcal{S} : L^2(0, T; L^2(\Omega)) \times \{p \in L^\infty(]0, T[\times \Omega, \mathbb{R}^d) \mid \|p\|_\infty < 1\} \to W(0, T)$$

is Fréchet-differentiable with derivative $z = D\mathcal{S}(u, p)(v, q)$ being the weak solution of the problem

$$\frac{\partial z}{\partial t} = \operatorname{div}(D_p^2 \nabla z^{\mathrm{T}}) + v + \operatorname{div}\left(\left(\frac{\partial D_p^2}{\partial p} q\right)\nabla y^{\mathrm{T}}\right) \quad \text{in }]0, T[\times \Omega \, ,$$

$$\frac{\partial z}{\partial_p \nu} = -\nu \cdot \left(\frac{\partial D_p^2}{\partial p} q\right)\nabla y^{\mathrm{T}} \qquad\qquad\quad \text{on }]0, T[\times \partial\Omega \, , \qquad (6.6)$$

$$z = 0 \qquad\qquad\qquad\qquad\qquad\qquad \text{on } \{0\} \times \Omega \, ,$$

with $y = \mathcal{S}(u, p)$ and $\frac{\partial D_p^2}{\partial p} q$ according to (6.5).

Proof. We will first compute the partial derivatives with respect to u and p. The mapping $u \mapsto \mathcal{S}(u, p)$ is affine linear and continuous, so the partial derivative $z_1 = \frac{\partial \mathcal{S}}{\partial u}(u, p)v$ is the weak solution of (2.4) with v as data, i.e. z_1 satisfies

$$\frac{\partial z_1}{\partial t} = \operatorname{div}(D_p^2 \nabla z_1^{\mathrm{T}}) + v \, , \qquad \frac{\partial z_1}{\partial_p \nu} = 0 \, , \qquad z_1(0) = 0$$

in the weak sense.

Now, let us compute $\frac{\partial \mathcal{S}}{\partial p}(u, p)$. Consider the mappings

$$E \in \operatorname{Iso}(W(0, T), L^2(\Omega) \times L^2(0, T; H^1(\Omega)^*))$$

which have an inverse

$$E^{-1} \in \mathrm{Iso}\big(L^2(\Omega) \times L^2(0,T;H^1(\Omega)^*),W(0,T)\big) .$$

The weak solution operator $u \mapsto \mathcal{S}(u,p)$ is such an inverse mapping, confer Remark 3.28. Since we know the derivative of the forward operator from Lemma 6.5, we are interested in the derivative of the mapping $E \mapsto E^{-1}$ on $\mathrm{Iso}\big(W(0,T),L^2(\Omega) \times L^2(0,T;H^1(\Omega)^*)\big)$. It is known that this mapping is continuously differentiable for each of such E with derivative $H \mapsto -E^{-1}HE^{-1}$. Here, in the notation of Lemma 6.5, $\mathcal{S}(\,\cdot\,,p) = \mathcal{E}(p)^{-1}(y_0,\,\cdot\,)$ so the chain rule yields that $H = \mathrm{D}\mathcal{E}(p)q$ when computing the derivative of the composed mapping. Finally, $\mathcal{S}(u,p) = \mathcal{E}(p)^{-1}(y_0,u)$, hence

$$\frac{\partial \mathcal{S}}{\partial p}(u,p)q = -\mathcal{S}_0\big((\mathrm{D}\mathcal{E}(p)q)\mathcal{S}(u,p),p\big) , \qquad (*)$$

remembering that $\mathrm{D}\mathcal{E}(p)q$ assigns a zero initial-value. By denoting $y = \mathcal{S}(u,p)$ and using that $\mathrm{D}\mathcal{E}(p)q$ is the weak form of a differential operator, one can rewrite this identity to $\frac{\partial \mathcal{S}}{\partial p}(u,p)q = z_2$ with z_2 being the weak solution of

$$\frac{\partial z_2}{\partial t} - \mathrm{div}\big(D_p^2 \nabla z_2^{\mathrm{T}}\big) = \mathrm{div}\Big(\Big(\frac{\partial D_p^2}{\partial p}q\Big)\nabla y^{\mathrm{T}}\Big),$$

$$\frac{\partial z_2}{\partial_p \nu} = -\nu \cdot \Big(\frac{\partial D_p^2}{\partial p}q\Big)\nabla y^{\mathrm{T}} \quad , \quad z_2(0) = 0 .$$

Thus, if we can show that the Fréchet derivative exists, then $\mathrm{D}\mathcal{S}(u,p)(v,q)$ indeed satisfies the claimed identity.

Having computed the partial derivatives with respect to u and p, we only need to verify that these derivatives are also continuous. Now $\frac{\partial \mathcal{S}}{\partial u}(u,p)v = \mathcal{S}_0(v,p)$, hence it is independent of u, so we only need to verify the continuity with respect to p, which is clear since even the partial derivative $\frac{\partial \mathcal{S}_0}{\partial p}(v,p)$ exists for all $\|p\|_\infty < 1$ (analog to the above). Next, examine the continuity of $\frac{\partial \mathcal{S}}{\partial p}$ with respect to u and p. Note that from $(*)$ follows that $\frac{\partial \mathcal{S}}{\partial p}$ is linear and continuous with respect to u, hence it suffices to examine the continuity with respect to p. But, again by $(*)$, (6.5) and Lemma 5.24, the derivative is a composition of continuous mappings with respect to p. Hence, both partial derivatives are continuous meaning that $\mathcal{S}$ is Fréchet-differentiable. ◀

Remark 6.10. The weak-solution mapping $\mathcal{S}$ is differentiable in vector fields satisfying $\|p\|_\infty < 1$ which corresponds to a non-degenerate solution $y \in W(0,T)$. Now one can ask if such a result also holds for degenerate solutions, i.e. for

$p \in \mathcal{Y}_d^*$ with $\|p\|_{\mathcal{Y}_d^*} \leq 1$. Unfortunately, the differentiability cannot be established with the above arguments.

The reason is, roughly speaking, that the derivative with respect to p differentiates y in the direction p which does not necessarily exist in $L^2(0, T; L^2(\Omega))$. On the other hand, the solution operator $\mathcal{S}(\,\cdot\,, p)$ does not smooth data in the direction of p and does not compensate the loss of smoothness. Therefore, the $\frac{\partial \mathcal{S}}{\partial p}(u, p)q$ is in general not a function in $L^2(0, T; L^2(\Omega))$.

6.2.2   The adjoint equations

In the following, we will deduce the adjoint of the derivative of the solution operator composed with the observation at the times $t_1, \ldots, t_N$ as described in the control problem (2.8). This turns out also to involve the solution of an evolution problem similar to (2.4), which we will call the adjoint problem.

In order to obtain these results, let us first define the observation operator.

Definition 6.11. Let σ fulfill Condition 5.21, $p \in L^\infty(]0, T[\times \Omega, \mathbb{R}^d)$ with $\|p\|_\infty \leq 1$ and $0 < t_1 < t_2 < \ldots < t_N = T$. The *observation operator* $\mathcal{O}$: $W_p(0, T) \to L^2(\Omega)^N$ at times $t_1, \ldots, t_N$ is defined as the mapping

$$\mathcal{O}(y) = \big(y(t_1), y(t_2), \ldots, y(t_N)\big) \ .$$

The following proposition is useful for determining the adjoint of the derivative of the solution operator with observation $\mathcal{O}\mathcal{S}$ with respect to u. Remember that the mapping $\mathcal{S}$ is linear and continuous with respect to u, see Theorem 3.25, thus the partial derivative $\frac{\partial \mathcal{S}}{\partial u}$ exists if one considers it between

$$\mathcal{S} : L^2(0, T; L^2(\Omega)) \times \{p \in L^\infty(]0, T[\times \Omega, \mathbb{R}^d) \mid \|p\|_\infty \leq 1\}$$
$$\to \mathcal{C}(0, T; L^2(\Omega)) \ .$$

Consequently, $\mathcal{O}\mathcal{S}$ is also differentiable with respect to u and $\frac{\partial \mathcal{O}\mathcal{S}}{\partial u}(u, p)v = \mathcal{O}\mathcal{S}_0(v, p)$ where

$$\frac{\partial \mathcal{O}\mathcal{S}_0}{\partial u}(u, p) \in \mathcal{L}\big(L^2(0, T; L^2(\Omega)), L^2(\Omega)^N\big) \ .$$

We are interested in the Hilbert-space adjoint for this mapping which can easily be obtained from the adjoint of $\mathcal{O}\mathcal{S}_p : \mathcal{V}_p^* \times L^2(\Omega) \to L^2(\Omega)^N$.

Proposition 6.12. *Let σ, p, $t_1, \ldots, t_N$ according to Definition 6.11 be given. Consider the weak solution operator of (2.4) mapping*

$$S_p : \mathcal{V}_p^* \times L^2(\Omega) \to W_p(0, T) \ .$$

Then, the adjoint of $\mathcal{OS}_p$ satisfies $(\mathcal{OS}_p)^*z = (v, v(0)) \in \mathcal{V}_p \times L^2(\Omega)$ where

$$v(t) = v_{n(t)}(t) \quad \text{with} \quad n(t) = n \quad \text{if } t \in [t_{n-1}, t_n[$$

and $v_n \in L^2(t_{n-1}, t_n; L^2(\Omega))$, $n = 1, \ldots, N$ are recursively defined by the weak solutions of the backwards equations

$$\frac{\partial v_n}{\partial t} + \operatorname{div}(D_p^2 \nabla v_n^{\mathrm{T}}) = 0 , \qquad \frac{\partial v_n}{\partial_p \nu} = 0 ,$$

$$v_n(t_n) = \begin{cases} z_n + v_{n+1}(t_n) & \text{if } 1 \le n < N \\ z_N & \text{if } n = N . \end{cases} \tag{6.7}$$

Proof. We will first verify that weak solutions of the backwards equations (6.7) with final values in $L^2(\Omega)$ exist for each fixed $1 \le n \le N$ by transforming the equation into a forward equation. This can be done by mapping $t \mapsto t_n - t$, letting $z \in L^2(\Omega)$, denoting by $p_n(t) = p(t_n - t)$ and considering the time interval $[0, t_n - t_{n-1}]$. Then, there exists a weak solution $\bar{v}_n \in W_{p_n}(0, t_n - t_{n-1})$ of the forward equation

$$\frac{\partial \bar{v}_n}{\partial t} - \operatorname{div}(D_{p_n}^2 \nabla \bar{v}_n^{\mathrm{T}}) = 0 , \qquad \frac{\partial \bar{v}_n}{\partial_p \nu} = 0 , \qquad \bar{v}_n(0) = z$$

due to Theorem 3.25. Substituting $v_n(t) = \bar{v}_n(t_n - t)$ with $t \in [t_{n-1}, t_n]$ swaps the sign of the derivative with respect to time, so

$$\frac{\partial v_n}{\partial t} + \operatorname{div}(D_{p_n}^2 \nabla v_n^{\mathrm{T}}) = 0 , \qquad \frac{\partial v_n}{\partial_p \nu} = 0 , \qquad v_n(t_n) = z$$

in the usual weak sense, which e.g. reads as: $v_n \in W_{p_n}(t_{n-1}, t_n)$, $v_n(t_n) = z$ with

$$\int_{t_{n-1}}^{t_n} \left\langle \frac{\partial v_n}{\partial t}(t), \bar{z}(t) \right\rangle_{V(t)^* \times V(t)} - \left\langle D_{p_n(t)} \nabla v_n^{\mathrm{T}}, D_{p_n(t)} \nabla \bar{z}(t) \right\rangle_{L^2} \, dt = 0 \quad (*)$$

for all $\bar{z} \in L^2(t_{n-1}, t_n; H^1(\Omega))$ (by density, also see (3.12)).

Finally, v is just the result of "gluing together" the components v_n on the respective intervals leading to a $v \in \mathcal{V}_p$ which is moreover piecewise continuous (with values in $L^2(\Omega)$) with possible jumps at $t_1, \ldots t_{N-1}$.

Now let a $z = (z_1, \ldots, z_N) \in L^2(\Omega)^N$ be given and denote by $y = S_p(u, y_0)$.

We have, by Proposition 3.16 and the fact that y is a weak solution of (2.4),

$$\langle y(t_n), v_n(t_n)\rangle_{L^2} - \langle y(t_{n-1}), v_n(t_{n-1})\rangle_{L^2}$$

$$= \int_{t_{n-1}}^{t_n} \langle \frac{\partial y}{\partial t}(t), v_n(t)\rangle_{V(t)^* \times V(t)} + \langle \frac{\partial v_n}{\partial t}(t), y(t)\rangle_{V(t)^* \times V(t)} \, dt$$

$$= \int_{t_{n-1}}^{t_n} -\langle D_{p(t)}\nabla y(t)^{\mathrm{T}}, D_{p(t)}\nabla v_n(t)^{\mathrm{T}}\rangle_{L^2} \, dt$$

$$+ \int_{t_{n-1}}^{t_n} \langle \frac{\partial v_n}{\partial t}(t), y(t)\rangle_{V(t)^* \times V(t)} + \langle u(t), v_n(t)\rangle_{V(t)^* \times V(t)} \, dt$$

$$= \int_{t_{n-1}}^{t_n} \langle u(t), v_n(t)\rangle_{V(t)^* \times V(t)} \, dt$$

since v_n is a weak solution of the backwards PDE according to $(*)$. Summing up yields

$$\sum_{n=1}^{N} \left(\langle y(t_n), v_n(t_n)\rangle_{L^2} - \langle y(t_{n-1}), v_n(t_{n-1})\rangle_{L^2} \right) = \int_0^T \langle u(t), v(t)\rangle_{V(t)^* \times V(t)} \, dt$$

and the left-hand side can be reordered to

$$\sum_{n=1}^{N} \left(\langle y(t_n), v_n(t_n)\rangle_{L^2} - \langle y(t_{n-1}), v_n(t_{n-1})\rangle_{L^2} \right) = \langle y(t_N), v_N(t_N)\rangle_{L^2}$$

$$+ \sum_{n=1}^{N-1} \langle y(t_n), v_n(t_n) - v_{n+1}(t_n)\rangle_{L^2} - \langle y(0), v_1(0)\rangle_{L^2} \ .$$

Taking the conditions on $v_n(t_n)$ and $y(0)$ into account finally implies

$$\sum_{n=1}^{N} \langle y_n(t_n), z_n\rangle_{L^2} = \int_0^T \langle u(t), v(t)\rangle_{V(t)^* \times V(t)} \, dt + \langle y_0, v(0)\rangle_{L^2}$$

which proves the formula for the adjoint mapping.   ◀

Remark 6.13. Note that the solution of the backwards equations in (6.7) on the intervals $[t_{n-1}, t_n]$ yields a function v_n which is in $\mathcal{C}(t_{n-1}, t_n; L^2(\Omega))$. In particular, there exist the left and right limits

$$\lim_{t \to t_{n-1}} v_n(t) = v_+(t_{n-1}) \ , \qquad \lim_{t \to t_n} v_n(t) = v_-(t_n) \ .$$

Hence, by joining all v_n to a v and setting $v_+(T) = 0$ we express the adjoint equation alternatively as

$$\frac{\partial v}{\partial t} + \operatorname{div}\left(D_p^2 \nabla v\right) = 0 \,, \qquad \frac{\partial v}{\partial_p \nu} = 0 \,, \tag{6.8}$$
$$v_-(t_n) = v_+(t_n) + z_n \qquad n = 1, \dots, N \,.$$

We can also formulate and prove a result which gives us the adjoint of the derivative of the control-to-observation mapping with respect to p. Note that since we consider

$$\mathcal{OS} : L^2\left(0, T; L^2(\Omega)\right) \times \{p \in L^\infty(]0, T[\times \Omega, \mathbb{R}^d) \mid \|p\|_\infty \leq 1\} \to L^2(\Omega)^N$$

we are seeking, for a certain (u, p), a linear and continuous mapping

$$\frac{\partial \mathcal{OS}}{\partial p}(u, p)^* : L^2(\Omega)^N \to L^\infty(]0, T[\times \Omega, \mathbb{R}^d)^*$$

which involves the dual of $L^\infty(]0, T[\times \Omega, \mathbb{R}^d)$. This space can be characterized as a space of certain additive measures (see [Yos80]), but fortunately it will turn out that the sought operator actually maps into $L^1(]0, T[\times \Omega, \mathbb{R}^d)$ regarded as a subspace of $L^\infty(]0, T[\times \Omega, \mathbb{R}^d)^*$. Since we only know the derivative of $\mathcal{S}$ with respect to p where $\|p\|_\infty < 1$, we again restrict ourselves to that case.

Proposition 6.14. *Let (u, p) be given such that $p \in L^\infty(]0, T[\times \Omega, \mathbb{R}^d)$ with $\|p\|_\infty < 1$ as well as $u \in L^2\left(0, T; L^2(\Omega)\right)$. Then the adjoint of the derivative of the solution/observation operator $\mathcal{OS}$ is given by*

$$\frac{\partial \mathcal{OS}}{\partial p}(u, p)^* z = \frac{\sigma'(|p|)}{|p|^2} \partial_p y \partial_p v \frac{p}{|p|} + \frac{\sigma(|p|)}{|p|^2}\left(\partial_p v P_p \nabla y^\mathrm{T} + \partial_p y P_p \nabla v^\mathrm{T}\right) \tag{6.9}$$

with $y = \mathcal{S}(u, p)$ and $\left(v, v(0)\right) = S_p^* z$ being the solution of the adjoint equations according to (6.8) and zero extension in 0. In particular, $\frac{\partial \mathcal{OS}}{\partial p}(u, p)^* : L^2(\Omega)^N \to L^1(]0, T[\times \Omega, \mathbb{R}^d) \subset L^\infty(]0, T[\times \Omega, \mathbb{R}^d)^*$.

Proof. Choose a $q \in L^\infty(]0, T[\times \Omega, \mathbb{R}^d)$ and let

$$((Dy)q)\bar{z} = \int_0^T \left\langle \left(\frac{\partial D_p^2}{\partial p} q\right)(t) \nabla y(t)^\mathrm{T}, \nabla \bar{z}(t)^\mathrm{T} \right\rangle \, dt$$

for all $\bar{z} \in V_p = L^2\left(0, T; H^1(\Omega)\right)$. By construction, each $(Dy)q$ is an element of V_p^*. According to Lemma 6.9, $\frac{\partial \mathcal{S}}{\partial p}(u, p)q = \bar{y}$ where $\bar{y} = -\mathcal{S}_0\left((Dy)q, p\right)$, so for a given $z \in L^2(\Omega)^N$ we can apply Proposition 6.12 which yields

$$\left\langle \frac{\partial \mathcal{OS}}{\partial p}(u, p)q, z \right\rangle_{L^2} = \langle \bar{y}, z \rangle_{L^2} = \langle -(Dy)q, v \rangle_{V^* \times V}$$

where $v \in \mathcal{V}_p$ is the solution of the adjoint equation according to (6.8) with data z. Together with the above and the identity (6.5) one gets (with coordinates t and x omitted)

$$- \int_0^T \left\langle \left(\frac{\partial D_p^2}{\partial p} q \right) \nabla y^{\mathrm{T}}, \nabla v^{\mathrm{T}} \right\rangle_{L^2} \, \mathrm{d}t$$

$$= \int_0^T \int_\Omega q \cdot \left(\frac{\sigma'(|p|)}{|p|^2} \partial_p y \partial_p v \frac{p}{|p|} + \frac{\sigma(|p|)}{|p|^2} \left(\partial_p v P_p \nabla y^{\mathrm{T}} + \partial_p y P_p \nabla v^{\mathrm{T}} \right) \right) \, \mathrm{d}x \, \mathrm{d}t$$

which means that the asserted identity for $\frac{\partial \mathcal{OS}}{\partial p}(u, p)^* z$ holds. Finally, we verify that the latter is an element of $L^1(]0, T[\times \Omega, \mathbb{R}^d)$. For this purpose we observe that (with zero extension where $p = 0$)

$$\frac{\partial_p y \partial_p v}{|p|^2} \in L^1(]0, T[\times \Omega) , \qquad \frac{\partial_p y P_p \nabla v^{\mathrm{T}}}{|p|}, \frac{\partial_p v P_p \nabla y^{\mathrm{T}}}{|p|} \in L^1(]0, T[\times \Omega, \mathbb{R}^d) ,$$

so the assumed property can be obtained if the remaining terms are in the spaces $L^\infty(]0, T[\times \Omega, \mathbb{R}^d)$ and $L^\infty(]0, T[\times \Omega)$, respectively. But remember that σ, σ' are bounded and $\sigma'(0) = 0$ by Condition 5.21, so by virtue of Lemma 5.24, one can easily see that indeed

$$\left\| \frac{\sigma'(|p|)}{|p|} p \right\|_\infty < \infty , \qquad \left\| \frac{\sigma(|p|)}{|p|} \right\|_\infty < \infty .$$

with continuous extension by 0 whenever $p = 0$, thus the products indeed make sense and yield (6.9). ◀

Remark 6.15. Note that the above result only holds for edge fields which satisfy $\|p\|_\infty < 1$, where the solution space is $W(0, T)$ and does not change in a L^∞-neighborhood of p. This is crucial for the calculations since, regarding (6.9), the directional derivative with respect to p is involved. One can now try to extend this identity to the case where $p \in L^\infty(]0, T[\times \Omega, \mathbb{R}^d)$ and $\|p\|_\infty = 1$ (or, more restrictive, $p \in \mathcal{Y}_d^*$, $\|p\|_{\mathcal{Y}_d^*} \le 1$) which implies that weak solutions of (2.4) are only in $W_p(0, T)$ and consequently may have jumps along p. Thus, even if $p \in \mathcal{Y}_d^*$, it is not possible to assure generally that $\partial_p y$ as well as $\partial_p v$ exists in $L^2(0, T; L^2(\Omega))$. Even worse, one of these weak directional derivatives can be a distribution which is not a function while $P_p \nabla y^{\mathrm{T}}$ or $P_p \nabla v^{\mathrm{T}}$ is only in $L^2(0, T; L^2(\Omega, \mathbb{R}^d))$, meaning that the product does not necessarily make sense (it would if e.g. $P_p \nabla y^{\mathrm{T}} \in L^2(0, T; H^1(\Omega, \mathbb{R}^d))$, which is not necessarily true). This is the reason that we only derive first-order necessary conditions for the case $\|p\|_\infty < 1$.

6.2.3 First-order necessary conditions

In the following, first-order necessary conditions for solutions of (6.4) in the situation of Theorem 6.6 and the case of L^2-penalization for u are derived. Unfortunately, we are only able to state these conditions if $\|p\|_\infty < 1$, i.e. where the parabolic equation does not degenerate (this is connected with the difficulties arising in differentiating $\mathcal{S}$ with respect to p, see also Remark 6.15). But first, we will collect some preliminary results regarding the derivative of the functionals in Lemma 6.5. For simplicity, we will only consider the quadratic case $r = 2$ in Φ_u and assume that $\{w_j\}$ in ℓ^∞, resulting in a smooth Φ_u.

Lemma 6.16. *In the situation of Theorem 6.6, consider the functionals*

$$\Psi_{u,p}(u,p) = \sum_{n=1}^{N} \frac{\left\| \left(\mathcal{O}\mathcal{S}(u,p)\right)_n - y_{\Omega,n} \right\|^2}{2}$$

with $0 < t_1 < \ldots < t_N = T$ and $y_\Omega \in L^2(\Omega)^N$,

$$\Phi_u(u) = \int_0^T \sum_{j=1}^{\infty} w_j \left| \langle u(t), \varphi_j \rangle \right|^2 \, \mathrm{d}t$$

with an orthonormal basis $\{\varphi_j\}$ of $L^2(\Omega)$, $\{w_j\}$ in ℓ^∞, $w_j \geq w_0 > 0$ and

$$\Phi_{p,0}(p) = \int_0^T \int_\Omega \sigma\big(|p(t,x)|\big) \, \mathrm{d}x \, \mathrm{d}t \; .$$

Then, $\Psi_{u,p}$, Φ_u and $\Phi_{p,0}$ are Fréchet-differentiable on

$$L^2\big(0,T;L^2(\Omega)\big) \times \{p \in L^\infty(]0,T[\times \Omega, \mathbb{R}^d) \mid \|p\|_\infty < 1\} \; .$$

The derivatives are, respectively,

$$\Psi'_{u,p}(u,p) = \left(v, \frac{\sigma'(|p|)}{|p|^2} \partial_p y \partial_p v \frac{p}{|p|} + \frac{\sigma(|p|)}{|p|^2} \big(\partial_p v P_p \nabla y^{\mathrm{T}} + \partial_p y P_p \nabla v^{\mathrm{T}} \big) \right) \quad (6.10a)$$

$$\Phi'_u(u) = 2 \sum_j w_j \langle u(\,\cdot\,), \varphi_j \rangle \varphi_j \quad (6.10b)$$

$$\Phi'_{p,0}(p) = \begin{cases} \dfrac{\sigma'(|p|)}{|p|} p & \text{where } p \neq 0 \\ 0 & \text{where } p = 0 \end{cases} \quad (6.10c)$$

where y and v denote the solutions of the primal equation (2.4) and adjoint equation (6.8) (with data $z = \mathcal{O}y - y_\Omega$), respectively.

Proof. First consider the functional $\Psi_{u,p}$, which is a composition of $\mathcal{OS}$ and $\Psi : L^2(\Omega)^N \to \mathbb{R}$ where

$$\Psi(y) = \sum_{n=1}^{N} \frac{\|y_n - y_{\Omega,n}\|^2}{2}$$

is Fréchet-differentiable with derivative

$$\langle \Psi'(y), z \rangle_{L^2} = \sum_{n=1}^{N} \langle y_n - y_{\Omega,n}, z_n \rangle_{L^2} .$$

Due to Lemma 6.9, $\mathcal{S}$ is differentiable on the claimed set, such that the derivative reads as

$$\Psi'_{u,p}(u,p)(v,q) = \sum_{n=1}^{N} \langle (\mathcal{OS}(u,p))_n - y_{\Omega,n}, \mathrm{D}(\mathcal{OS})(u,p)(v,q) \rangle_{L^2}$$

or, equivalently, denoting by $y = \mathcal{S}(u,p)$,

$$\Psi'_{u,p}(u,p) = \mathrm{D}(\mathcal{OS})(u,p)^*(\mathcal{O}y - y_\Omega) .$$

Since the control-to-observation mapping $\mathcal{OS}$ is linear and continuous with respect to the first argument, $\frac{\partial \mathcal{OS}}{\partial u}(u,p) = \mathcal{OS}_p(\,\cdot\,,0)$ is also a linear and continuous mapping, just depending on p. According to Proposition 6.12, we have $(\mathcal{OS}_p)^*(\mathcal{O}y - y_\Omega) = (v, v(0))$ where v solves the backwards parabolic equation (6.8) with the observation error $\mathcal{O}y - y_\Omega$ as data. So $\frac{\partial \mathcal{OS}}{\partial u}(u,p)^*(\mathcal{O}y - y_\Omega) = v$ in $L^2(0,T;L^2(\Omega))$, since we do not control the initial value y_0 of the evolution equation and have to discard $v(0)$. Eventually, Proposition 6.14 yields that

$$\frac{\partial \mathcal{OS}}{\partial p}(u,p)^*(\mathcal{O}y - y_\Omega) = \frac{\sigma'(|p|)}{|p|^2}\partial_p y \partial_p v \frac{p}{|p|} + \frac{\sigma(|p|)}{|p|^2}\left(\partial_p v P_p \nabla y^{\mathrm{T}} + \partial_p y P_p \nabla v^{\mathrm{T}}\right)$$

which can be regarded as an element of $L^1(]0,T[\times \Omega, \mathbb{R}^d)$.

Turning towards Φ_u and $\Phi_{p,0}$, standard differentiation rules give

$$\Phi'_u(u) = 2 \sum_j \omega_j \langle u(\,\cdot\,), \varphi_j \rangle \varphi_j$$

while $\Phi_{p,0}$ is a composition of the superposition $p \mapsto \sigma(|p|)$ mapping between the spaces $L^\infty(]0,T[\times \Omega, \mathbb{R}^d) \to L^\infty(]0,T[\times \Omega)$ and the integration over $]0,T[\times \Omega$. With the help of Lemma 5.24, this leads to

$$\Phi'_{p,0}(p)q = \int_0^T \int_\Omega \sigma'(|p(t,x)|) \frac{p(t,x)}{|p(t,x)|} \cdot q(t,x) \, \mathrm{d}x \, \mathrm{d}t$$

meaning that

$$\Phi'_{p,0}(p) = \frac{\sigma'(|p|)}{|p|} p$$

in points where $p \neq 0$ and 0 by continuous extension of $\frac{\sigma'(|p|)}{|p|}$ in points where $p = 0$, giving an element of the space $L^1(]0, T[\times \Omega, \mathbb{R}^d)$ regarded as a subspace of $L^\infty(]0, T[\times \Omega, \mathbb{R}^d)^*$. ◀

Beside the smooth parts, the penalization of p in (6.4) moreover involves non-smooth functionals. On the one hand, one has the convex restrictions $\|p\|_{\mathcal{Y}_d^*} \leq 1$ and $p \in \mathcal{P}_\partial$ expressed with indicator functionals, while on the other hand, we penalized with the sum of $\mathrm{tv}^*(p)$ and the $\mathrm{ess\,sup}_{t \in]0,T[} \mathrm{TV}(\nabla p(t))$ which is lower semi-continuous and positively one-homogeneous. These restrictions/functionals are mainly put in for reasons of regularization, as without them, we could not ensure the existence of minimizers, especially for the degenerate case $\|p\|_\infty = 1$. But the degenerate case is exactly the problem when proving first-order necessary conditions, since there is a lack of differentiability of the solution operator with respect to p. So, by assuming that for an optimal pair (u^*, p^*) the vector field p^* satisfies $\|p^*\|_\infty < 1$, we also can say that there was no necessity of regularizing so much. And since we originally aimed at solving (2.7), we go back to this functional for a moment and assume that (u^*, p^*) with $\|p\|_\infty < 1$ is an optimal solution of (2.7). For this setting, the first-order necessary conditions turn out to be a set of nice equations. After that, we derive extensions which take the non-smooth parts into account, introducing some additional terms of the first-order necessary conditions.

The following proposition establishes first-order necessary conditions for the non-degenerate and unregularized case (2.7), i.e. essentially, the objective functional (6.4) without the restriction $\|p\|_{\mathcal{Y}_d^*} \leq 1$ and the total-variation functionals as well as $\omega_j = \frac{1}{2}$ is considered. Additionally, the optimal p^* has to satisfy $\|p^*\|_\infty < 1$ and the space $L^\infty(]0, T[\times \Omega, \mathbb{R}^d)$ is chosen for p.

Proposition 6.17. *Let the situation of Lemma 6.16 be given, $q = 2$ and $\omega_j = \frac{1}{2}$ but consider the optimization problem*

$$\min_{\substack{u \in L^2(0,T;L^2(\Omega)) \\ p \in L^\infty(]0,T[\times\Omega,\mathbb{R}^d) \\ \|p\|_\infty \leq 1}} \sum_{n=1}^{N} \frac{\|y(t_n) - y_{n,\Omega}\|_2^2}{2} + \frac{\lambda_u}{2} \int_0^T \|u(t)\|_2^2 \, dt$$

$$+ \lambda_p \int_0^T \int_\Omega \sigma\big(|p(t,x)|\big) \, dx \, dt \, , \quad (6.11)$$

where $y = \mathcal{S}(u,p)$. Assume that a solution (u^*, p^*) of this control problem is such that $\|p^*\|_\infty < 1$.

Then, (u^*, p^*) have to satisfy the following optimality system:

$$\frac{\partial y^*}{\partial t} = \operatorname{div}\left(D_{p^*}^2 \nabla y^{*\mathrm{T}}\right) + u^* \;, \quad \frac{\partial y^*}{\partial_{p^*}\nu} = 0 \;, \quad y^*(0) = y_0$$

$$primal\ equation$$

$$\frac{\partial v^*}{\partial t} = -\operatorname{div}\left(D_{p^*}^2 \nabla v^{*\mathrm{T}}\right) \;, \qquad \frac{\partial v^*}{\partial_{p^*}\nu} = 0 \;,$$

$$v_-^*(t_n) = v_+^*(t_n) + y^*(t_n) - y_{n,\Omega} \qquad n = 1,\dots,N$$

$$adjoint\ equation$$

$$\begin{aligned}
u^* &= -\lambda_u v^* & in\ L^2\left(0,T;L^2(\Omega)\right) \\
\partial_{p^*} v^* P_{p^*} \nabla y^{*\mathrm{T}} &= -\partial_{p^*} y^* P_{p^*} \nabla v^{*\mathrm{T}} & in\ L^1\left(]0,T[\times\Omega,\mathbb{R}^d\right) \\
\partial_{p^*} y^* \partial_{p^*} v^* &= -\lambda_p |p^*|^2 & in\ L^1\left(]0,T[\times\Omega\right)
\end{aligned}$$

$$necessary\ optimality\ conditions$$

(6.12)

Proof. Note that (6.11) amounts to minimizing the objective functional $F = \Psi_{u,p} + \lambda_u \Phi_u + \lambda_p \Phi_{p,0}$ according to Lemma 6.16. Applying this lemma, we conclude that F is differentiable at the minimizer and regarded in the space $L^2\left(0,T;L^2(\Omega)\right) \times L^\infty(]0,T[\times\Omega,\mathbb{R}^d)$. Hence, it is necessary that $F'(u^*,p^*) = 0$ in the respective dual space. Splitting the components of F' into the parts with respect to u and p as well as using the formulas (6.10), this amounts to

$$v^* + \lambda_u u^* = 0$$

$$\frac{\sigma'(|p^*|)}{|p^*|^2} \partial_{p^*} y^* \partial_{p^*} v^* \frac{p^*}{|p^*|} + \frac{\sigma(|p^*|)}{|p^*|^2}\left(\partial_{p^*} v^* P_{p^*} \nabla y^{*\mathrm{T}} + \partial_{p^*} y^* P_{p^*} \nabla v^{*\mathrm{T}}\right)$$

$$+\lambda_p \frac{\sigma'(|p^*|)}{|p^*|} p^* = 0$$

with $y^* = \mathcal{S}(u^*,p^*)$ and v^* denoting the weak solution of the adjoint equation (6.8) with data $z^* = \mathcal{O}y^* - y_\Omega$ (of course, in points where $p^* = 0$, the respective functions are continuously extended by 0 again). Note that the equation for u^* has to be interpreted in the space $L^2\left(0,T;L^2(\Omega)\right)$ while the equation for p^* can be stated in $L^1(]0,T[\times\Omega,\mathbb{R}^d)$, since both $\frac{\partial \Psi_{u,p}}{\partial p}$ and $\frac{\partial \Phi_{p,0}}{\partial p}$ are contained in $L^1(]0,T[\times\Omega,\mathbb{R}^d)$, see Lemma 6.16 as well as Proposition 6.14.

This observation moreover allows to consider the equation for p^* pointwise almost everywhere. Then, since σ fulfills Condition 5.21, for almost every point where $p^* = 0$, the equation yields no information. Therefore, we consider points where $p^* \neq 0$ in the following. Note that for each $p \in \mathbb{R}^d$ and $p \neq 0$, the vectors p and each $P_p q$ with $q \in \mathbb{R}^d$ are orthogonal. Hence one can split the equation almost everywhere into parts with respect to span$\{p\}$ and rg(P_p). The first-order necessary condition for almost every point where $p^* \neq 0$ then becomes

$$\left(|p^*|^{-2} \partial_{p^*} y^* \partial_{p^*} v^* + \lambda_p\right) \frac{\sigma'(|p^*|)}{|p^*|} p^* = 0$$

$$\frac{\sigma(|p^*|)}{|p^*|^2} \left(\partial_{p^*} v^* P_{p^*} \nabla y^{*T} + \partial_{p^*} y^* P_{p^*} \nabla v^{*T}\right) = 0 \ .$$

In almost every point $(t, x) \in \,]0, T[\times \Omega$ where $p^*(t, x) \neq 0$, this is equivalent to (remember that σ is strictly monotone increasing)

$$\partial_{p^*} y^*(t, x) \partial_{p^*} v^*(t, x) = -\lambda_p |p^*(t, x)|^2$$

$$\partial_{p^*} v^*(t, x) P_{p^*} \nabla y^{*T}(t, x) = -\partial_{p^*} y^*(t, x) P_{p^*} \nabla v^{*T}(t, x)$$

where both equations are also fulfilled if $p^*(t, x) = 0$. Going back from the pointwise a.e. formulation to the respective function space gives the optimality system as stated in (6.12).　　　◀

Remark 6.18.

(a) The parts of the first-order optimality conditions (6.12) which are hard to solve are clearly the equations involving p^*. Once a p^* of a pair (u^*, p^*) fulfilling these conditions is known, the $u^* \in L^2\left(0, T; L^2(\Omega)\right)$ can be obtained by solving a linear equation:

$$y^* = S_{p^*}(u^*, y_0) \quad , \quad v^* = (\mathcal{O}S_{p^*})_1^*(\mathcal{O}y^* - y_\Omega) \quad , \quad v^* + \lambda_u u^* = 0$$

where $(\mathcal{O}S_{p^*})_1^*$ denotes taking the first component of the adjoint of $\mathcal{O}S_{p^*}$, see Proposition 6.12. This is equivalent to

$$\left(\lambda_u I + (\mathcal{O}S_{p^*})_1^* \mathcal{O}S_{p^*}(\,\cdot\,, 0)\right) u^* = (\mathcal{O}S_{p^*})_1^*\left(y_\Omega - \mathcal{O}S_{p^*}(0, y_0)\right) \ ,$$

a linear equation in $L^2\left(0, T; L^2(\Omega)\right)$. It is easy to see that the linear operator involved is continuous and positive definite, hence by the Lax-Milgram theorem we know that there is a solution operator which is moreover linear and continuous. By plugging this into the first-order optimality system, the problem can be reduced to solving an equation only for p^*, which is unfortunately even more complicated.

(b) Observe that, regardless of the data $y_0, \ldots, y_N$, there is always a u^* for which $(u^*, 0)$ satisfies (6.12), confer the above. This amounts to isotropic diffusion in each point $(t, x) \in \,]0, T[\, \times \Omega$ which is certainly not always a solution of (6.11) for general data $y_0, \ldots, y_N$. So, the first-order necessary conditions are in general not sufficient of optimality.

Beside the smooth parts, the penalization of p in (6.4) moreover involves non-smooth functionals. On the one hand, one has the convex restrictions $\|p\|_{\mathcal{Y}_d^*} \le 1$ and $p \in \mathcal{P}_\partial$ expressed by indicator functionals, while on the other hand, we penalized with the sum of $\operatorname{tv}^*(p)$ and the $\operatorname{ess\,sup}_{t \in]0,T[} \operatorname{TV}(\nabla p(t))$ which is lower semi-continuous and positively one-homogeneous. The subgradient for indicator functional $I_{\{\|p\|_{\mathcal{Y}_d^*} \le 1\}}$ therefore consists of vectors which can be interpreted as the respective set of normal vectors and while the latter functional has a nice expression involving some closed and convex polar sets. Therefore, we aim at describing the respective subgradient in terms of these notions.

Note that in order to be complete, we also have to investigate the subgradient of the functional $I_{\mathcal{P}_\partial}$ which has its origin in the condition that only vector fields which do not degenerate at the boundary are allowed. But, again since the resulting first-order necessary conditions can only be formulated for $\|p\|_\infty < 1$ and the parameter $M_\partial \in \,]0,1[$ which is involved in the definition of $\mathcal{P}_\partial$ can be arbitrarily close to 1, we can also assume that an optimal p^* satisfies $1 > M_\partial > \|p^*\|_\infty$, so the condition does not apply in a neighborhood of p^* and thus has no influence on the first-order necessary conditions. We will therefore drop the requirement $p \in \mathcal{P}_\partial$ in the following.

Lemma 6.19. *The subgradient of $I_{\{\|p\|_{\mathcal{Y}_d^*} \le 1\}}$ in $\mathcal{Y}_d^*$ at a $p \in \mathcal{Y}_d^*$ with $\|p\|_{\mathcal{Y}_d^*} \le 1$ is given by*

$$\partial I_{\{\|p\|_{\mathcal{Y}_d^*} \le 1\}}(p) = \{q \in \mathcal{Y}_d^{**} \mid \langle q, p \rangle = \|q\|_{\mathcal{Y}_d^{**}}\}$$

in all other points, it is empty.

Moreover, the subgradient of

$$\Phi_{\mathrm{TV}}(p) = \mu_1 \operatorname{tv}^*(p) + \mu_2 \operatorname*{ess\,sup}_{t \in]0,T[} \operatorname{TV}(\nabla p(t))$$

as a functional in $\mathcal{Y}_d^*$ is given by

$$\partial \Phi_{\mathrm{TV}}(p) = \{q \in C_{\mathrm{TV}}^{00} \mid \langle q, p \rangle = \Phi_{\mathrm{TV}}(p)\}$$

where C_{TV}^{00} is the (right) bipolar (in $\mathcal{Y}_d^{**}$) of the set $C_{\mathrm{TV}} \subset \mathcal{Y}_d$ defined by

$$C_{\mathrm{TV}} = \{q \in \mathcal{Y}_d \mid q(t,x) = q_1'(t)q_2(x), \ q_1 \in C_0^1(]0,T[), \ \|q_1\|_\infty \le \mu_1,$$
$$q_2 \in Y_d, \ \|q_2\|_{Y_d} \le 1\}$$
$$+ \{q \in \mathcal{Y}_d \mid q = \operatorname{div} \operatorname{div} \bar q, \ \bar q \in L^1(0,T; C_0^2(\Omega, \mathbb{R}^{d \times d \times d})),$$
$$\int_0^T \|\bar q(t)\|_\infty \, dt \le \mu_2\}, \quad (6.13)$$

where $(\operatorname{div}\operatorname{div}\bar{q})_k = \sum_{l=1}^{d} \frac{\partial}{\partial x_l}\left(\sum_{m=1}^{d} \frac{\partial \bar{q}_{k,l,m}}{\partial x_m}\right).$

Proof. That the subgradient of $I_{\{\|p\|_{\mathcal{Y}_d^*} \leq 1\}}$ is given by the claimed formula is a consequence of the Fenchel identity in convex analysis, see [ET76]. A $q \in \mathcal{Y}_d^{**}$ satisfies $q \in \partial I_{\{\|p\|_{\mathcal{Y}_d^*} \leq 1\}}(p)$ if and only if

$$\langle q, p \rangle = I_{\{\|p\|_{\mathcal{Y}_d^*} \leq 1\}}(p) + I^*_{\{\|p\|_{\mathcal{Y}_d^*} \leq 1\}}(q)$$

where $I^*_{\{\|p\|_{\mathcal{Y}_d^*} \leq 1\}}$ denotes the conjugate functional of $I_{\{\|p\|_{\mathcal{Y}_d^*} \leq 1\}}$, i.e.

$$I^*_{\{\|p\|_{\mathcal{Y}_d^*} \leq 1\}}(q) = \sup_{\|p\|_{\mathcal{Y}_d^*} \leq 1} \langle q, p \rangle = \|q\|_{\mathcal{Y}_d^{**}}.$$

This proves the first part.

To prove the subgradient identity for Φ_{TV}, one can again employ the Fenchel identity which states that a $q \in \partial \Phi_{\mathrm{TV}}(p)$ if and only if

$$\langle q, p \rangle = \Phi_{\mathrm{TV}}(p) + \Phi^*_{\mathrm{TV}}(q).$$

The desired statement will be implied by proving that $\Phi^*_{\mathrm{TV}} = I_{C_{\mathrm{TV}}^{00}}$ with C_{TV} as defined in (6.13). But the conjugate functional of Φ_{TV} is given by

$$\Phi^*_{\mathrm{TV}}(q) = \sup_{p \in \mathcal{Y}_d^*} \langle q, p \rangle - \Phi_{\mathrm{TV}}(p)$$

which only attains 0 and ∞, since Φ_{TV} is positively one-homogeneous. Thus, it is an indicator functional and it is easy to see that

$$\Phi^*_{\mathrm{TV}}(q) = I_{C_{\mathrm{TV}}^{**}}(q) \ , \quad C_{\mathrm{TV}}^{**} = \{q \in \mathcal{Y}_d^{**} \mid \langle q, p \rangle \leq \Phi_{\mathrm{TV}}(p) \text{ for all } p \in \mathcal{Y}_d^*\}.$$

By the one-homogeneity of the dual pairing, this means

$$C_{\mathrm{TV}}^{**} = \{q \in \mathcal{Y}_d^{**} \mid \langle q, p \rangle \leq 1 \text{ for all } \Phi_{\mathrm{TV}}(p) \leq 1\} = \{\Phi_{\mathrm{TV}}(p) \leq 1\}^0 ,$$

i.e. C_{TV}^{**} is the polar set of $\{\Phi_{\mathrm{TV}}(p) \leq 1\}$. So, it suffices to show that $C_{\mathrm{TV}}^0 = \{\Phi_{\mathrm{TV}}(p) \leq 1\}$, or equivalently, $I^*_{C_{\mathrm{TV}}} = \Phi_{\mathrm{TV}}$. For this purpose, we introduce

$$C_{\mathrm{tv}^*} = \{q \in \mathcal{Y}_d \mid q(t,x) = q_1'(t)q_2(x) \ , \ q_1 \in C_0^1(]0,T[), \|q_1\|_\infty \leq 1 \ ,$$
$$q_2 \in Y_d \ , \ \|q_2\|_{Y_d} \leq 1\}$$
$$C_{\mathrm{TV}\nabla} = \{q \in \mathcal{Y}_d \mid q = \operatorname{div}\operatorname{div}\bar{q} \ , \ \bar{q} \in L^1(0,T;C_0^2(\Omega,\mathbb{R}^{d \times d \times d})) \ ,$$
$$\int_0^T \|\bar{q}(t)\|_\infty \, \mathrm{d}t \leq 1\}$$

and observe that $I^*_{C_{TV}} = \mu_1 I^*_{C_{tv^*}} + \mu_2 I^*_{C_{TV}\nabla}$. Our aim is to obtain $I^*_{C_{tv^*}} = tv^*$ as well as $I^*_{C_{TV}\nabla}(p) = \text{ess sup}_{t\in]0,T[}\ TV(\nabla p(t))$. Recall that the total variation of a measurable function $p_1 :]0,T[\to \mathbb{R}$ can also be written as

$$TV(p_1) = \sup\left\{ \int_0^T q_1'(t)p_1(t)\ dt \ \Big| \ q_1 \in C_0^1(]0,T[)\ ,\ \|q_1\|_\infty \le 1 \right\}$$

so, according to Definition 5.4,

$$tv^*(p) = \sup_{\substack{q_2\in Y_d \\ \|q_2\|_{Y_d}\le 1}} \sup_{\substack{q_1\in C_0^1(]0,T[) \\ \|q_1\|_\infty\le 1}} \int_0^T q_1'(t) \int_\Omega p(t,x)\cdot q_2(x)\ dx\ dt$$

$$= \sup_{q\in C_{tv^*}} \langle p,\,q\rangle = I^*_{C_{tv^*}}(p)\ .$$

Proving the identity for $I^*_{C_{TV}\nabla}$ is technically a little more complicated but resembles the proof of the $L^1 \times L^\infty$-duality. First observe that for a $p \in Y_d^*$ and $q = \text{div div}\ \bar{q}$ with $\bar{q} \in C_0^2(]0,T[\times \Omega, \mathbb{R}^{d\times d\times d})$ we have

$$\langle p,\,q\rangle = \int_0^T \int_\Omega p(t,x)\cdot \text{div div}\ \bar{q}(t,x)\ dx\ dt$$

$$= -\int_0^T \int_\Omega \nabla p(t,x)\cdot \text{div}\ \bar{q}(t,x)\ dx\ dt \le \int_0^T TV(\nabla p(t))\|\bar{q}(t)\|_\infty\ dt$$

$$\le \text{ess sup}_{t\in]0,T[} TV(\nabla p(t))$$

so we have to construct a sequence for which the supremum is attained. Note that the difficulty here is the lack of a sign function, i.e. we cannot guarantee the existence of a mapping $\text{sgn} : Y_d^* \to Y_d$ such that $\langle p,\,\text{sgn}(p)\rangle = \|p\|_{Y_d^*}$ since Y_d is not reflexive.

Choose a sequence $\{q_{2,l}\}$ in Y_d (and corresponding $\bar{q}_l$ in $C_0^2(\Omega, \mathbb{R}^{d\times d\times d})$) which is dense in

$$\{q \in Y_d \mid q = \text{div div}\ \bar{q}\ ,\ \bar{q} \in C_0^2(\Omega, \mathbb{R}^{d\times d\times d})\ ,\ \|\bar{q}_l\|_\infty \le 1\}\ .$$

This is possible since $C_0^2(\Omega, \mathbb{R}^{d\times d\times d})$ is separable and div div maps into Y_d with dense image. Moreover, without loss of generality one can always assume $q_{2,1} = 0$. Now construct the following sequence of measurable $v_m :]0,T[\to \mathbb{R}$:

$$v_m(t) = \sup_{l=1,\ldots,m} \int_\Omega p(t)\cdot q_{2,l}\ dx$$

Note that $\{v_m\}$ is a monotone increasing sequence of a.e. non-negative functions. Define the sets

$$S_{m,0} = \emptyset \quad , \quad S_{m,l} = \{t \in]0, T[\mid \langle p(t), q_{2,l} \rangle = v_m(t)\} \backslash S_{m,l-1} .$$

Since the supremum is taken over finitely many numbers, the definition

$$\tilde{q}_m(t, x) = \sum_{l=1}^{m} \chi_{S_{m,l}}(t) q_{2,l}(x)$$

yields that $\langle p(t), \tilde{q}_m(t) \rangle = v_m(t)$ for each $m \geq 1$ and almost every $t \in]0, T[$.

Consider ess $\sup_{t \in]0,T[} v_m(t) = \|v_m\|_\infty$ which is also monotone increasing and converging to ess $\sup_{t \in]0,T[} TV(\nabla p(t))$ a.e. by construction. Of course, for each m there is a $q_{1,m} \in L^1(]0, T[)$ such that $\|q_{1,m}\|_1 \leq 1$ and $\langle q_{1,m}, v_m \rangle \to$ ess $\sup_{t \in]0,T[} TV(\nabla p(t))$ as a consequence of the $L^1 \times L^\infty$-duality.

Set $q_m(t, x) = q_{1,m}(t) \tilde{q}_m(t, x)$ for we can verify that for a.e. $t \in]0, T[$ we have $q_m(t) = \text{div div}(q_{1,m}(t) \bar{q}_l)$ for some $1 \leq l \leq m$. Moreover, each $\bar{q}_l$ satisfies $\|\bar{q}_l\|_\infty \leq 1$, hence $q_m \in C_{TV\nabla}$ and

$$\int_0^T \int_\Omega q_m(t, x) \cdot p(t, x) \, dx \, dt = \int_0^T q_{1,m}(t) v_m(t) \, dx \to \underset{t \in]0,T[}{\text{ess sup}} \; TV(\nabla p(t))$$

as $m \to \infty$, what was to show. ◀

Finally, with the help of the preceding lemma, it is possible to give a description of the first-order necessary conditions for minimizers of (6.4) for the case where $\|p^*\|_\infty < M_\partial < 1$.

Theorem 6.20. *In the situation of Theorem 6.6 and $r = 2$ as well as $\omega_j = \frac{1}{2}$, assume that there is an optimal solution (u^*, p^*) of (6.4) such that $\|p^*\|_\infty < M_\partial < 1$. Then, (u^*, p^*) satisfies the following optimality system:*

$$\frac{\partial y^*}{\partial t} = \text{div}\left(D_{p^*}^2 \nabla y^{*T}\right) + u^* , \quad \frac{\partial y^*}{\partial_{p^*} \nu} = 0 , \quad y^*(0) = y_0$$

primal equation

$$\frac{\partial v^*}{\partial t} = -\text{div}\left(D_{p^*}^2 \nabla v^{*T}\right) , \quad \frac{\partial v^*}{\partial_{p^*} \nu} = 0 ,$$

$$v_-^*(t_n) = v_+^*(t_n) + y^*(t_n) - y_{n,\Omega} \qquad n = 1, \ldots, N$$

adjoint equation

$$u^* = -\lambda_u v^* \quad in\ L^2\big(0, T; L^2(\Omega)\big)$$

$$q_{\parallel}^* = \frac{\sigma'(|p^*|)}{|p^*|^2}\big(\partial_{p^*} y^* \partial_{p^*} v^* + \lambda_p |p^*|^2\big)\frac{p^*}{|p^*|}$$

$$q_{\perp}^* = \frac{\sigma(|p^*|)}{|p^*|^2}\big(\partial_{p^*} v^* P_{p^*} \nabla y^{*\mathrm{T}} + \partial_{p^*} y^* P_{p^*} \nabla v^{*\mathrm{T}}\big)$$

$$q_I^* \in \mathcal{Y}_d^{**}: \quad \langle q_I^*, p^* \rangle = \|q_I^*\|_{\mathcal{Y}_d^{**}}$$

$$q_{\mathrm{TV}}^* \in C_{\mathrm{TV}}^{00}: \quad \langle q_{\mathrm{TV}}^*, p^* \rangle = \mu_1 \operatorname{tv}^*(p^*) + \mu_2 \operatorname*{ess\,sup}_{t \in]0,T[} \operatorname{TV}\big(\nabla p^*(t)\big)$$

$$-q_{\parallel}^* - q_{\perp}^* = q_I^* + q_{\mathrm{TV}}^* \quad in\ \mathcal{Y}_d^{**}$$

necessary optimality conditions

Proof. First note that the objective functional F can be written as the sum of a smooth part $F_1 = \Psi_{u,p} + \lambda_u \Phi_u + \lambda_p \Phi_{p,0}$ according to Lemma 6.16 and a non-smooth convex part $F_2 = I_{\{\|p\|_{\mathcal{Y}_d^*} \leq 1\}} + \Phi_{\mathrm{TV}} + I_{\mathcal{P}_\partial}$ according to Lemma 6.19. Now, if one has a minimizer (u^*, p^*) with $\|p^*\|_\infty < M_\partial < 1$, the minimization property yields, for each $(u, p) \in L^2\big(0, T; L^2(\Omega)\big) \times \mathcal{Y}_d^*$,

$$0 \leq \Big(F_1\big(u^* + t(u - u^*), p^* + t(p - p^*)\big) - F_1(u^*, p^*)\Big) + F_2\big(p^* + t(p - p^*)\big) - F_2(p^*)$$

$$\leq \Big(F_1\big(u^* + t(u - u^*), p^* + t(p - p^*)\big) - F_1(u^*, p^*)\Big) + t\big(F_2(p) - F_2(p^*)\big)$$

for $t \in]0, 1]$ by convexity of F_2. Diving by t and letting $t \to 0$ then gives, since F_1 is differentiable (see Lemma 6.16),

$$0 \leq F_1'(u^*, p^*)(u - u^*, p - p^*) + F_2(p) - F_2(p^*)$$

$$\Leftrightarrow \quad F_2(p^*) - \frac{\partial F_1}{\partial u}(u^*, p^*)(u - u^*) - \frac{\partial F_1}{\partial p}(u^*, p^*)(p - p^*) \leq F_2(p)$$

for all $(u, p) \in L^2\big(0, T; L^2(\Omega)\big) \times \mathcal{Y}_d^*$, which is in turn equivalent to

$$\frac{\partial F_1}{\partial u}(u^*, p^*) = 0 \quad , \quad -\frac{\partial F_1}{\partial p}(u^*, p^*) \in \partial F_2(p^*) .$$

Again, with the notation $y^* = \mathcal{S}(u^*, p^*)$ and v^* denoting the weak solution of the adjoint problem (6.8) with data $\mathcal{O}y^* - y_\Omega$, the equation $\frac{\partial F_1}{\partial u}(u^*, p^*) = 0$ amounts to

$$u^* + \lambda_u v^* = 0 ,$$

see also Proposition 6.17 and (6.10).

To give a description of $-\frac{\partial F_1}{\partial p}(u^*, p^*) \in \partial F_2(p^*)$, we first examine the subgradient of F_2: Observe that $I_{\{\|p\|_{\mathcal{Y}_d^*} \leq 1\}}$ as well as $I_{\mathcal{P}_\partial}$ is continuous in 0 as they are indicator functionals for neighborhoods of 0, respectively. Furthermore, $\Phi_{\text{TV}}(0) = 0 < \infty$, hence we can apply the identity $\partial F_2 = \partial I_{\{\|p\|_{\mathcal{Y}_d^*} \leq 1\}} + \partial \Phi_{\text{TV}} + \partial I_{\mathcal{P}_\partial}$, also see [Sho96]. In particular, since $\|p^*\|_\infty < M_\partial$, $\partial I_{\mathcal{P}_\partial}(p^*) = \{0\}$, so with the help of Lemma 6.19 we obtain that $\partial F_2(p^*) = \{q_I^* + q_{\text{TV}}^*\}$ with

$$q_I^* \in \mathcal{Y}_d^{**}: \quad \langle q_I^*, p^* \rangle = \|q_I^*\|_{\mathcal{Y}_d^{**}},$$
$$q_{\text{TV}}^* \in C_{\text{TV}}^{00}: \quad \langle q_{\text{TV}}^*, p^* \rangle = \mu_1 \operatorname{tv}^*(p^*) + \mu_2 \operatorname*{ess\,sup}_{t \in]0,T[} \operatorname{TV}\big(\nabla p^*(t)\big) \qquad (*)$$

where C_{TV}^{00} is the bipolar of C_{TV} according to (6.13). Now, by examining (6.10), one can easily convince oneself that $\frac{\partial F_1}{\partial p}(u^*, p^*) = q_{\|}^* + q_\perp^*$ with

$$q_{\|}^* = \frac{\sigma'(|p^*|)}{|p^*|^2}\big(\partial_{p^*} y^* \partial_{p^*} v^* + \lambda_p |p^*|^2\big)\frac{p^*}{|p^*|},$$
$$q_\perp^* = \frac{\sigma(|p^*|)}{|p^*|^2}\big(\partial_{p^*} v^* P_{p^*} \nabla y^{*\mathrm{T}} + \partial_{p^*} y^* P_{p^*} \nabla v^{*\mathrm{T}}\big).$$

The condition $-\frac{\partial F_1}{\partial p}(u^*, p^*) \in \partial F_2(p^*)$ then amounts to the existence of q_I^* and q_{TV}^* according to $(*)$, such that $-q_{\|}^* - q_\perp^* = q_I^* + q_{\text{TV}}^*$. This constitutes the stated necessary conditions. ◀

Chapter 7

Numerical realization

This chapter is concerned with the problem of treating the optimal control problem (2.8) numerically. This task includes many aspects which have to be taken into account. First, there are two common approaches to the numerical treatment of infinite-dimensional optimization problems: "first discretize, then optimize" and "first optimize, then discretize". The former means that the infinite-dimensional problem is reduced to a discretized finite-dimensional problem which is then optimized by a finite-dimensional numerical optimization algorithm, while in the latter, an infinite-dimensional optimization algorithm, such as gradient descent or sequential quadratic programming, is implemented for which the involved continuous objects (e.g. the derivative of the functional, the solutions of linear subproblems) are discretized whenever there is the need to. Unfortunately, such algorithms often involve the derivative of the objective functional which does not necessarily exist for the problem (2.8), see Section 6.2. Therefore, we will follow the "first discretize, then optimize" approach.

Since the optimal control problem is an optimization problem with the degenerate partial differential equation (2.4), one has to deal with the discretization of this PDE in addition to the discretization of the control variables. As we have shown in the previous chapters, the solution space $W_p(0, T)$ depends on p. Moreover, it is not a Bochner space, i.e. the time-variant space $\mathcal{V}_p$ containing functions $y(t)$ may involve $V_{p(t)}$ which are different for each $t \in [0, T]$. For reasons of simplicity at this point, we use finite differences to discretize the PDE, in contrast to a Galerkin approach such as finite elements on triangular meshes, for instance. This leads to a discretized solution operator which can examined with respect to the data u and p quite easily. Also, we restrict ourselves to $d = 2$ and rectangular domains $]0, M_1[\times]0, M_2[$ which amounts to the case of images.

Furthermore, as the use of finite differences for the discretization of (2.4) suggests, we will also approximate the controls u and p on an equidistant point mesh. This will automatically imply a bounded approximation of the Jacobian ∇p, making the part $\|\nabla p\|_\infty \leq M_x^{-1}$, which is a consequence of $\|p\|_{y_d^*} \leq 1$ (see Propositions 5.2 and 4.3), unnecessary. Additionally, the requirement that p is non-degenerate at the boundary and the terms $\mathrm{tv}^*(p)$ as well as $\mathrm{ess\,sup}_{t\in]0,T[} \, \mathrm{TV}(\nabla p(t))$ in the penalization for p will also be dropped, since they only serve as a device leading to the well-posedness of the infinite-dimensional problem and becomes redundant in the discretized case.

Finally, note that the following is only intended to show up the possibility of computing approximate solutions for the problem (2.8). Convergence as the step-size h goes to 0 will not be considered, as such an analysis would be beyond the scope of this work. Moreover, rather simple discretizations of low order are chosen as higher-order schemes would imply more complex and more technical computations which may obstruct the ideas of the implementation.

7.1 Discretization of the PDE with finite differences

The following section describes how linear degenerate parabolic equations of the type (2.4) can be discretized and solved with the help of finite differences. It begins with setting up the grid as well as the notation for the discretized variables y, u, p. Note that in the following, y_k is often denoting the discrete solution of the PDE and not related to the images which have to be controlled.

Definition 7.1. Let $d = 2$, $T > 0$, $\Omega = \,]0, M_1[\, \times \,]0, M_2[$ with $M_1, M_2 > 0$. Choose $0 < h < \min\{M_1, M_2\}$ and set $\tau = T/K$ for a $K \in \mathbb{N}$, $K \geq 1$. Then define the space of discretized functions on the rectangle Ω as

$$R_h = \{\{y_{i,j}\} \mid y_{i,j} \in \mathbb{R} \,,\, 0 < ih < M_1 \,,\, 0 < jh < M_2\}$$

where $y_{i,j}$ corresponds to the value of y at $(ih, jh) \in \Omega$. Denote by I and J the largest integers for which $ih < M_1$ and $jh < M_2$, respectively. A scalar product and norm on R_h is given by

$$\langle y, z \rangle_h = h^2 \sum_{j=1}^{J} \sum_{i=1}^{I} y_{i,j} z_{i,j} \quad , \quad \|y\|_h = \left(h^2 \sum_{j=1}^{J} \sum_{i=1}^{I} y_{i,j}^2 \right)^{1/2} . \tag{7.1a}$$

Additionally, we agree to use the following absolute value and scalar product for $y \in R_h^{d'}$ with $d' \geq 1$:

$$|y_{i,j}| = \left(\sum_{l=1}^{d'} y_{l,i,j}^2 \right) \quad , \quad \langle y, z \rangle_h = h^2 \sum_{j=1}^{J} \sum_{i=1}^{I} \sum_{l=1}^{d'} y_{l,i,j} z_{l,i,j} . \tag{7.1b}$$

The norm is then defined, as usual, as $\|y\|_h = \sqrt{\langle y, y \rangle_h}$.

Furthermore, define the space of discretized functions of the time/space rectangle $[0, T] \times \Omega$ as

$$\mathcal{R}_{\tau,h} = \{\{y_k\} \mid y_k \in R_h , \ k = 0, \dots, K\}$$

with the above τ. A y_k then corresponds to the discretized function y evaluated at $k\tau \in [0, T]$. The maximum norms for $y \in \mathcal{R}_{\tau,h}$ and $y \in \mathcal{R}_{\tau,h}^{d'}$ (which is given in analogy) are defined as

$$\|y\|_\infty = \max_{k,i,j} |y_{k,i,j}| . \tag{7.1c}$$

We also associate a scalar product with $\mathcal{R}_{\tau,h}$ and $\mathcal{R}_{\tau,h}^{d'}$:

$$\langle y, z \rangle_{\tau,h} = \tau \sum_{k=0}^{K} \langle y, z \rangle_h . \tag{7.1d}$$

In the following, we suppose that the parameters h, τ, I, J, K of the above definition are given. Then, the discretized solutions of (2.4) are sought in the space $\mathcal{R}_{\tau,h}$. The data of this equation $(\bar{y}_0, u, p)$ (we denote by $\bar{y}_0$ the initial value to avoid confusion with the discretized initial time-step y_0) can be split into the part $(\bar{y}_0, u)$ which belongs to the right-hand side of the equation and acts linearly and the part p which amounts to the control of the coefficients via the diffusion tensor D_p^2. Following the above approach, it makes sense to discretize $\bar{y}_0 \in R_h$ as well as $u_k \approx u(\tau k)$ with $u_k \in R_h$. We combine these two vectors to one $(\bar{y}_0, u) \in \mathcal{R}_{\tau,h}$ where

$$(\bar{y}_0, u)_k = \begin{cases} \bar{y}_0 & \text{if } k = 0 \\ u_k & \text{if } k \geq 1 . \end{cases}$$

Hence, we can also say that the initial condition $u_0 = \bar{y}_0$ is satisfied and speak of the right-hand side $u \in \mathcal{R}_{\tau,h}$. This motivates the following definition.

Definition 7.2. Let a discrete grid according to Definition 7.1 be given. Then define the space $\mathcal{U}_{\tau,h} = \mathcal{R}_{\tau,h}$ equipped with the scalar product

$$\langle y, z \rangle_{\tau,h,0} = \langle y_0, z_0 \rangle_h + \tau \sum_{k=1}^{K} \langle y_k, z_k \rangle_h . \tag{7.2}$$

Next, we derive a discretization of the differential operator mapping $y \mapsto -\operatorname{div} D_p^2 \nabla y^{\mathrm{T}}$. It will be useful to know the square root of the diffusion tensor, see (3.3), hence we restate the respective expression for the case where $d = 2$. For $p \in \mathbb{R}^2$ with $|p| \leq 1$ we have

$$
\begin{aligned}
D_p &= I - \left(1 - \sqrt{1 - \sigma(|p|)}\right) \frac{p}{|p|} \otimes \frac{p}{|p|} \\
&= \frac{1}{|p|^2} \begin{pmatrix} p_2^2 + \sqrt{1 - \sigma(|p|)} p_1^2 & |p|^2 - \left(1 - \sqrt{1 - \sigma(|p|)}\right) p_1 p_2 \\ |p|^2 - \left(1 - \sqrt{1 - \sigma(|p|)}\right) p_1 p_2 & p_1^2 + \sqrt{1 - \sigma(|p|)} p_2^2 \end{pmatrix},
\end{aligned}
\tag{7.3}
$$

which can also be applied pointwise for $p \in R_h^2$.

For a $y \in R_h$, we will utilize the most basic finite-difference approximation scheme for the gradient:

$$
(\nabla_h y)_{i,j}^{\mathrm{T}} = \frac{1}{h} \begin{pmatrix} y_{i+1,j} - y_{i,j} \\ y_{i,j+1} - y_{i,j} \end{pmatrix}
\tag{7.4}
$$

where we set $y_{I+1,j} = y_{I,j}$ and $y_{i,J+1} = y_{i,j}$, which is the usual way to incorporate the boundary conditions $\frac{\partial y}{\partial_p \nu} = \nabla y \cdot D_p^2 \nu = 0$ on $\partial \Omega$ (see [EBY99], for example). The discrete negative divergence $- \mathrm{div}_h$ will be defined as the adjoint of the discrete gradient operator which reads as

$$
(- \mathrm{div}_h z)_{i,j} = \frac{\bar{z}_{1,i-1,j} - \bar{z}_{1,i,j} + \bar{z}_{2,i,j-1} - \bar{z}_{2,i,j}}{h},
$$

$$
\bar{z}_{1,i,j} = \begin{cases} 0 & \text{if } i \in \{0, I\} \\ z_{1,i,j} & \text{if } 1 \leq i \leq I - 1 \end{cases}, \quad \bar{z}_{2,i,j} = \begin{cases} 0 & \text{if } j \in \{0, J\} \\ z_{2,i,j} & \text{if } 1 \leq i \leq J - 1 . \end{cases}
$$

Also denoting by D_p the linear operator assigning $z \mapsto D_p z$ pointwise for $z \in R_h^2$ and a fixed $p \in R_h^2$, we can set $A_{p,h} y = D_p \nabla_h y^{\mathrm{T}}$ which maps $R_h \to R_h^2$ and gives

$$
- \mathrm{div}_h D_p^2 \nabla_h y^{\mathrm{T}} = A_{p,h}^* A_{p,h} y
\tag{7.5}
$$

which we take as an approximation for $- \mathrm{div}(D_p^2 \nabla y^{\mathrm{T}})$. (Note that the factor h^2 in the definition of the scalar product in R_h and R_h^2 does not matter, one can take the usual adjoint.)

In the following step, our aim is to derive an approximation for the homogeneous equation $\frac{\partial y}{\partial t} - \mathrm{div} D_p^2 \nabla y^{\mathrm{T}} = 0$. For this purpose, we will use the discretization (7.5) as well as the approximation $\frac{\partial y}{\partial t}(\tau(k+1)) \approx \frac{1}{\tau}(y_{k+1} - y_k)$ for $k = 0, \ldots, K - 1$, leading to an implicit scheme with respect to time. Now, let $p \in R_{\tau,h}^2$ with $|p| \leq 1$ for each component. Then, the discretized version of (2.4) with $u = 0$ reads as follows:

$$
\frac{y_{k+1} - y_k}{\tau} - \mathrm{div}_h D_{p_{k+1}}^2 \nabla_h y_{k+1}^{\mathrm{T}} = 0 \quad, \quad y_0 = u_0 \quad, \quad k = 0, \ldots, K - 1 .
$$

With (7.5), this can be reformulated to give the following recursive system of linear equations:

$$y_0 = u_0 \quad , \quad (I + \tau A_{p_{k+1},h}^* A_{p_{k+1},h})y_{k+1} = y_k \quad , \quad k = 0, \ldots, K-1 \,.$$

Thus, solving the discrete problem according to the above definition amounts to the iteration (writing A_{k+1} short hand for $A_{p_{k+1},h}$)

$$y_0 = u_0$$
$$y_{k+1} = (I + \tau A_{k+1}^* A_{k+1})^{-1} y_k$$

which can be performed since, for each $y \in R_h$,

$$\langle (I + \tau A_{k+1}^* A_{k+1})y, \, y \rangle_h = \|y\|_2^2 + \tau \|A_{k+1} y\|_2^2 \geq \|y\|_2^2 \,,$$

meaning in particular that the linear operator is injective and thus invertible on the finite-dimensional space R_h.

By denoting

$$S_{k+1} = (I + \tau A_{k+1}^* A_{k+1})^{-1} \quad , \quad S_{k'}^k = \begin{cases} S_{k+1} S_k \ldots S_{k'+1} & \text{for } k' \leq k \\ I & \text{for } k' > k \end{cases} ,$$

we can write the solution operator for the homogeneous equation as

$$y_0 = u_0 \quad , \quad y_{k+1} = S_0^k y_0 \quad , \quad k = 0, \ldots, K-1 \,.$$

The solution of the inhomogeneous equation, i.e. $u \in R_{\tau,h}$ as right-hand side then can be derived from a discrete version of the variation of constants formula, giving

$$y_{k+1} = S_0^k \left(u_0 + \tau \sum_{k'=0}^{k} (S_0^{k'})^{-1} u_{k'+1} \right) = S_0^k u_0 + \tau \sum_{k'=0}^{k} S_{k'+1}^k u_{k'+1} \tag{7.6}$$

In block-matrix notation, this also reads as

$$\begin{pmatrix} y_0 \\ y_1 \\ y_2 \\ y_3 \\ \vdots \\ y_K \end{pmatrix} = \begin{pmatrix} I & 0 & 0 & 0 & \cdots & 0 \\ S_0^0 & \tau I & 0 & 0 & & 0 \\ S_0^1 & \tau S_1^1 & \tau I & 0 & & 0 \\ S_0^2 & \tau S_1^2 & \tau S_2^2 & \tau I & \cdots & 0 \\ \vdots & & & & \ddots & \vdots \\ S_0^{K-1} & \tau S_1^{K-1} & \tau S_2^{K-1} & \tau S_3^{K-1} & \cdots & \tau I \end{pmatrix} \begin{pmatrix} u_0 \\ u_1 \\ u_2 \\ u_3 \\ \vdots \\ u_K \end{pmatrix} . \tag{7.7}$$

An iterative formulation for (7.6) can also be derived easily. Since it will be used for the numerical computations, we will employ it as the basis of the definition of discrete solutions.

Definition 7.3. Let a grid according to Definition 7.1 and $u \in \mathcal{U}_{\tau,h}$, $p \in \mathcal{R}_{\tau,h}^2$ be given.

A $y \in \mathcal{R}_{\tau,h}$ is defined to be a *discrete solution* of the *discrete problem* (2.4) if

$$y_0 = u_0 \, ,$$
$$y_{k+1} = (I + \tau A_{k+1}^* A_{k+1})^{-1} y_k + \tau u_{k+1} \tag{7.8}$$

with $k = 0, \ldots, K-1$ and $A_{k+1} y = D_{p_{k+1}} \nabla_h y^{\mathrm{T}}$ according to (7.3) and (7.4).

Remark 7.4.

(a) The numerical solution of the linear equations in (7.8) can usually be done with iterative solvers for symmetric positive definite matrices such as a preconditioned conjugate gradient method (confer [SB93], for example). As it is common for finite-difference discretizations for PDEs, the forward operation only involves the identity, taking the discrete gradient, a pointwise multiplication with a $\mathbb{R}^{2\times2}$-matrix as well as its adjoint, which can be implemented quite easily and fast.

(b) The data of the discrete equation is given by $u_0, \ldots, u_K$ on the one hand, but only by $p_1, \ldots, p_K$ on the other hand. The parameter p_0 is neglected, since an implicit time-discretization scheme has been chosen. This still makes sense because one has to perform $K-1$ steps in time to get to the K-th step, with each step depending on one p_k, yielding at most $K-1$ different discrete edge fields.

For the solution of the optimization problem associated with the PDE, we also need to know how the adjoint of the solution operator can be computed. With the block-matrix representation (7.7), however, this adjoint solution operator is easy to obtain.

Proposition 7.5. *The adjoint of the discrete solution operator for (2.4) according to Definition 7.3 mapping between the spaces $\mathcal{U}_{\tau,h} \to \mathcal{R}_{\tau,h}$ (with scalar products (7.2) and (7.1d)) is given by the iteration:*

$$v_K = \tau z_K$$
$$v_k = (I + \tau A_{k+1}^* A_{k+1})^{-1} v_{k+1} + \tau z_k \tag{7.9}$$

for $k = 0, \ldots, K-1$. In particular, for each $z \in \mathcal{R}_{\tau,h}$ the above $v \in \mathcal{U}_{\tau,h}$ satisfies

$$\langle y, z \rangle_{\tau,h} = \langle u, v \rangle_{\tau,h,0}$$

whenever y is a solution of the discrete problem with data u.

Proof. First note how the adjoints with respect to $\mathcal{R}_{\tau,h} \to \mathcal{R}_{\tau,h}$ and $\mathcal{R}_{\tau,h} \to \mathcal{U}_{\tau,h}$ are related. Let y, u satisfy (7.8) and $z, \bar{v}$ such that $\langle y, z \rangle_h = \langle u, \bar{v} \rangle_h$. Set v such that $v_0 = \tau \bar{v}_0$ and $v_k = \bar{v}_k$. Then:

$$\langle y, z \rangle_{\tau,h} = \langle u, \bar{v} \rangle_{\tau,h} = \langle u_0, \tau \bar{v}_0 \rangle_h + \tau \sum_{k=1}^{K} \langle u_k, \bar{v}_k \rangle_h = \langle u, v \rangle_{\tau,h,0} \,,$$

thus it suffices to know the adjoint of the discrete solution operator with respect to $\mathcal{R}_{\tau,h} \to \mathcal{R}_{\tau,h}$.

The latter can easily be derived from the block-matrix notation (7.7):

$$\begin{pmatrix} \tau^{-1} v_0 \\ v_1 \\ v_2 \\ v_3 \\ \vdots \\ v_K \end{pmatrix} = \begin{pmatrix} I & \mathcal{S}_0^{0*} & \mathcal{S}_0^{1*} & \mathcal{S}_0^{2*} & \cdots & \mathcal{S}_0^{K-1*} \\ 0 & \tau I & \tau \mathcal{S}_1^{1*} & \tau \mathcal{S}_1^{2*} & & \tau \mathcal{S}_1^{K-1*} \\ 0 & 0 & \tau I & \tau \mathcal{S}_2^{2*} & & \tau \mathcal{S}_2^{K-1*} \\ 0 & 0 & 0 & \tau I & \cdots & \tau \mathcal{S}_3^{K-1*} \\ \vdots & & & & \ddots & \vdots \\ 0 & 0 & 0 & 0 & \cdots & \tau I \end{pmatrix} \begin{pmatrix} z_0 \\ z_1 \\ z_2 \\ z_3 \\ \vdots \\ z_K \end{pmatrix} . \quad (*)$$

where $v \in \mathcal{R}_{\tau,h}$ is the image of the adjoint solution operator under $z \in \mathcal{R}_{\tau,h}$. The $\mathcal{S}_{k'}^{k*}$ satisfy

$$\mathcal{S}_{k'}^{k*} = \mathcal{S}_{k'+1}^{*} \mathcal{S}_{k'+2}^{*} \cdots \mathcal{S}_{k+1}^{*} = \mathcal{S}_{k'+1} \mathcal{S}_{k'+2} \cdots \mathcal{S}_{k+1}$$

since the $\mathcal{S}_k$ are self-adjoint by definition. The block-matrix in $(*)$ admits a structure which is similar to (7.7) and can therefore be translated back into the (backwards) iteration

$$v_K = \tau z_K$$
$$v_k = (I + \tau A_{k+1}^{*} A_{k+1})^{-1} v_{k+1} + \tau z_k$$

where $k = 0, \ldots, K - 1$, which is the claimed representation. ◀

Remark 7.6.

(a) Comparing (7.8) and (7.9) yields that the adjoint solution operator admits the same structure as the solution operator itself, with the difference that it is iterating backwards and takes weighted "end-time" data τz_K in contrast to the unweighted initial data u_0. Such a result is in accordance with the adjoint equations derived in Subsection 6.2.2.

(b) If one only considers the evaluation of the discrete solution at the end-time, i.e. $u \mapsto y_K$, then the adjoint is given as follows. For a $\bar{z} \in R_h$ define

$z \in \mathcal{R}_{\tau,h}$ by setting $z_K = \tau^{-1}\bar{z}$ and $z_k = 0$ for $k = 0,\ldots,K-1$. The $v \in \mathcal{U}_{\tau,h}$ according to (7.9) then satisfies

$$\langle u,\, v \rangle_{\tau,h,0} = \langle y,\, z \rangle_{\tau,h} = \langle y_K,\, \bar{z} \rangle_h\ ,$$

meaning that v is indeed the adjoint associated with the discrete solution of (2.4) with end-time evaluation. The iteration

$$\begin{aligned} v_K &= \bar{z} \\ v_k &= (I + \tau A_{k+1}^* A_{k+1})^{-1} v_{k+1} \end{aligned} \tag{7.10}$$

for $k = 0,\ldots,K-1$ then corresponds exactly to the solution of the backwards equation with end-time value $\bar{z}$ and homogeneous right-hand side.

(c) It is notable that other discretizations with respect to the time-variable are not necessarily preserved when the adjoint is taken. For instance, performing a fully implicit time-step scheme, i.e.

$$y_{k+1} = (I + \tau A_{k+1}^* A_{k+1})^{-1}(y_k + \tau u_k)$$

yields a "backwards equation" which does not correspond to such an implicit time-step scheme.

7.2   The discrete optimization problem

After having computed a discretization of the PDE (2.4), we are now able to derive a corresponding discretized version of (2.8) and to compute first-order necessary conditions for optimality. These conditions can be used for developing a gradient descent algorithm for the computation of stationary points. Again, for reasons of simplicity, we only consider the case $N = 1$, i.e. the interpolation problem for two images $\bar{y}_0$ and $\bar{y}_T$ in the interval $[0,T]$. Moreover, we choose a quadratic penalty functional for u and, as already mentioned, a pointwise penalization of p.

Definition 7.7. Let a grid according to Definition 7.1 be given and let $\bar{y}_0, \bar{y}_T \in \mathcal{R}_h$, σ according to Condition 5.21 and $\lambda_u, \lambda_p > 0$. Then, the *discretized optimization problem* associated with (2.8) is to solve

$$\min_{\substack{u \in \mathcal{U}_{\tau,h}, u_0 = \bar{y}_0 \\ p \in \mathcal{R}_{\tau,h}^2, \|p\|_\infty \le 1}} \frac{\|\bar{y}_T - y_K\|_2^2}{2} + \lambda_u \|u\|_2^2 + \lambda_p \tau h^2 \sum_{k=0}^{K} \sum_{i,j=1}^{I,J} \sigma(|p_{k,i,j}|) \tag{7.11}$$

with $y \in \mathcal{R}_{\tau,h}$ being the discrete solution of (2.4) with data u and p according to (7.8). A minimizing pair $(u^*, p^*) \in \mathcal{U}_{\tau,h} \times \mathcal{R}_{\tau,h}^2$ is then called a *discrete solution* of (2.8).

Remark 7.8.

(a) The proof of existence of solutions for (7.11) is much easier than in the continuous case (2.8). We will see in the following that the discrete solution operator of the PDE is continuously differentiable as are the functionals in (7.11) with respect to y, u and p. This means in particular that the objective functional is lower semi-continuous. It is moreover obvious that a minimizing sequence has to be bounded and hence admits a convergent subsequence for which the limit has to be a minimizer (since the set of feasible u, p is closed in $\mathcal{U}_{\tau,h} \times \mathcal{R}_{\tau,h}^2$).

(b) As already noted in Remark 7.4, the discrete solution operator of the degenerate evolution equation only depends on $p_1, \ldots, p_K$ and not on p_0, thus we can assume in the following that always $p_0 = 0$.

The next step towards a numerical algorithm for the solution of the discrete problem is to derive first-order necessary conditions. The differentiability of the discrete solution operator according to Definition 7.3 is examined first.

Proposition 7.9. *The operator* $\mathcal{S} : \mathcal{U}_{\tau,h} \times \mathcal{R}_{\tau,h}^2 \to \mathcal{R}_{\tau,h}$ *assigning each* (u, p) *with* $p_0 = 0$ *the solution* y *according to* (7.8), *is continuously differentiable with partial derivatives*

$$\frac{\partial \mathcal{S}}{\partial u}(u, p)v = \mathcal{S}(v, p) \quad , \quad \frac{\partial \mathcal{S}}{\partial p}(u, p)q = \mathcal{S}((Dy)q, p) \tag{7.12a}$$

with

$$(Dy)q = \mathrm{div}_h\left(\frac{\partial D_p^2}{\partial p}(p)q\right)\nabla_h \mathcal{S}(u, p)^{\mathrm{T}} \tag{7.12b}$$

$$\frac{\partial D_p^2}{\partial p}(p)q = \begin{cases} -(q \cdot p)\frac{\sigma'(|p|)}{|p|}\frac{p}{|p|} \otimes \frac{p}{|p|} - \sigma(|p|)\frac{p}{|p|} \otimes \frac{q}{|p|}P_p & p \neq 0 \\ \qquad\qquad -\sigma(|p|)P_p\frac{q}{|p|} \otimes \frac{p}{|p|} & \\ 0 & p = 0 \end{cases} \tag{7.12c}$$

$$P_p = I - \frac{p}{|p|} \otimes \frac{p}{|p|}$$

where ∇_h *on* $\mathcal{R}_{\tau,h}$ *is given by* $(\nabla_h y)_k = \nabla_h y_k$ *(and an analog definition for* div_h*) and* $\frac{\partial D_p^2}{\partial p}$ *has to be understood in the pointwise sense.*

Proof. First observe that the solution operator $\mathcal{S}$ is continuous. This can be seen, for example, in the representation (7.7), revealing that $\mathcal{S}$ is a composition of the continuous mapping $p \mapsto D_p$ according to (7.3), linear mappings, matrix multiplication and inversion which are all continuous on their respective domains of definition.

Now compute the partial derivatives. Observe that $\mathcal{S}(u, p)$ is linear with respect to u which already proves the formula for $\frac{\partial \mathcal{S}}{\partial u}$. Regarding the derivative with respect to p, we first verify that for $p, q \in \mathbb{R}^2$ the identity (7.12c) holds. Also note that the derivative is continuous according to the assumptions on σ (Condition 5.21, also confer Lemma 5.24). By exploiting the pointwise structure, the analog statements can be derived for $p, q \in R_h^2$ as well as $p, q \in \mathcal{R}_{\tau,h}^2$.

Furthermore, investigate the matrix-valued mapping

$$S : R_h^2 \to \mathcal{L}(R_h, R_h) \quad , \quad p \mapsto (I - \tau \operatorname{div}_h D_p^2 \nabla_h^{\mathrm{T}})^{-1}$$

according to (7.5), which is also continuously differentiable with derivative

$$\frac{\partial S}{\partial p}(p)q = \tau S(p)\Big(\operatorname{div}_h\Big(\frac{\partial D_p^2}{\partial p}(p)q\Big)\nabla_h^{\mathrm{T}}\Big)S(p)$$

$$= S(p)A(p,q)S(p)$$

with $A(p,q) = \operatorname{div}_h\big(\frac{\partial D_p^2}{\partial p}(p)q\big)\nabla_h^{\mathrm{T}}$ and recalling that the derivative of the matrix inversion $Q \mapsto Q^{-1}$ at some Q is $H \mapsto -Q^{-1}HQ^{-1}$.

Now, taking $p, q \in \mathcal{R}_{\tau,h}^2$ and introducing the short-hand notation $S_{k+1} = S(p_{k+1})$ and $A_{k+1} = A(p_{k+1}, q_{k+1})$, we want to plug this into the formula (7.6). But first, observe that for $0 \leq l \leq K - 1$ and $k \geq l$

$$\frac{\partial}{\partial p_{l+1}}\Big(\prod_{k''=k'}^{k} S_{k''+1}\Big)q_{l+1} = \tau\Big(\prod_{k''=l}^{k} S_{k''+1}\Big)A_{l+1}\Big(\prod_{k''=k'}^{l} S_{k''+1}\Big)$$

from which follows, with $y = \mathcal{S}(u, p)$ and the help of (7.6), that

$$\Big(\frac{\partial \mathcal{S}}{\partial p_{l+1}}(u,p)q_{l+1}\Big)_k = \frac{\partial}{\partial p_{l+1}}\Big(\prod_{k'=0}^{k} S_{k'+1}u_0 + \tau \sum_{k'=0}^{k}\prod_{k''=k'}^{k} S_{k''+1}u_{k'+1}\Big)q_{l+1}$$

$$= \tau\Big(\prod_{k''=l}^{k} S_{k''+1}\Big)A_{l+1}\Big(\prod_{k''=0}^{l} S_{k''+1}u_0 + \tau \sum_{k'=0}^{l}\prod_{k''=k'}^{l} S_{k''+1}u_{k'+1}\Big)$$

$$= \tau\Big(\prod_{k''=l}^{k} S_{k''+1}\Big)A_{l+1}y_{l+1} \ ,$$

while for $l > k$, we have

$$\Big(\frac{\partial \mathcal{S}}{\partial p_{l+1}}(u,p)q_{l+1}\Big)_k = 0 \ .$$

Finally, the desired partial derivative reads as

$$\left(\frac{\partial \mathcal{S}}{\partial p}(u,p)q\right)_k = \sum_{l=0}^{K-1}\left(\frac{\partial \mathcal{S}}{\partial p_{l+1}}(u,p)q_{l+1}\right)_k = \tau\sum_{l=0}^{k-1}\left(\prod_{k''=l}^{k} S_{k''+1}\right)A_{l+1}y_{l+1} \ .$$

This amounts to the iteration

$$z_0 = 0$$

$$z_{k+1} = S(p_{k+1})z_k + \tau\operatorname{div}_h\left(\frac{\partial D_p^2}{\partial p}(p_{k+1})q_{k+1}\right)\nabla_h y_{k+1}^{\mathrm{T}}$$

for $k = 0,\ldots,K-1$, which is, in turn, the discrete solution operator according to (7.8) with a special right-hand side. Thus, one can say that

$$\frac{\partial \mathcal{S}}{\partial p}(u,p)q = \mathcal{S}\big((\mathrm{D}y),p\big)$$

with $\mathrm{D}y$ according to (7.12b), finally showing (7.12a).

It remains to show that both partial derivatives are continuous. This will imply the differentiability as well as the desired statement. But looking at (7.12a), one can see that the partial derivatives are a composition of the continuous $\mathcal{S}$ and, again, linear operations, matrix multiplication as well as (7.12c), which has already been identified as continuous, yielding the stated continuity. ◀

We are now able to differentiate the functional in (7.11) and to compute the derivative. These computations are very similar to the ones performed in Section 6.2, giving exactly a discrete version of (6.12), but this time valid for all $p \in \mathcal{R}_{\tau,h}^2$.

Lemma 7.10. *Consider the situation of Definition 7.7. The functionals*

$$\Psi_{u,p}(u,p) = \frac{\left\|\big(\mathcal{S}(u,p)\big)_K - \bar{y}_T\right\|_2^2}{2},$$

$$\Phi_u(u) = \lambda_u\|u\|_2^2 \quad, \quad \Phi_p(p) = \lambda_p\tau h^2\sum_{k=0}^{K}\sum_{i,j=1}^{I,J}\sigma(|p_{k,i,j}|) \tag{7.13a}$$

are continuously differentiable as mappings $\mathcal{U}_{\tau,h} \times \mathcal{R}_{\tau,h}^2 \to \mathbb{R}$, $\mathcal{U}_{\tau,h} \to \mathbb{R}$ and $\mathcal{R}_{\tau,h}^2 \to \mathbb{R}$, respectively, with derivatives which can be identified with

$$\Psi_{u,p}'(u,p) = \left(v, \frac{\sigma'(|p|)}{|p|^2}\partial_p y\partial_p v\frac{p}{|p|} + \frac{\sigma(|p|)}{|p|^2}\big(\partial_p v P_p\nabla_h y^{\mathrm{T}} + \partial_p y P_p\nabla_h v^{\mathrm{T}}\big)\right)$$

$$\Phi_u'(u) = 2\lambda_u u$$

$$\Phi_p'(p) = \lambda_p\frac{\sigma'(|p|)}{|p|}p$$

$$\tag{7.13b}$$

via the scalar products (7.1d) and (7.2), where

$$y = \mathcal{S}(u,p) \quad , \quad v = \mathcal{S}^*(y_K - \bar{y}_T, p)$$

with $\mathcal{S}^*(\bar{z}, p) \in \mathcal{U}_{\tau,h}$ denoting the result of the iteration performed in (7.10) and

$$\partial_p y = p \cdot \nabla_h y^{\mathrm{T}} \quad , \quad \partial_p v = p \cdot \nabla_h v^{\mathrm{T}} \quad , \quad P_p = I - \tfrac{p}{|p|} \otimes \tfrac{p}{|p|}$$

in a pointwise sense. Moreover, the above identities have to be understood in the sense that in points where $p = 0$, the respective terms are always extended by 0.

Proof. According to Proposition 7.9 and (7.12), the gradient of $\Psi_{u,p}$ applied to $(\bar{u}, \bar{p}) \in \mathcal{U}_{\tau,h} \times \mathcal{R}^2_{\tau,h}$ reads as

$$\Psi'_{u,p}(u,p)(\bar{u},\bar{p}) = \Big\langle \Big(\frac{\partial \mathcal{S}}{\partial u}(u,p)\bar{u} + \frac{\partial \mathcal{S}}{\partial p}(u,p)\bar{p}\Big)_K, \, y_K - \bar{y}_T \Big\rangle_h$$

$$= \langle \bar{u}, \, \mathcal{S}^*(y_K - \bar{y}_T, p)\rangle_{\tau,h,0} + \langle \mathcal{S}((\mathrm{D}y)\bar{p}, p), \, y_K - \bar{y}_T\rangle_h$$

$$= \langle \bar{u}, \, v\rangle_{\tau,h,0} + \langle (\mathrm{D}y)\bar{p}, \, v\rangle_{\tau,h,0} \,,$$

see also Remark 7.6 and the definition of v. With (7.12b), one can further see that

$$\langle (\mathrm{D}y)\bar{p}, \, v\rangle_{\tau,h,0} = \Big\langle \mathrm{div}_h \Big(\frac{\partial D_p^2}{\partial p}(p)\bar{p}\Big)\nabla_h y^{\mathrm{T}}, \, v\Big\rangle_{\tau,h,0}$$

$$= \Big\langle -\Big(\frac{\partial D_p^2}{\partial p}(p)\bar{p}\Big)\nabla_h y^{\mathrm{T}}, \, \nabla_h v^{\mathrm{T}}\Big\rangle_{\tau,h}$$

which, with (7.12c) and considered in each point (with indices dropped), turns out to be

$$-\Big(\frac{\partial D_p^2}{\partial p}(p)\bar{p}\Big)\nabla_h y^{\mathrm{T}} \cdot \nabla_h v^{\mathrm{T}} = \bar{p} \cdot \Big(\frac{\sigma'(|p|)}{|p|^2}(p \cdot \nabla_h y^{\mathrm{T}})(p \cdot \nabla_h v^{\mathrm{T}})\frac{p}{|p|}$$

$$+ \frac{\sigma(|p|)}{|p|^2}(p \cdot \nabla_h v^{\mathrm{T}})P_p \nabla_h y^{\mathrm{T}} + \frac{\sigma(|p|)}{|p|^2}(p \cdot \nabla_h y^{\mathrm{T}})P_p \nabla_h v^{\mathrm{T}}\Big)$$

if $p \neq 0$ and 0 else. Consequently, with the notation introduced above,

$$\langle (\mathrm{D}y)\bar{p}, \, v\rangle_{\tau,h,0}$$

$$= \Big\langle \bar{p}, \, \frac{\sigma'(|p|)}{|p|^2}\partial_p y \partial_p v \frac{p}{|p|} + \frac{\sigma(|p|)}{|p|^2}(\partial_p v P_p \nabla_h y^{\mathrm{T}} + \partial_p y P_p \nabla_h v^{\mathrm{T}})\Big\rangle_{\tau,h} \,,$$

proving the representation of $\Psi'_{u,p}$.

The differentiability of Φ_u and Φ_p as well as the formulas for the gradients can be obtained from standard differentiation rules and the properties of σ, see Lemma 5.24.   ◀

Remark 7.11. It is easy to see that the gradient of the objective functional F in (7.11) reading as

$$F(u,p) = \Psi_{u,p}(u,p) + \Phi_u(u) + \Phi_p(p)$$

is given by

$$\frac{\partial F}{\partial u}(u,p) = \frac{\partial \Psi_{u,p}}{\partial u}(u,p) + \Phi'_u(u) \quad , \quad \frac{\partial F}{\partial p}(u,p) = \frac{\partial \Psi_{u,p}}{\partial p}(u,p) + \Phi'_p(p)$$

which can be computed with the help of (7.13b).

Finally, we can write first-order necessary conditions for optimality of discrete solutions of the minimization problem according to Definition 7.7.

Proposition 7.12. *In the situation of Definition 7.7, a pair $(u^*, p^*) \in \mathcal{U}_{\tau,h} \times \mathcal{R}^2_{\tau,h}$ with $p_0^* = 0$ and $\|p^*\|_\infty \leq 1$ which is a minimizer of (7.11) satisfies:*

$$y_0^* = \bar{y}_0$$
$$y_{k+1}^* = (I + \tau A_{k+1}^* A_{k+1})^{-1} y_k^* + \tau u_{k+1}^* \qquad (7.14a)$$
discrete primal equation

$$v_K^* = y_K^* - \bar{y}_T$$
$$v_k^* = (I + \tau A_{k+1}^* A_{k+1})^{-1} v_{k+1}^* \qquad (7.14b)$$
discrete adjoint equation

$$v^* + 2\lambda_u u^* = 0$$
$$\partial_{p^*} v^* P_{p^*} \nabla_h y^{*T} + \partial_{p^*} y^* P_{p^*} \nabla_h v^{*T} = 0$$
$$\partial_{p^*} v^* \partial_{p^*} y^* + \lambda_p |p^*|^2 = 0 \quad \vee \quad |p^*| = 1 \qquad (7.14c)$$
discrete necessary optimality conditions

where (7.14c) has to be interpreted pointwise.

Proof. According to Remark 7.11, the objective functional F associated with (7.11) is differentiable. Hence, if (u^*, p^*) with $p_0^* = 0$ and $\|p^*\|_\infty \leq 1$ is optimal, then we have for all (u,p) with $p_0 = 0$ and $\|p\|_\infty \leq 1$ that

$$\left\langle \frac{\partial \Psi_{u,p}}{\partial u}(u^*, p^*) + \Phi'_u(u^*), \, u - u^* \right\rangle_{\tau,h,0}$$

$$+ \left\langle \frac{\partial \Psi_{u,p}}{\partial p}(u^*, p^*) + \Phi'_p(p^*), \, p - p^* \right\rangle_{\tau,h} \geq 0 \, .$$

Plugging $p = p^*$ and all $u \in \mathcal{U}_{\tau,h}$ into the above inequality yields, with (7.13b), that $v^* + 2\lambda_u u^* = 0$ holds where v^* is the solution of the adjoint equation according to (7.14b).

Likewise, setting $u = u^*$ and testing with all feasible p yields

$$\frac{\partial \Psi_{u,p}}{\partial p}(u^*, p^*) + \Phi'_p(p^*) = \mu p^* \quad , \quad \begin{cases} \mu = 0 & \text{where } |p^*| < 1 \\ \mu \leq 0 & \text{where } |p^*| = 1 \,. \end{cases}$$

Multiplying this equation pointwise with P_p from the left then gives, again with taking (7.13b) into account,

$$\frac{\sigma(|p^*|)}{|p^*|^2} \left(\partial_{p^*} v^* P_{p^*} \nabla_h y^{*\mathrm{T}} + \partial_{p^*} y^* P_{p^*} \nabla_h v^{*\mathrm{T}} \right) = 0$$

implying the second necessary condition since $\sigma(|p^*|)/|p^*|^2 = 0$ if only if $p^* = 0$, see Condition 5.21. Finally, the third condition follows from taking the pointwise inner product with p^*,

$$\frac{\sigma'(|p^*|)}{|p^*|} \left(\partial_{p^*} y^* \partial_{p^*} v^* + \lambda_p |p^*|^2 \right) = \mu |p^*|^2 \,,$$

and noting that $0 \leq |p^*| < 1$ implies the one possibility of the third condition since $\mu = 0$ in that case. If $|p^*| = 1$, then the equation contains no further information and can therefore be omitted. ◀

Remark 7.13. Regarding the first-order necessary conditions, one can see that they always can be satisfied by $p = 0$ and a suitable u yielding a solution which may be undesired since this corresponds to a maximum amount of diffusion which is not necessarily optimal. Thus, the set of pairs (u, p) satisfying the first-order necessary conditions is indeed strictly larger than the set of optimal solutions.

Moreover, as one can conclude by examining the derivatives (7.13b), a step along a gradient direction will not change points where $p = 0$, hence we have to choose an initial guess with preferably $p \neq 0$ in each discrete point. Furthermore, if $|p| = 1$ for one point, a step in the direction of the gradient will be perpendicular to p (since $\sigma'(1) = 0$), meaning that $|p| \geq 1$ in the next step, forcing a gradient descent algorithm to project back to $|p| = 1$. Thus, an initial guess for p should be chosen such that $0 < |p| < 1$ is satisfied for all discrete points.

7.3 A gradient descent algorithm

Finally, we have all ingredients for obtaining a gradient descent algorithm for finding points fulfilling the first-order necessary conditions stated in Proposition 7.12. We will use the classical gradient projection algorithm with Goldstein-Armijo step-size rule which is quite easy to implement. We will first give an outline of the abstract algorithm for the minimization of $F(u, p)$ over the convex set $X = \mathcal{U}_{\tau,h} \times \{\|p\|_\infty \leq 1\} \subset \mathcal{U}_{\tau,h} \times \mathcal{R}^2_{\tau,h}$.

Algorithm 7.14.

1. Choose the parameters $\alpha_0 > 0$, $\beta \in \,]0, 1[$, $\vartheta \in \,]0, 1[$ and an initial guess $(u^0, p^0) \in X$.

2. Let the approximation $(u^n, p^n) \in X$ be given. Set $m = 0$.

3. For some $m \geq 0$, compute

$$
(u^{n,m}, p^{n,m}) = P_X\left(u^n - \beta^m \alpha_0 \frac{\partial F}{\partial u}(u^n, p^n), p^n - \beta^m \alpha_0 \frac{\partial F}{\partial p}(u^n, p^n)\right),
$$

   where P_X denotes the orthogonal projection onto X.

4. If

$$
F(u^n, p^n) - F(u^{n,m}, p^{n,m}) < \vartheta \left\langle \frac{\partial F}{\partial u}(u^n, p^n), u^n - u^{n,m} \right\rangle_{\tau,h,0}
$$
$$
+ \vartheta \left\langle \frac{\partial F}{\partial p}(u^n, p^n), p^n - p^{n,m} \right\rangle_{\tau,h},
$$

   then set $m := m + 1$ and return to Step 3.

5. Otherwise, set $(u^{n+1}, p^{n+1}) = (u^{n,m}, p^{n,m})$, $n := n + 1$ and continue with Step 2 or stop if some tolerance/stationary point is reached.

In the following, it is summarized what has to be done to implement the algorithm numerically.

Algorithm 7.15.

1. • Choose the parameters $\alpha_0 > 0$, $\beta \in {]}0, 1[$, $\vartheta \in {]}0, 1[$ and set the variable $n = 0$.

    • Generate an initial guess (u^0, p^0) according to Remark 7.13.

    • Compute the discrete solution y^0 of the primal equation associated with (u^0, p^0) according to the iteration (7.14a).

    • Calculate F_0 by plugging (y^0, u^0, p^0) into the functional in (7.11).

2. • Initialize the variable $m = 0$.

    • Obtain the derivative $(\frac{\partial F}{\partial u}, \frac{\partial F}{\partial p})$ of F at (u^n, p^n) as follows: Compute the solution v^n of the adjoint equation associated with y^n according to the iteration (7.14b).

    • Set

    $$\mathrm{D}_u^n = v^n + 2\lambda_u u^n$$

    $$\mathrm{D}_p^n = \frac{\sigma'(|p^n|)}{|p^n|^2} \left(\partial_{p^n} y^n \partial_{p^n} v^n + |p^n|^2 \lambda_p \right) \frac{p^n}{|p^n|}$$

    $$+ \frac{\sigma(|p^n|)}{|p^n|^2} \left(\partial_{p^n} v^n P_{p^n} \nabla_h y^{nT} + \partial_{p^n} y^n P_{p^n} \nabla_h v^{nT} \right).$$

3. • Set

    $$u^{n,m} = u^n - \beta^m \alpha_0 \mathrm{D}_u^n \quad , \quad p^{n,m} = P_{\{\|p\|_\infty \leq 1\}} (p^n - \beta^m \alpha_0 \mathrm{D}_p^n)$$

    where $P_{\{\|p\|_\infty \leq 1\}}(p) = \frac{p}{\max\{1, |p|\}}$ (pointwise).

    • Compute the discrete solution $y^{n,m}$ of the primal equation with data $(u^{n,m}, p^{n,m})$ according to (7.14a).

4. • Plug $(y^{n,m}, u^{n,m}, p^{n,m})$ into the functional in (7.11) to obtain the value $F_{n,m}$.

    • Compute $s_{n,m}$ according to

    $$s_{n,m} = \vartheta \left(\langle \mathrm{D}_u^n, u^n - u^{n,m} \rangle_{\tau,h,0} + \langle \mathrm{D}_p^n, p^n - p^{n,m} \rangle_{\tau,h} \right).$$

    • If $F_n - F_{n,m} < s_{n,m}$, then increase the variable $m := m + 1$ and continue with Step 3.

5. Otherwise, set $(y^{n+1}, u^{n+1}, p^{n+1}, F_{n+1}) = (y^{n,m}, u^{n,m}, p^{n,m}, F_{n,m})$, increase the variable $n := n + 1$ and continue with Step 2 or stop if some appropriate criterion is satisfied.

Since this algorithm is a descent algorithm in finite dimensions for which the iterates are contained in a compact set, there will be at least a subsequence of $\{(u^n, p^n)\}$ which converges to a stationary point for the functional in (7.11). For details involving convergence of optimization algorithms in finite dimensions, see [Lue89], for example.

7.4 Numerical examples

This section finally shows some examples of approximate solutions computed with the introduced gradient descent algorithm. Its purpose is not to provide a full numerical discussion, it can more be seen as an illustration of how the model developed in Chapter 2 behaves in certain situations and what it is able to accomplish. Computations were performed for a set of sample images where each pair features different characteristics: a pair of artificially created abstract images, a mammography image and a photograph. We do not go into detail with respect to the actual implementation, but it is worth mentioning that the computations were carried out on a single desktop PC which was possible due to the small size of the images and the fact that only short sequences were produced. In the following, the generated movie y is depicted as a frame sequence. Moreover, the control associated with the state y is shown frame-wise. While y is represented by gray-scale images, the control variable u, which indicates sources or sinks leading to contrast changes, is represented by an adapted color map. Sources appear as red regions while sinks are depicted blue. The edge field p is also encoded with colors depending on the absolute value and angle for each point. Each angle is represented by a unique fully-saturated hue while each absolute valued between 0 and 1 is assigned an intensity (also known as value in the HSV color-model).

Example 7.16. The method was first tested with two artificially created images of relatively small size showing static and varying features, see Figure 7.1. Note the differences between the two images: On the one hand, the rectangle has different size and position. On the other hand, the location of the gradient in the background is different. Due to this essential differences, the associated interpolation problem is potentially a hard test and should also demonstrate which effects the underlying model is capable to produce and where its limits are. To get an impression of the parameters used for the computations, also see Figure 7.1.

You can see the state y associated with a stationary point (u, p) calculated by Algorithm 7.15 in Figure 7.2. The controls u and p are shown in Figures 7.3 and 7.4, respectively. Note that only a selection of the 50 frames of u

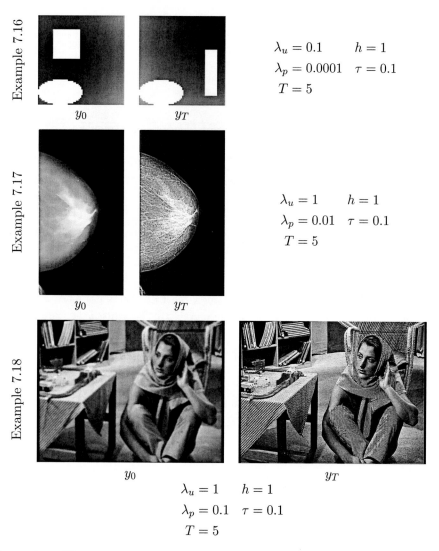

Figure 7.1: The test images as well as the problem-dependent parameters of Examples 7.16–7.18. The top row depicts the data of Example 7.16, while the middle and the bottom rows show the data of Example 7.17 and 7.18, respectively.

and p as well as the 51 frames of y (which includes the initial value y_0) are depicted. (The different numbering results from the fact that y has one more frame than u and p.) As one can see in the sequence y, the interpolation is such that the rectangle in y_0 dissolves smoothly while the rectangle in y_T is created from a blurry rectangular object which becomes sharper in the end. Both can be interpreted as an interpolation of the background and the rectangle across different scales. Moreover, the ellipse in the corner seems to "bleed" in the beginning but it edges stay sharp during the movie, while the gradient in the background is also interpolated in a smooth way. All this observations are in accordance with the goals of the model developed in Chapter 2.

Let us also discuss the control variables (u, p) computed by Algorithm 7.15. The parameter λ_u, which can be interpreted as a penalization for u, is relatively high resulting in a solution which does not vary too much with respect to the time-variable. As expected, u is negative in the region where the rectangle in y_0 vanishes, positive where the rectangle in y_T is created and approximately zero in the ellipse.

Furthermore, the penalization introduced for p is relatively small, so p indicates many "edges", even in regions where y is smooth. Visually, these falsely detected edges do not seem to have much influence on the interpolation goals we have in mind, since, as we can see in Figure 7.2, the model yields reasonable interpolating sequences.

Example 7.17. To test the algorithm and method on some medical data, we performed computations for a mammography image which has been digitally enhanced to emphasize fine and coarse images, respectively. The sample image was taken from the Digital Database for Screening Mammography of the University of South Florida (again, see Figure 7.1). The problem was posed such that it fades from coarse to fine, potentially the harder problem. Furthermore, the parameters λ_u and λ_p are chosen relatively high, such that only mild contrast changes and a minimal amount of edges are supposed to appear.

You can see the outcome of the optimization process in the subsequent figures: The interpolating movie is depicted in Figure 7.5, while the corresponding controls u and p can be found in Figures 7.6 and 7.7, respectively. Again, the underlying PDE produces a movie which interpolates and reveals intermediate scales one after another. Note that no essential contrast changes are necessary to fade from one image to the other. This is reflected in the control variable u which is approximately zero most of the time. To the end, however, it becomes non-zero in order to allow the formation of the very fine details. The edge field p behaves similar: With increasing level of detail, more and more edges appear such that the already created structures are not blurred by the diffusion process.

Example 7.18. In this example, it is illustrated how the model acts on photographs and how this is different from linear interpolation. A sample image is chosen which contains typical features of a natural image: There are clearly defined objects with edges, for which some have a smooth texture (the women's face and arms as well as the carpet, for instance). Other objects feature a fine oscillating structure (the women's trousers, the table cloth and the chair, for instance). The goal was to find a movie which interpolates between a version where the oscillations were removed to a version with exaggerated edges.

You can see the optimized interpolating movie in Figure 7.8. As expected, some intermediate details appear first while the smooth parts remain smooth. Then, the high oscillations are created and eventually, the value y_T is approximately attained. In Figure 7.9, a magnified detail of this movie is compared to the linear interpolation of y_0 and y_T. The differences become most apparent in the middle of the interpolation: In the movie generated by the optimization process, not every detail is revealed and there are still some smooth regions which will become more oscillatory later. In contrast to this, all details are already visible in the images resulting from linear interpolation.

The numerical experiments can be summarized as follows. It seems like modeling the problem of interpolating two images across intermediate scales with an optimal control problem in which both the source term u as well as the diffusion tensor D_p^2 are controlled, leads to reasonable results when implemented numerically and tested with real images. It moreover becomes clear how this interpolation actually behaves. It indeed interpolates the scales: coarse features are able to evolve smoothly and naturally into fine features and vice versa. This process is mainly diffusion-driven, hence the impression of a natural transition is given. Furthermore, we again emphasize the fact that the diffusion process is allowed to degenerate which leads to the formation and preservation of edges. This is the main reason that the generated sequences are perceived as images. However, diffusion does not model transportation of mass which can, for example, result in the movement of objects. This becomes clear in Example 7.16, where the rectangles are not moving and rather dissolving into the background and appearing from the background, respectively. Hence, the optimization procedure does not capture all effects one perceives as natural but at least the effects which were originally modeled.

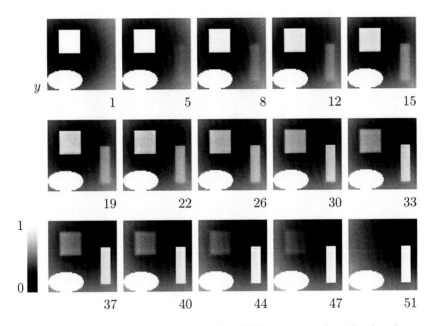

Figure 7.2: The discrete solution y of the PDE associated with the discrete control for the artificial test images of Example 7.16.

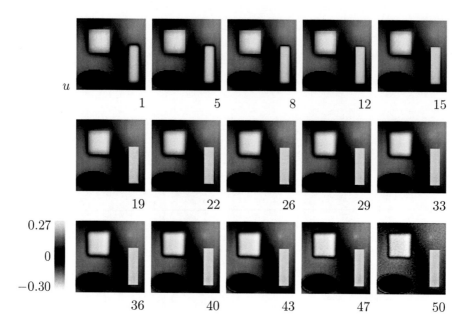

Figure 7.3: The discrete optimized sources and sinks u of Example 7.16.

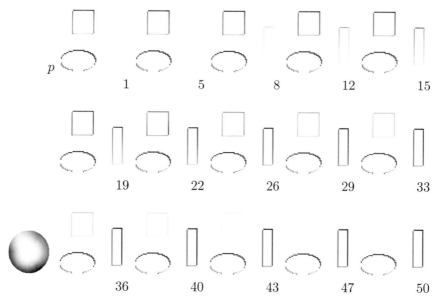

Figure 7.4: The discrete optimized edge field p of Example 7.16.

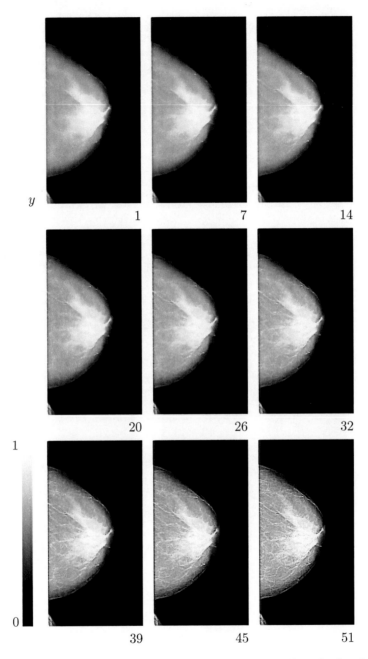

Figure 7.5: The discrete solution y of the PDE associated with the discrete control for the mammography images of Example 7.17.

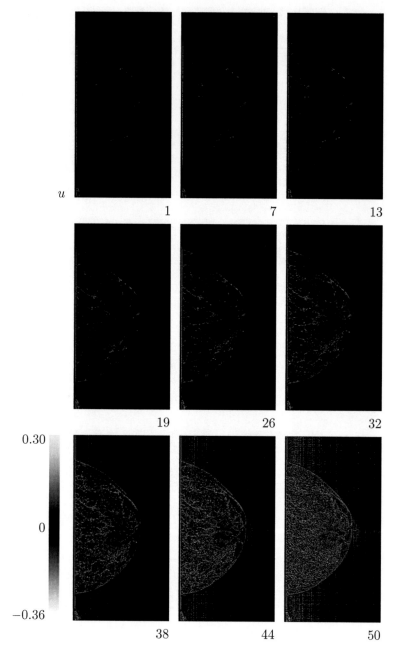

Figure 7.6: The discrete optimized sources and sinks u of Example 7.17.

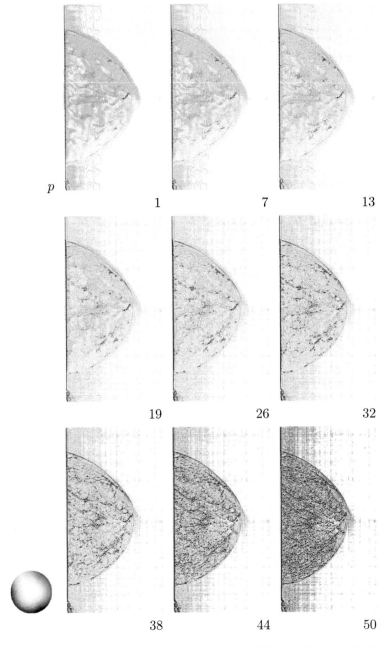

Figure 7.7: The discrete optimized edge field p of Example 7.17.

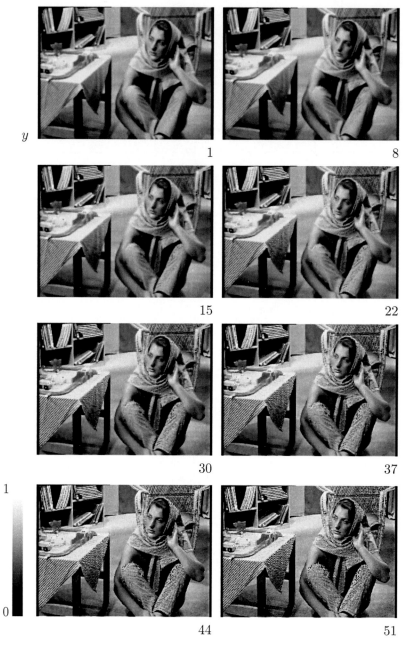

Figure 7.8: The discrete solution y of the PDE associated with the discrete optimized control for the photograph of Example 7.18.

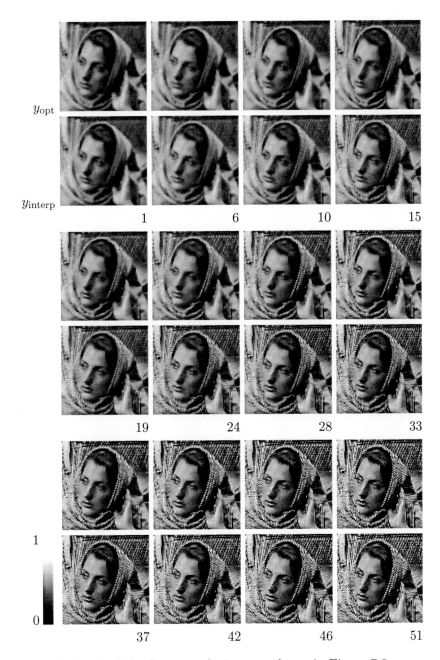

Figure 7.9: A detail of the discrete solution y_{opt} shown in Figure 7.8 compared with the corresponding linear interpolation y_{interp}.

Chapter 8

Summary

The image-processing problem of finding a multiscale interpolation of some given images admitting both coarse and fine scales can be modeled by the optimal control problem

$$\min_{\substack{u \in L^2 \\ |p| \le 1}} \sum_{n=1}^{N} \frac{\|y(t_n) - y_n\|^2}{2} + \lambda_u \|u\|_2^2 + \lambda_p \int_0^T \int_\Omega \sigma\big(|p(t,x)|\big) \ \mathrm{d}x \ \mathrm{d}t + \tilde{\Phi}_p(p) \ ,$$

$$\frac{\partial y}{\partial t} = \mathrm{div}\big(D_p^2 \nabla y^{\mathrm{T}}\big) + u \quad \text{in }]0,T[\times \Omega \ ,$$

$$\frac{\partial y}{\partial_p \nu} = 0 \qquad\qquad\qquad \text{on }]0,T[\times \partial\Omega \ ,$$

$$y = y_0 \qquad\qquad\qquad \text{on } \{0\} \times \Omega$$

with the anisotropic degenerate diffusion tensor

$$D_p^2 = \left(I - \sigma(|p|) \frac{p}{|p|} \otimes \frac{p}{|p|} \right) \ .$$

Such a model involves a class of degenerate parabolic evolution equations which is controlled in the right-hand side u as well as in the edge field p. The latter enters the diffusion tensor of the equation in a non-linear way and allows the process to degenerate which in turn results in preservation and formation of edges, a desirable effect in the context of images.

Mathematical analysis of the above problem has been carried out in this thesis. The analysis includes studying the class of degenerate parabolic equations associated with the control problem. Existence and uniqueness of solutions has been proven in Chapter 3 (confer Theorem 3.25) for the slightly more general problem (3.1) and for non-smooth coefficients (see Conditions 3.1 and 3.2). Appropriate solution spaces V_p, $\mathcal{V}_p$ and $W_p(0,T)$, which are the closures of Sobolev spaces, have been introduced to obtain existence. The uniqueness holds in $W_p(0,T)$, a space which is potentially smaller than $\bar{W}_p(0,T)$, the natural space for weak solutions in the sense of Definition 3.17. All $W_p(0,T)$, however, are, in a certain sense, weak compactly embedded in $C^*\bigl(0,T;L^2(\Omega)\bigr)$, see Theorem 3.40, so one is able to obtain a weakly-convergent subsequence of solutions even if the data u and p both are varied and the solution spaces are changed.

The important question of whether such subsequences of solutions converge to the corresponding solution if the data also converges weakly leads to the close study of the solution spaces which has been carried out in Chapters 4 and 5. Chapter 4 examines the stationary case for which the results can be carried over to the time-variant case in Chapter 5. Notions of weak weighted and directional derivatives, $w\nabla y$ and $\partial_q y$, have been introduced to examine the solution spaces of the PDE with respect to p. The results hold for weights and directions in Y_1^* and Y_d^* which consists of, roughly speaking, $H^{1,\infty}$-functions, see Section 4.1. It has been proven in Propositions 4.19 and 4.25, respectively, that the weak weighted derivative is a meaningful extension to the classical weak derivative, while the weak directional derivative can be interpreted locally as taking the partial derivative in one direction and applying an appropriate coordinate transform. As it has been pointed out in Section 4.3, these constructions satisfy certain weak closedness conditions which to turn out to be crucial for showing existence of solutions of the optimal control problem. The important property that smooth functions are dense in the associated Sobolev spaces $H^r_{w,\partial q_1,\dots,\partial q_K}$ (a result also known as $H = W$) has been established in Theorem 4.43.

In Chapter 5, the notions and results have been extended to the time-variant setting in which it has been applied to PDEs of the type (2.4). The additional time-variable introduces some difficulties arising from a lack of compactness. These problems have been handled in Section 5.1 with the help of the notion of bounded semivariation. As a result, weak closedness properties can be carried over with this additional assumption while density of smooth functions still holds, see Section 5.2. The theory on weighted and directional Sobolev spaces has then been applied to the degenerate parabolic equation (2.4) under the condition that $p \in \mathcal{Y}_d^*$ with $\|p\|_{\mathcal{Y}_d^*} \le 1$ which, roughly speaking, states that the weak gradient of p with respect to the space coordinates is still bounded. This condition and the requirement that p is non-degenerate at the boundary, see Definition 5.32, has been used in Sections 5.3 and 5.4 to establish uniqueness of solutions in $\bar{W}_p(0,T)$ as well as an alternate weak formulation for the PDE (Theorem 5.35) and to derive some boundedness and regularity results, respectively.

The main purpose of examining the weak weighted and directional derivative and the associated spaces has to been to provide a tool which can be utilized to prove that the minimization problem (2.8) admits a solution. This has been done in Chapter 6. In Section 6.1, existence results valid for arbitrary dimensions have been established. The minimization problem has been regularized with the functional $\tilde{\Phi}_p$ in order to meet the requirements ensuring the applicability of weighted and directional Sobolev spaces (i.e. $\|p\|_{\mathcal{Y}_d^*} \leq 1$ and p non-degenerate at the boundary) and the weak closedness properties:

$$\tilde{\Phi}_p(p) = I_{\{\|p\|_{\mathcal{Y}_d^*} \leq 1\}}(p) + \mu_1 \operatorname{tv}^*(p) + \mu_2 \operatorname*{ess\,sup}_{t \in]0,T[} \operatorname{TV}\big(\nabla p(t)\big) \ .$$

Here, on the one hand, the regularization with respect to the time-variable t has been carried out with the tv*-functional while, on the other hand, the amount of compactness necessary with respect to the space-variable x has been enforced with a TV-like penalization on the gradient with respect to x. Roughly speaking, with this regularization, existence of a minimizer has been proven in Theorem 6.6. Moreover, the optimization problem (2.8) has also been investigated with respect to first-order necessary conditions for optimality in Section 6.2. Differentiability properties of the solution operator with respect to both u and p have been subject to investigation. There, the major drawback is that the differentiability for the degenerate (and interesting) case $\|p\|_\infty = 1$ could not be established. Nevertheless, the adjoint equation can be formulated for the general case (cf. Proposition 6.12) where first-order necessary conditions have been stated in case the optimal solution leads to a non-degenerate PDE. Both the unregularized functional ($\tilde{\Phi}_p = 0$, see (6.12)) as well as the regularized functional have been considered (see Theorem 6.20). As can be seen there, the influence of $\tilde{\Phi}_p$ can be expressed with additional terms involving dual norms and polar sets. Again, these conditions are only necessary for minimizers for which $\|p\|_\infty < 1$ holds (provided that M_∂ is close enough to 1).

The minimization problem has been discretized with finite differences in Chapter 7. A first-discretize-then-optimize approach has been chosen. In this situation, analysis can be carried out for the discretized problem, including derivation of an adjoint equation (Proposition 7.5) and of discrete first-order optimality conditions (7.14) for each discrete p fulfilling $\|p\|_\infty \leq 1$. In the discrete case (7.11), the situation is different from the continuous case since finite-difference derivatives always exist. Therefore, also the regularization term $\tilde{\Phi}_p$ can be dropped completely in order to get a well-posed problem. Such considerations have been presented in Section 7.2 and led to a simple numerical algorithm for finding stationary points, see Section 7.3. Eventually, this algorithm was tested in Section 7.4 on some sample images, showing that the desired effects can indeed be captured by the model.

To give a outlook, there are many starting points for further investigation: First, one can do a further examination of the class of degenerate parabolic equations (2.4). It would be interesting to know if there are more regularity results. For example, in the spirit of Proposition 5.40, the existence of "mixed derivatives" $\partial^*_{q_{k'}} \partial_{q_k} y$ for data $u \in L^2(0, T; L^2(\Omega))$ could be examined. By duality, there also would be the possibility to allow for weaker data u. Another direction within the theory of PDEs is to examine degenerate stationary problems of the type

$$- \operatorname{div}\left(D_p^2 \nabla y^{\mathrm{T}}\right) = u \quad \text{in } \Omega \quad , \quad \frac{\partial y}{\partial_p \nu} = 0 \quad \text{on } \partial\Omega$$

for $d \geq 2$ and in particular derive inequalities of Poincaré-Wirtinger type. For $d = 1$, there are counterexamples showing that such inequalities do not exist for general degeneracies but for $d \geq 2$ there is always at least one direction with "full" diffusion, so there is hope that,

$$\|P_X y\|_2 \leq C \|D_p \nabla y^{\mathrm{T}}\|_2 \quad , \quad X = \{y \in L^2(\Omega) \mid D_p \nabla y^{\mathrm{T}} = 0\}$$

(with P_X denoting the orthogonal projection onto X) can be established.

Regarding weighted Sobolev spaces, there moreover seems to be some sort of "gap" between the class of A_p-weights of Muckenhoupt and the class Y_1^*: The A_p-weights can be less smooth than the Y_1^*-weights but are not allowed to vanish on an non-null set in the sense of Lebesgue. Obtaining similar results (such as weak closedness as well as $H = W$) for weighted (and directional) Sobolev spaces with more general weights would also be interesting. Moreover, it could be possible to prove $W_p(0, T) = \bar{W}_p(0, T)$ without or at least with weaker conditions than the requirement that p is non-degenerate at the boundary. This would also allow for less regularization in the optimal control problem. Furthermore, differentiability of the solution operator S with respect to p and u was only established for the non-degenerate case. It seems that it is necessary to weaken the norm of the space the operator S is mapping into. Finding such a space and examining differentiability S with respect to this space may be subject to further investigation. Another approach would be to find an appropriate weakened notion of differentiability the operator S satisfies. Both cases would permit first-order necessary conditions in the continuous case for the interesting case where $\|p\|_\infty = 1$ and possibly permit the application of abstract numerical optimization algorithms for which convergence rates are known.

Finally, improving the numerical algorithms can also be studied. It is well-known that gradient-type descent algorithms converge rather slowly, in comparison to methods of Newton-type. A different aspect in numerical analysis would be to develop finite-element discretizations of the degenerate parabolic equation (2.4) such that approximation rates can be established. In view of the solution spaces $W_p(0, T)$ and its characterization by weighted and directional

Sobolev spaces, one can, for instance, think of special discrete spaces for p for which conforming finite-elements on a triangular mesh can be defined.

To conclude, analyzing and solving the optimal control problem associated with degenerate parabolic equations for which the degeneracy can be controlled is an extensive mathematical task. Although this thesis tries to provide a thorough treatment, give insight into the nature of this problem and to suggest solutions, there are still some open questions which would be interesting to answer.

Appendix A

Notation

A.1 Symbols

d	The dimension of the problem. Usually, $d = 2$
Ω	A bounded domain of $\mathbb{R}^d$, i.e. Ω is open, connected and bounded
t	The time variable, $t \in [0, \infty[$
x	The space variable, $x \in \mathbb{R}^d$
N	The number of control images in the optimal control problem (2.8)
y_n	The control images
t_n, T	The control points in time and the end-time point, respectively ($T = t_N$)
y	The solution of the degenerate parabolic equation (2.4)
u	The distributed control parameter in (2.4)
p	The edge field in (2.4)
$p_1, \ldots, p_d$	The diffusion directions in the generalized equation (3.1)
σ	The edge-intensity function $[0, 1] \to [0, 1]$ in (2.4), see also Conditions 3.1 and 5.21
D_p^2	The diffusion tensor associated with the vector field p (see (2.3))
$D_{p_1, \ldots, p_d}^2$	The diffusion tensor associated with the diffusion directions $p_1, \ldots, p_d$ (see (3.2a))
$\Phi_u, \Phi_p, \tilde{\Phi}_p$	Penalization functionals for u, p and $p_1, \ldots, p_d$

λ_u, λ_p	Regularization parameters
M_x^{-1}	Bound for the derivative of p with respect to space
M_∂	Bound for $|p|$ at the boundary of Ω, see Definition 5.32
$S_{p_1,\dots,p_d}(u, y_0)$	Linear solution operator for (3.1) with sources u, initial value y_0 and associated fixed diffusion directions $p_1, \dots, p_d$
$\mathcal{S}(u, p)$	Solution operator for (2.4) associated with the optimal control problem (2.8), i.e. with sources u and edge field p, but fixed y_0
$\mathcal{S}_0(u, p)$	Same as $\mathcal{S}(u, p)$ but with initial value $y_0 = 0$
$V_{p_1,\dots,p_d}, V_p$	The abstract stationary spaces for associated with (3.1) (and (2.4)), see Definition 3.7
$\mathcal{V}_{p_1,\dots,p_d}, \mathcal{V}_p$	The abstract solution spaces associated with (3.1) (and (2.4)), see Definition 3.12
$W_{p_1,\dots,p_d}(0, T)$, $W_p(0, T)$	The abstract time-regular solution spaces associated with (3.1) (and (2.4)), see Definition 3.14
Y_k, Y_{div}, $Y_k^*, Y_{\mathrm{div}}^*$	Spaces of admissible time-invariant weights/directions and its duals, respectively, see Definitions 4.1 and 4.12 as well as Section 4.1
$\mathcal{Y}_k, \mathcal{Y}_{\mathrm{div}}$, $\mathcal{Y}_k^*, \mathcal{Y}_{\mathrm{div}}^*$	Spaces of admissible time-variant weights/directions and its duals, respectively, see Definition 5.1 as well as Section 5.1
$H^r_{w, \partial q_1, \dots, \partial q_K}$	The weighted and directional Sobolev space associated with $w \in Y_1^*$ and $q_1, \dots, q_K \in Y_{\mathrm{div}}^*$, see Definition 4.26 as well as Chapter 4
$\mathcal{H}^r_{w, \partial q_1, \dots, \partial q_K}$	The time-variant weighted and directional Sobolev space associated with $w \in \mathcal{Y}_1^*$ and $q_1, \dots, q_K \in \mathcal{Y}_{\mathrm{div}}^*$, see Definition 5.13 as well as Section 5.2

A.2   Functions, operators and sets

A^{T}	Matrix transposition $(A^{\mathrm{T}})_{i,j} = A_{j,i}$
$y \cdot z$	The standard inner product of two row or column vectors $y, z \in \mathbb{R}^k \ : \ y \cdot z = \sum_{i=1}^k y_i z_i$
$y \otimes z$	The standard outer product of the row or column vectors $y \in \mathbb{R}^k$ and $z \in \mathbb{R}^l$ $(y \otimes z)_{i,j} = y_i z_j$

$|y|$ The Euclidean absolute value (resp. Frobenius norm for matrices)

$$y \in \mathbb{R} \quad : \quad |y| = \max\{y, -y\}$$

$$y \in \mathbb{R}^k \quad : \quad |y| = \sqrt{\textstyle\sum_{i=1}^k y_i^2}$$

$$y \in \mathbb{R}^{k \times l} \quad : \quad |y| = \sqrt{\textstyle\sum_{i=1}^k \sum_{j=1}^l y_{i,j}^2} \text{ etc.}$$

$|k|$ The absolute value of a multi-index $k \in \mathbb{N}^d$

$$|k| = \textstyle\sum_{i=1}^d k_i$$

∇y The (weak) Jacobian/gradient of y

$$y : \mathbb{R}^k \supset U \to \mathbb{R}^l \quad : \quad (\nabla y)_{i,j} = \frac{\partial y_j}{\partial x_i}$$

Note: ∇y is a row vector if $y : U \to \mathbb{R}$.

$\nabla^m y$ The m-th derivative of y (m-linear mapping, symmetric)

Note: $(\nabla y)(h_1, \ldots, h_k)$ is $\nabla^m y$ applied to $h_1, \ldots, h_k$ (and therefore a $(m - k)$-linear mapping)

div y The divergence of the vector field/matrix y

$$y \in \mathbb{R}^{k \times d} \quad : \quad (\text{div } y)_i = \sum_{j=1}^d \frac{\partial y_{i,j}}{\partial x_j}$$

Δy The Laplacian of a function/vector field y

$$\Delta y = \text{div } \nabla y$$

$\langle y, z \rangle$ The scalar product of y and z associated with a (pre)-Hilbert space H

$\langle y, z \rangle_{L^2}$ The L^2 scalar product of functions $y, z \in L^2(\Omega)$

$$\langle y, z \rangle_{L^2} = \textstyle\int_\Omega y(x) z(x) \, \mathrm{d}x$$

$\|y\|$ The norm of a y which is element of some normed space X

$\|y\|_r$ The r-Lebesgue norm of a real-valued, vector-valued or matrix valued function on Ω

$\|y\|_{m,\infty}$ The supremum norm up to the m-th derivative of y

$$\|y\|_{m,\infty} = \max_{k=0,\ldots,m} \|\nabla^k y\|_\infty$$

$\mathrm{TV}(y)$ The total variation of a function $y : U \to \mathbb{R}^k$ where U is an open subset of $\mathbb{R}^d$

$\mathrm{tv}^*(y)$ The total semivariation of a function $y :]0, T[\to X$ for some Banach space X

I_C The indicator functional associated with a $C \subset X$, where X is usually a Banach space

$$I_C(y) = \begin{cases} 0 & \text{if } y \in C \\ \infty & \text{if } y \notin C. \end{cases}$$

$K \subset\subset \Omega$ The set K is a compact subset of Ω

A.3   Vector and function spaces

X^*	The dual of a normed space X
$X^\sim$	The closure of a metric/normed space X with respect to its topology
$X \hookrightarrow Y$	The normed space X is continuously embedded in the normed space Y
$X \hookleftarrow Y$	The normed space X is compactly embedded in the space Y
$\mathcal{C}^m(\Omega)$ $\mathcal{C}^m(\Omega, \mathbb{R}^k)$ $\mathcal{C}^m(\Omega, \mathbb{R}^{k \times l})$	The linear space of functions on Ω (with values in $\mathbb{R}$, $\mathbb{R}^k$ and $\mathbb{R}^{k \times l}$, respectively) which are up to m times continuously differentiable
$\mathcal{C}^m(\overline{\Omega})$ $\mathcal{C}^m(\overline{\Omega}, \mathbb{R}^k)$ $\mathcal{C}^m(\overline{\Omega}, \mathbb{R}^{k \times l})$	The normed space of functions on Ω (with values in $\mathbb{R}$, $\mathbb{R}^k$ and $\mathbb{R}^{k \times l}$, respectively) which are up to m times continuously differentiable and whose derivatives can be extended to the compact set $\overline{\Omega}$ $$\|y\|_{m,\infty} = \max_{\mu=0,\ldots,m} \|\nabla^\mu y\|_\infty$$
$\mathcal{C}^m(0, T; X)$	The space of functions on $[0, T]$ with up to m continuous derivatives with values in the metric space X
$\mathcal{C}^\infty(\Omega)$	The linear space of functions for which each (classical) derivative exists
$\mathcal{C}^\infty(\overline{\Omega})$	The linear space of arbitrarily smooth functions on Ω for which each derivative can be extended to the compact set $\overline{\Omega}$
$\mathcal{C}_0^\infty(\Omega)$	The linear space of arbitrarily smooth functions with compact support in Ω
$L^r(\Omega)$ $L^r(\Omega, \mathbb{R}^k)$ $L^r(\Omega, \mathbb{R}^{k \times l})$	The Lebesgue space of exponent r with values in $\mathbb{R}$, $\mathbb{R}^k$ and $\mathbb{R}^{k \times l}$ respectively and Euclidean vector norm resp. Frobenius matrix norm $$\|y\|_r = \left(\int_\Omega |y(x)|^r \, \mathrm{d}x \right)^{1/r} \quad , \quad \|y\|_\infty = \operatorname*{ess\,sup}_{x \in \Omega} |y(x)|$$
$L^r(0, T; X)$	The Bochner space of exponent $1 \leq r \leq \infty$ over $]0, T[$ with values in the Banach space X $$\|y\|_r = \left(\int_0^T \|y(t)\|_X^r \, \mathrm{d}t \right)^{1/r} \quad , \quad \|y\|_\infty = \operatorname*{ess\,sup}_{t \in]0, T[} \|y(t)\|_X$$

$H^m(\Omega)$ The Sobolev space of up to m times weakly differentiable func-
$H^m(\Omega, \mathbb{R}^k)$ tions with square integrable derivatives and values in $\mathbb{R}$, $\mathbb{R}^k$ and
$H^m(\Omega, \mathbb{R}^{k\times l})$ $\mathbb{R}^{k\times l}$, respectively

$$\|y\|_{H^m} = \left(\int_\Omega \sum_{k=0}^m |\nabla^k y(x)|^2 \, \mathrm{d}x \right)^{1/2}$$

$$\langle y, z \rangle_{H^m} = \int_\Omega \sum_{k=0}^m \nabla^k y(x) \cdot \nabla^k z(x) \, \mathrm{d}x$$

$H^{m,r}(\Omega)$ The Sobolev space of up to m times weakly differentiable func-
$H^{m,r}(\Omega, \mathbb{R}^k)$ tions with derivative in $L^r(\Omega)$ and values in $\mathbb{R}$, $\mathbb{R}^k$ and $\mathbb{R}^{k\times l}$,
$H^{m,r}(\Omega, \mathbb{R}^{k\times l})$ respectively

$\mathcal{L}(X, Y)$ The space of bounded linear mappings between the Banach spaces $X \to Y$

$\mathrm{Iso}(X, Y)$ The set of linear isomorphisms between the Banach spaces $X \to Y$

Bibliography

[Ada62]   John Frank Adams. Vector fields on spheres. *The Annals of Mathematics*, 75(3):603–632, 1962.

[ADK99]   Gilles Aubert, Rachid Deriche, and Pierre Kornprobst. Computing optical flow via variational techniques. *SIAM Journal on Applied Mathematics*, 60(1):156–182, 1999.

[AE56]   Richard F. Arens and James Eells. On embedding uniform and topological spaces. *Pacific Journal of Mathematics*, 6(3):397–403, 1956.

[AF03]   Robert A. Adams and John J. F. Fournier. *Sobolev spaces*, volume 140 of *Pure and Applied Mathematics*. Academic Press, 2nd edition, 2003.

[AK02]   Gilles Aubert and Pierre Kornprobst. *Mathematical Problems in Image Processing: Partial Differential Equations and the Calculus of Variations*, volume 147 of *Applied Mathematical Sciences*. Springer, 2002.

[Alt99]   Hans Wilhelm Alt. *Lineare Funktionalanalysis*. Springer, 3rd edition, 1999.

[Ama90]   Herbert Amann. *Ordinary Differential Equations: An Introduction to Nonlinear Analysis*, volume 13 of *de Gruyter Studies in Mathematics*. De Gruyter, 1990.

[AQ06]   Herbert Amann and Pavol Quittner. Optimal control problems governed by semilinear equations with low regularity data. *Advances in Differential Equations*, 11(1):1–33, 2006.

[AV94]   Robert Acar and Curtis R. Vogel. Analysis of bounded variation penalty methods for ill-posed problems. *Inverse Problems*, 10(6):1217–1229, 1994.

[AZ90]   Jürgen Appell and Petr Petrovich Zabrejko. *Nonlinear Superpo-sition Operators*, volume 95 of *Cambridge Tracts in Mathematics*. Cambridge University Press, 1990.

[Bel05]   Aziz Belmiloudi. Optimal control problems of nonlinear degener-ate parabolic differential systems with logistic time-varying delays. *IMA Journal of Mathematical Control and Information*, 22(1):88–108, 2005.

[BK89]   H. Thomas Banks and Karl Kunisch. *Estimation techniques for distributed parameter systems*. Birkhäuser, 1989.

[Bro81]   Chaim Broit. *Optimal registration of deformed images*. PhD thesis, University of Pennsylvania, 1981.

[Can84]   John Rozier Cannon. *The one-dimensional heat equation*, volume 23 of *Encyclopedia of mathematics and its applications*. Addison-Wesley, 1984.

[CDLL98]   Antonin Chambolle, Ronald A. DeVore, Nam-yong Lee, and Bradley J. Lucier. Nonlinear wavelet image processing: Variational problems, compression, and noise removal through wavelet shrink-age. *IEEE Transactions on Image Processing*, 7(3):319–335, 1998.

[CER90]   David Colton, Richard Ewing, and William Rundell, editors. *Inverse Problems in Partial Differential Equations*. Society for Industrial and Applied Mathematics, 1990.

[CFY95]   Eduardo Casas, Luis A. Fernández, and Jiongmin Yong. Opti-mal control of quasilinear parabolic equations. *Proceedings of the Royal Society of Edinburgh, Section A: Mathematics*, 125(3):545–565, 1995.

[CK97]   Andrej Cherkaev and Robert Kohn, editors. *Topics in the Mathematical Modelling of Composite Materials*, volume 31 of *Progress in Nonlinear Differential Equations and Their Applica-tions*. Birkhäuser, 1997.

[CKP98]   Eduardo Casas, Karl Kunisch, and Cecilia Pola. Some applica-tions of BV functions in optimal control and calculus of variations. *ESAIM: Proceedings*, 4:83–96, 1998.

[CL97]   Antonin Chambolle and Pierre-Louis Lions. Image recovery via total variation minimization and related problems. *Numerische Mathe-matik*, 76(2):167–188, 1997.

[CLMC92] Francine Catté, Pierre-Louis Lions, Jean-Michel Morel, and Tomeu Coll. Image selective smoothing and edge detection by nonlinear diffusion. *SIAM Journal on Numerical Analysis*, 29(1):182–193, 1992.

[CLSW98] Francis H. Clarke, Yuri S. Ledyaev, Ronald J. Stern, and Peter R. Wolenski. *Nonsmooth Analysis and Control Theory*, volume 178 of *Graduate Texts in Mathematics*. Springer, 1998.

[CS05] Tony Chan and Jianhong Shen. *Image Processing And Analysis: Variational, PDE, Wavelet, And Stochastic Methods*. Society for Industrial and Applied Mathematics, 2005.

[Dav05] Guy David. *Singular Sets of Minimizers for the Mumford-Shah Functional*, volume 233 of *Progress in Mathematics*. Birkhäuser, 2005.

[DiB93] Emmanuele DiBenedetto. *Degenerate Parabolic Equations*. Springer, 1993.

[DL90] Robert Dautray and Jacques-Louis Lions. *Mathematical Analysis and Numerical Methods for Science and Technology*, volume 5. Springer, 1990.

[DR00] Jérôme Droniou and Jean-Pierre Raymond. Optimal pointwise control of semilinear parabolic equations. *Nonlinear Analysis*, 39(2):135–156, 2000.

[DS57] Nelson Dunford and Jacob T. Schwartz. *Linear Operators. Part I: General Theory*, volume VII of *Pure and Applied Mathematics*. Wiley Interscience, 1957.

[DT05] Ingrid Daubechies and Gerd Teschke. Variational image restoration by means of wavelets: Simultaneous decomposition, deblurring, and denoising. *Applied and Computational Harmonic Analysis*, 19(2):1–16, 2005.

[DU77] Joseph Diestel and John Jerry, Jr. Uhl. *Vector Measures*, volume 15 of *Mathematical Surveys and Monographs*. American Mathematical Society, 1977.

[EBY99] Gwynne A. Evans, Jonathan M. Blackledge, and Peter D. Yardley. *Numerical Methods for Partial Differential Equations*. Springer Undergraduate Mathematics Series. Springer, 1999.

[EG79]   Murray Eisenberg and Robert Guy. A proof of the hairy ball theorem. *The American Mathematical Monthly*, 86(7):571–574, 1979.

[EG92]   Lawrence Craig Evans and Ronald F. Gariepy. *Measure Theory and Fine Properties of Functions*. CRC Press, 1992.

[ET76]   Ivar Ekeland and Roger Temam. *Convex Analysis and Variational Problems*, volume 1 of *Studies in Mathematics and its Applications*. North-Holland, 1976.

[Eva98]   Lawrence Craig Evans. *Partial Differential Equations*, volume 19 of *Graduate Studies in Mathematics*. American Mathematical Society, 1998.

[FY99]   Angelo Favini and Atsushi Yagi. *Degenerate Differential Equations in Banach Spaces*, volume 215 of *Textbooks and Monographs in Pure and Applied Mathematics*. Marcel Dekker, 1999.

[GGZ74]   Herbert Gajewski, Konrad Gröger, and Klaus Zacharias. *Nichtlineare Operatorgleichungen und Operatordifferentialgleichungen*, volume 38 of *Mathematische Lehrbücher und Monographien*. Akademie-Verlag, 1974.

[Giu84]   Enrico Giusti. *Minimal Surfaces and Functions of Bounded Variation*, volume 80 of *Monographs in Mathematics*. Birkhäuser, 1984.

[Goo70]   Kent Robert Goodrich. A Riesz representation theorem. *Proceedings of the American Mathematical Society*, 24(3):629–636, 1970.

[Gri85]   Pierre Grisvard. *Elliptic Problems in Nonsmooth Domains*. Pitman, 1985.

[GT98]   David Gilbarg and Neil S. Trudinger. *Elliptic Partial Differential Equations of Second Order*. Number 224 in Grundlehren der mathematischen Wissenschaften. Springer, 2nd edition, 1998.

[Gut90]   Semion Gutman. Identification of discontinuous parameters in flow equations. *SIAM Journal on Control and Optimization*, 28(5):1049–1060, 1990.

[Hoc89]   Reinhard Hochmuth. *Regularitätsresultate für ein Randwertproblem einer nicht-hypoelliptischen linearen partiellen Differentialgleichung*. PhD thesis, Freie Universität Berlin, 1989.

[HS81]   Berthold K. P. Horn and Brian G. Schunck. Determining optical
          flow. *Artificial Intelligence*, 17:185–203, 1981.

[Kil94]   Tero Kilpeläinen. Weighted Sobolev spaces and capacity. *Annales
          Academiæ Scientiarum Fennicæ*, 19:95–113, 1994.

[Kil97]   Tero Kilpeläinen. Smooth approximation in weighted Sobolev
          spaces. *Commentationes Mathematicae Universitas Carolinae*,
          38(1):29–35, 1997.

[Kuf80]   Alois Kufner. *Weighted Sobolev Spaces*, volume 31 of *Teubner-Texte
          zur Mathematik*. Teubner, 1980.

[KV84]   Robert V. Kohn and Michael Vogelius. Determining conductivity
          by boundary measurements. *Communications on Pure and Applied
          Mathematics*, 37:289–298, 1984.

[KV85]   Robert V. Kohn and Michael Vogelius. Determining conductivity
          by boundary measurements. II: Interior results. *Communications
          on Pure and Applied Mathematics*, 38:643–667, 1985.

[Lin94]   Tony Lindeberg. *Scale-Space Theory in Computer Vision*. Kluwer
          Academic Publishers, 1994.

[Lio61]   Jacques-Louis Lions. *Équations différentielles opérationnelles et
          problèmes aux limites*, volume 111 of *Die Grundlehren der math-
          ematischen Wissenschaften in Einzeldarstellungen mit besonderer
          Berücksichtigung der Anwendungsgebiete*. Springer, 1961.

[Lio68]   Jacques-Louis Lions. *Contrôle optimal de systèmes gouvernés par
          des équations aux dérivées partielles*. Dunod, 1968.

[LR91]   Patricia K. Lamm and I. Gary Rosen. An approximation theory
          for the estimation of parameters in degenerate Cauchy problems.
          Journal of Mathematical Analysis and Applications, 162(1):13–48,
          1991.

[LSU68]   O. A. Ladyženskaja, V. A. Solonnikov, and N. N. Ural'ceva. *Linear
          and Quasi-linear Equations of Parabolic Type*, volume 23 of *Transla-
          tions of Mathematical Monographs*. American Mathematical Society,
          1968.

[Lue89]   David G. Luenberger. *Linear and Nonlinear Programming*. Addison-
          Wesley, 2nd edition, 1989.

[LY95]   Suzanne M. Lenhart and Jiongmin Yong. Optimal control for degen-
          erate parabolic equations with logistic growth. *Nonlinear Analysis,
          Theory, Methods & Applications*, 25(7):681–698, 1995.

[Mey01]   Yves Meyer. *Oscillating Pattern in Image Processing and Nonlin-
          ear Evolution Equations: The Fifteenth Dean Jacqueline B. Lewis
          Memorial Lectures*, volume 22 of *University Lecture Series*. Ameri-
          cal Mathematical Society, 2001.

[MM79]   Moshe Marcus and Victor J. Mizel. Complete characterization of
          functions which act, via superposition, on Sobolev spaces. *Trans-
          actions of the American Mathematical Society*, 251:181–218, July
          1979.

[Mod04]   Jan Modersitzki. *Numerical Methods for Image Registration*. Oxford
          University Press, 2004.

[MS64]   Norman G. Meyers and James Serrin. $H = W$. *Proceedings of
          the National Academy of Sciences of the United States of America*,
          51(6):1055–1056, 1964.

[MS85]   David Mumford and Jayant Shah. Boundary detection by minimiz-
          ing functionals. In *Proceedings of the IEEE Computer Society Con-
          ference on Computer Vision and Pattern Recognition*, pages 22–26,
          1985.

[MS89]   David Mumford and Jayant Shah. Optimal approximations by piece-
          wise smooth functions and associated variational problems. *Com-
          munications on Pure and Applied Mathematics*, 42:577–685, 1989.

[Mur77]   François Murat. Contre-exemples pour divers problèmes où le
          contrôle intervient dans les coefficients. *Annali di Matematica Pura
          ed Applicata*, 112(1):49–68, 1977.

[OR73]   O. A. Oleĭnik and E. V. Radkevič. *Second Order Equations With
          Nonnegative Characteristic Form*. American Mathematical Society,
          1973.

[PM90]   Pietro Perona and Jitendra Malik. Scale-space and edge detection
          using anisotropic diffusion. *IEEE Transactions of Pattern Analysis
          and Machine Intelligence*, 12(7):629–639, 1990.

[ROF92]   Leonid I. Rudin, Stanley J. Osher, and Emad Fatemi. Nonlinear to-
          tal variation based noise removal algorithms. *Physica D: Nonlinear
          Phenomena*, 60(1–4):259–268, 1992.

[SB93]   Josef Stoer and Roland Bulirsch. *Introduction to Numerical Analysis*, volume 12 of *Texts in Applied Mathematics*. Springer, 2nd edition, 1993.

[Sho96]   Ralph Edwin Showalter. *Monotone Operators in Banach Space and Nonlinear Partial Differential Equations*, volume 49 of *Mathematical Surveys and Monographs*. American Mathematical Society, 1996.

[SNFJ97]   Jon Sporring, Mads Nielsen, Luc M. Florack, and Peter Johansen, editors. *Gaussian Scale-Space Theory*, volume 8 of *Computational Imaging and Vision*. Kluwer Academic Publishers, 1997.

[Tri95]   Hans Triebel. *Interpolation Theory, Function Spaces, Differential Operators*. Johann Ambrosius Barth, 2nd edition, 1995.

[Trö05]   Fredi Tröltzsch. *Optimale Steuerung partieller Differentialgleichungen*. Vieweg, 2005.

[Tur00]   Bengt Ove Turesson. *Nonlinear Potential Theory and Weighted Sobolev Spaces*, volume 1736 of *Lecture Notes in Mathematics*. Springer, 2000.

[VO04]   Luminita Aura Vese and Stanley J. Osher. Image denoising and decomposition with total variation minimization and oscillatory functions. *Journal of Mathematical Imaging and Vision*, 20(1–2):7–18, 2004.

[Wal96]   Wolfgang Walter. *Gewöhnliche Differentialgleichungen*. Springer, 6th edition, 1996.

[Wei98]   Joachim Weickert. *Anisotropic diffusion in image processing*. Teubner, 1998.

[Wit83]   Andrew P. Witkin. Scale-space filtering. In *Proceedings of the Eighth International Joint Conference on Artificial Intelligence, August 8–12 1983, Karlsruhe, Germany*, volume 2, pages 1019–1022, 1983.

[Yos80]   Kôsaku Yosida. *Functional Analysis*. Springer, 6th edition, 1980.

[Zhi98]   Vasilii Vasilievich Zhikov. Weighted Sobolev spaces. *Sbornik: Mathematics*, 189(8):1139–1170, 1998.